1 MONTH OF
FREE
READING

at

www.ForgottenBooks.com

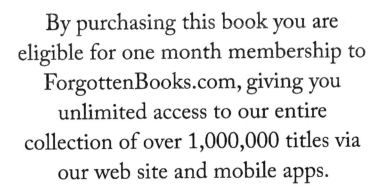

By purchasing this book you are eligible for one month membership to ForgottenBooks.com, giving you unlimited access to our entire collection of over 1,000,000 titles via our web site and mobile apps.

To claim your free month visit:

www.forgottenbooks.com/free854349

ISBN 978-0-428-97298-1
PIBN 10854349

A COURSE O

·LECTURES

IN

Natural Philofophy.

By the late

RICHARD HELSHAM, M.D.

Profeſſor of PHYSICK and NATURAL PHILOSOPHY
in the Univerſity of DUBLIN.

PUBLISHED BY

BRYAN ROBINSON, M.D.

The FOURTH EDITION

LONDON:

Printed for J. NOURSE, oppoſite *Katherine-Street* in the
Strand, Bookſeller in Ordinary to his MAJESTY.

M.DCC.LXVII.

THE
PREFACE.

THAT the reader may be duly prepared for the perufal of the following Treatife, it will be neceffary that he firft acquaint himfelf with the genuine Method and Rules of Philofophizing, as they have been delivered by Sir ISAAC NEWTON.

His Method of Philofophizing is thus laid down in his *Opticks* *.

    " As in mathematicks, fo in natural philo-
" fophy, the inveftigation of difficult things
" by way of *Analyfis*, ought ever to precede
" the method of compofition. This *Analyfis*
" confifts in making experiments and obfer-
" vations, and in drawing general conclufions
" from them by induction, and admitting of
" no objections againft the conclufions, but
" fuch as are taken from experiments or other
" certain truths. And although the arguing
" from experiments and obfervations by in-
" duction be no demonftration of general
" conclufions; yet it is the beft way of ar-
" guing which the nature of things admits
" of, and may be looked upon as fo much
" the ftronger, by how much the induction

* Opt. p. 380.

A 2   " is

" is more general. And if no exception oc-
" cur from *Phænomena*, the conclusion may
" be pronounced generally. But if at any
" time afterwards, any exceptions shall occur
" from experiments, it may then be pro-
" nounced with such exceptions as shall occur.
" By this way of *Analysis*, we may proceed
" from compounds to ingredients, and from
" motions to the forces producing them;
" and in general from effects to their causes,
" and from particular causes to more general
" ones, till the argument ends in the most
" general. This is the method of *Analysis*:
" And the *Synthesis* consists in assuming the
" causes discovered, and established as prin-
" ciples, and by them explaining the *Phæ-*
" *nomena* proceeding from them, and proving
" the explanations."

His Rules of Philosophizing, delivered in
his *Principles* *, are these four.

R U L E I.

" *More causes of natural things are not to be*
" *admitted, than are both true and sufficient*
" *for explaining their phænomena.*

" Thus Philosophers say; nature does no-
" thing in vain, and in vain that is done by
" more causes, which can be done by fewer:
" For nature is simple, and delights not in
" superfluous causes of things."

* Philos. Natur. Princip. Mathemat. p. 387.

R U L E

RULE II.

" *Of natural effects therefore of the same*
" *kind the same causes are to be assigned, as far*
" *as it can be done.*

" As of respiration in a man and in a
" beast; of the descent of stones in *Europe*
" and in *America*; of light in a culinary fire
" and in the sun; of the reflexion of light in
" the earth and in the planets."

RULE III.

" *The qualities of bodies which cannot be*
" *increased and diminished, and which agree*
" *to all bodies in which experiments can be*
" *made, are to be reckoned as qualities of all*
" *bodies whatsoever.*

" For the qualities of bodies are not known
" but by experiments; and therefore, as ma-
" ny are to be reckoned general as generally
" agree with experiments, and those which
" cannot be diminished cannot be taken
" away. Certainly dreams are not to be de-
" vised at pleasure contrary to the tenor of
" experiments; nor must we depart from the
" analogy of nature, since she is wont to be
" simple, and always consonant to herself.
" The extension of bodies is not known but
" by the senses, nor is it perceived in all bo-
" dies: but because it agrees to all bodies
" which are perceiveable, it is affirmed of all

A 3 " what

" whatfoever. We experience many bodies
" to be hard. But the hardnefs of the whole
" arifes from the hardnefs of the parts, and
" thence with good reafon we conclude the
" undivided parts not only of thofe bodies
" which are perceived, but alfo of all others,
" to be hard. We gather all bodies to be
" impenetrable, not by reafon, but by fenfe.
" We find the bodies we handle to be im-
" penetrable, and thence conclude impene-
" trability to be a property of all bodies what-
" foever. That all bodies are moveable, and
" by certain forces (which I call *vires iner-*
" *tiæ*) perfevere in motion or reft, we gather
" from thefe fame properties in bodies which
" are feen. Extenfion, hardnefs, impenetra-
" bility, mobility, and *vis inertiæ* of the
" whole, arife from the extenfion, hardnefs,
" impenetrability, mobility, and *vis inertiæ*
" of the parts ; and thence we conclude that
" all the leaft parts of all bodies are extend-
" ed, and hard, and impenetrable, and move-
" able, and endued with *vires inertiæ*. And
" this is the foundation of all Philofophy.
" Farther we know from the *Phænomena*, that
" the parts of bodies which are divided
" and mutually contiguous to one another,
" may be feparated from one another; and it
" is certain from mathematicks, that the un-
" divided parts may by reafon be diftin-
　　　　　　　　　　　　　　　" guifhed

" guished into less parts. But whether those
" parts distinct, and not yet divided, can
" by the powers of nature be divided and
" separated from one another, is uncertain.
" But if it should appear, even by one single
" experiment, that by breaking a hard and
" solid body, any undivided particle suffered
" a division; we might conclude by the
" force of this Rule, that not only the di-
" vided parts were separable, but that the un-
" divided parts might be divided *in infinitum.*

Lastly, If it be universally evident by
" experiments and astronomical observations,
" that all bodies round the earth gravitate
" towards the earth, and that in proportion
" to the quantity of matter in each, and that
" the moon gravitates towards the earth in
" proportion to its quantity of matter, and in
" like manner our sea gravitates towards the
" moon, and that all the planets mutually
" gravitate towards one another, and that
" there is a similar gravity of comets to-
" wards the sun; we must pronounce by
" this Rule, that all bodies gravitate mutu-
" ally towards one another. For the argu-
" ment from the *Phænomena* will be stronger
" for an universal gravity, than for the im-
" penetrability of bodies, concerning which
" in the heavenly bodies we have no experi-
" ment, no observation at all."

RULE

RULE IV.

" *In experimental Philosophy propositions*
" *collected from the* Phænomena *by induction,*
" *are to be deemed, notwithstanding contrary*
" Hypotheses, *either accurately or very nearly*
" *true, till other* Phænomena *occur, by which*
" *they may be rendered either more accurate*
" *or liable to exceptions.*

" This ought to be done, left arguments
" of induction ſhould be deſtroyed by *Hy-*
" *potheses.*"

This Method and theſe Rules, have been
carefully obſerved by our Author in theſe
LECTURES; which, from the clearneſs and
diffuſiveneſs of the ſtile, and the eaſy and
juſt manner of reaſoning, are, in my opinion, better fitted for the inſtruction of youth,
than any thing which I have ſeen on this
ſubject.

I have added a few *Problems* by way of
APPENDIX.

THE

THE
CONTENTS.

LECTURE I.

The CONTENTS.

LECTURE XV.

APPENDIX.

( )

LECTURE I.

Of ATTRACTION.

AS natural philofophy is a fcience in its own nature entertaining and delightful, and withal conducive in many inftances to the eafe and convenience of life; it is not to be wondered that there have been men in all ages who have laid themfelves out in the improvement and cultivation of it. But it is a matter of no fmall furprize to think how inconfiderable a progrefs the knowledge of nature had made in former ages, when compared with the vaft improvements it has received from the numberlefs difcoveries of later times; infomuch, that fome of the branches of natural philofophy, which at this day is almoft compleat in all its parts, were utterly unknown before the laft century. If we look into the reafon of this, we fhall find it to be chiefly owing to the wrong meafures that were taken by philofophers of former ages in their purfuits after natural knowledge: for they difregarding experiments, the only fure foundation whereon to build a rational philofophy, bufied themfelves in framing hypothefes, for the folution of natural appearances, which as they were creatures of the brain, without any foundation in nature, were generally fpeaking fo lame and defective, as in many cafes not to anfwer thofe very phænomena for whofe fakes they had been contrived. Whereas the philofophers of later times, laying afide thofe falfe lights, as being of no other ufe than to mifguide the underftanding

in

in its fearches into nature, betook themfelves to experiments and obfervations; and from thence collected the general powers and laws of nature; which with a proper application, and the affiftance of mathematical learning, enabled them to account for moft of the properties and operations of bodies; and to folve many difficulties in the natural appearances, which were utterly inexplicable on the foot of hypothefes. By this means has natural philofophy within the compafs of one century been brought out of the greateft darknefs and obfcurity into the cleareft light; and this has been chiefly owing to the unparalleled abilities, and indefatigable induftry of that great and accurate philofopher Sir ISAAC NEWTON; who to his great honour has in his principles of natural philofophy, and his incomparable treatife of light and colours, cleared more difficulties, and difclofed more and more important truths relating to nature, than are to be met with in the voluminous writings of all that went before him. To illuftrate fome of thefe truths by experiments is the defign of this courfe, which confifts of four parts. In the firft are confidered folid bodies and their properties. In the fecond water and watery fluids. In the third the elaftic fluid of air. And in the laft the fubtile fluid of light. But before I proceed to thefe particulars, it will be neceffary to fay fomething concerning certain principles, forces, or powers, wherewith all parts of matter of what kind foever, fo far as experience reaches, feem to be endued; and whereby they act upon one another for producing a great part of the phænomena of nature.

Such is firft that power whereby the minute particles of matter do in fome circumftances tend towards one another, which is commonly called attraction; the caufe whereof is in a great meafure unknown, tho' the thing itfelf is manifeft from experiments. For if two polifhed plates of brafs be laid

laid one upon another, having their contiguous sides
smeared with oil, they will cohere in *vacuo*, and
with such firmness, that when they are suspended,
the force of gravity in the lower plate will not
suffice to separate and pull them asunder.

That the cohesion of these plates is to be attributed to the mutual attractions of their contiguous
parts, cannot I think admit of a doubt, since the
pressure of the outward air on their external surfaces, (to whose force this effect might otherwise
have been attributed) is in this case taken off.

The use of the oil is to fill up the minute cavities in the surfaces, and by so doing to prevent the
lodgment of air between the plates; which upon
the removal of the outward air would expand itself
by reason of its elasticity, and thereby force the
plates asunder.

The forementioned attraction is in like manner
collected from the following experiments.

If two plane polished plates of glass be laid together, so that their sides be parallel, and at a very
small distance from one another; and their lower
edges be dipped in water; the water will rise up
between them, and the less the distance of the
glasses is, the greater will the height be to which
the water rises. If the distance be about the hundredth part of an inch, it will rise to the height of
about an inch; and if the distance be greater or less
in any proportion, the height will be reciprocally
proportional to the distance very nearly.

The reason why the water ascends between the
plates is, that those parts of the surfaces of the
glasses which lie next above the surface of the water, and are contiguous thereto, attract the water,
and by that means cause it to ascend; and this
ascent continues till the weight of the elevated
water becomes equal to the force of the attracting
surfaces, and then the motion ceases, the water
tending as much downward by the force of its own
gravity,

Exp. 2.

gravity, as it doth upward by the attraction of the glasses.

The reason why the water rises to heights which are inversly as the distances of the glasses, is this: the absolute attractive force of the glasses, whereby the water is raised, continues unvaried whatever be the distance of the glasses; for the height and length of the glass surfaces, whose attractions influence the ascent of the water, are always the same, and consequently the attractive force must be so too; and for that reason will constantly support the same weight of water; but the quantity and consequently the weight of the elevated water will always be the same, if its height be reciprocally as its base, that is, in this case, as the distance of the plates; for the length of the base being equal to the length of the plates, it continues unvaried; and therefore the base will ever be as its breadth, that is, as the interval between the plates.

Exp. 3.   If the glass plates instead of being set parallel to one another, be made to meet at one of their ends, and kept at a little distance at the other; and their lower edges be then dipped in water, spirit of wine, or any other convenient liquor; the inward sides of the plates being first moistened with a clean cloth dipped in the liquor; the liquor will rise between the plates; and the upper surface of the elevated liquor will form a curve, the heights of whose several points above the surface of the stagnating liquor will be to one another reciprocally as their perpendicular distances from the concourse of the plates.

Fig. 1.   the plates. For the illustration of which, let AE be the surface of the stagnating liquor wherein the lower edges of the plates are immersed, A H the concourse of the plates, and F, G, I, K, L the curve formed by the surface of the elevated liquor; from any points in the curve as G, I, K, L taken at pleasure, let fall the right lines GB, IC, KD, LE perpendicular to A E, and those lines will express the

heights

heights of the refpective points of the curve above the furface of the ftagnant liquor; whilft A B, A C, A D, A E denote the perpendicular diftances of the fame points from the concourfe of the glaffes; now thefe heights and diftances are to one another in a reciprocal proportion: for if we fuppofe the lines G B, I C, K D, L E to be fo many pillars of liquor confifting of four fides, two of which are terminated by the plates, and the other two by the contiguous liquor; and if thofe fides which lie next the plates be of an equal but exceedingly fmall breadth in all the pillars, then will the attracting furfaces of the plates which fupport thofe pillars be likewife equal, and confequently the quantities fupported, that is, the pillars muft be fo too. But in order to have them equal, their heights muft be reciprocally proportional to their bafes; which bafes inafmuch as they are fuppofed to be equally broad muft be as their lengths, that is, as the intervals between the glaffes in thofe parts where the pillars are taken; and therefore the heights of the pillars muft be reciprocally as the intervals between the plates; but from the nature of fimilar triangles the intervals between the glaffes at different diftances from the concourfe are to one another directly as thofe diftances; whence it follows, that the heights of the pillars are to one another reciprocally as their refpective diftances from the concourfe of the plates; that is, if G B be double of I C, then is A C double of A B.

From what has been faid it is plain that the curve formed by the upper furface of the elevated liquor muft be an hyperbola; for from the nature of the hyperbola the external ordinates are reciprocally as the abfciffæ; wherefore if A B, A C, A D, A E be taken for the abfciffæ; then will B G, C I, D K, E L, be the refpective ordinates; and confequently the curve which paffes through the points G, I, K, L is an hyperbola.

2

·As

As water or any other proper fluid afcends between polifhed plates of glafs by the force of their
attractions; fo does it likewife in flender pipes of glafs
open at both ends; for if fuch tubes be dipped at
one end into water, fpirit of wine, or any other
convenient fluid, the liquor will rife within the pipes
to a confiderable height, and this experiment (as
alfo thofe before made) fucceeds in the very fame
manner in *vacuo*, as in the open air, for the liquor
conftantly afcends to the fame height in both.

That the afcent of liquor in thefe fmall tubes,
as alfo between polifhed plates of glafs, is to be attributed to fome power in the glafs ftrongly acting
on the liquor, and not to the preffure either of the
ftagnating liquor or incumbent atmofphere, is evident from this confideration; that as much of the
liquor remains fufpended in the pipes, and between
the plates, when they are lifted out of the ftagnating fluid, either in *vacuo* or the open air, as was
elevated above the furface of the fluid, while they
were immerfed therein: and therefore whatever caufe
concurred to the elevating of the liquor while the
plates and pipes were therein immerfed, and expofed
to the air; the fame contributes as powerfully to keep
it up, when the ftagnating liquor is removed, and
the preffure of the atmofphere taken off, and confequently muft be fome power inherent in the glafs.

The heights to which the liquor rifes in flender
pipes, are to one another reciprocally as the diameters. For the power which raifes the liquor in a flender pipe, being the attractive force of that part only
of the internal concave furface which lies next
above the liquor, and conftitutes a ring of an indefinitely fmall height, which height is ever the fame
whatever be the diameter of the ring, becaufe the
diftance to which the attractive force of glafs
reaches is unvaried; and the attractive force of fuch
an annular furface being as the number of attracting
parts whereof it is compofed, that is, as the furface,
                                                which

which becaufe its height is given is as the periphery, that is, as the diameter, the attractive force of
the pipe muſt be as the diameter. Wherefore if in
comparing the forces of two fuch pipes we make
F to denote the attractive force of the larger, and f
the attractive force of the fmaller, and alfo D and
d to denote their diameters; we fhall have this analogy, viz. F : f :: D : d, that is, the force of the
larger pipe is to that of the fmaller as the diameter of the larger to the diameter of the fmaller:
but thefe forces are likewife to one another in the
fame ratio with the quantities of liquor which they
keep fufpended, for they continue to elevate the liquor till fuch time as the weights, and confequently
the quantities of liquor drawn up, become a balance
to the attracting forces. Wherefore if H be put
for the height of the liquor in the pipe, whofe diameter is D, and h for its height in the pipe whofe
diameter is d; then will H multiplied into the fquare
of D be as the quantity of liquor in the larger pipe;
and h multiplied into the fquare of d as the quantity of liquor in the fmaller pipe; whence we have
this fecond analogy, F : f :: H x D² : h x d²; and
by fubftituting D and d in the room of F and f,
to which they are proportional, as appears from
the firſt analogy, we fhall have D : d :: HD² : hd²;
and then multiplying extreams and means, and
throwing off fimilar quantities, we fhall have
HD = hd, and by refolving this equation into an
analogy, we fhall have H : h :: d : D, that is, the
height to which the liquor rifes in the larger pipe is
to the height to which it rifes in the fmaller, as
the diameter of the fmaller pipe to that of the larger;
fo that the heights of the liquor are reciprocally
proportional to the diameters of the pipes.

By virtue of this attractive force, wherewith fmall
pipes are endued, plants receive nourifhment from
the earth; the flender tubes whereof their roots are
compofed, fucking in various juices according to

their

their different natures and conftitutions. From the fame attractive force it is that fponges take in water ; and that water afcends in loaf fugar, when any part of it is dipped therein ; thofe parts of the fugar which lie next above the water attracting, and thereby raifing the fame. And here it muft be obferved that the water rifes by the action of thofe particles alone which are contiguous to, and lie next above the furface of the elevated water; thofe particles which are at any the leaft fenfible diftance above the water being too far removed to influence the water by their attractions : and what has been thus obferved of fugar, is likewife true of polifhed plates, flender pipes, and every other attracting body, by virtue of whofe attractions fluids are raifed. For if thofe parts of attracting furfaces which are at any fenfible diftance above the furface of the fluid, do in any meafure contribute to the afcent ; it is evident that the fluid *ceteris paribus* muft rife to a greater height when the attracting furfaces are continued to a confiderable height above the elevated fluid, than when they terminate at a very little diftance above the fame. But the

Exp. 5. contrary appears from experiment. For if two polifhed plates of glafs fet parallel to one another at the diftance of about the hundredth part of an inch, be immerfed in water fo far that only an inch and one tenth be fuffered to remain above the water, the water will rife up between them to the height of about an inch ; and if the furface of the ftagnating water be then depreffed by drawing off fome of the water, the elevated water will likewife defcend between the plates fo as ftill to preferve the height of about an inch and no more.

Exp. 6. If a polifhed plate of glafs be laid parallel to the horizon, and another plate of the fame kind be laid thereon, fo as that they may touch at one of their ends, and be kept at a very fmall diftance at the other : being firft moiftened on their inward fides

. with

with a clean cloth or feather dipped in oil of o-
ranges; and if a drop of the oil be placed between the plates at that end where they are at some
distance from each other, so as that it may be touched by both the plates, it will begin to move towards the concourse of the glasses, and will continue
to go on with an accelerated motion till it arrives
at the concourse. And if during the motion of the
drop, that end of the glasses where they meet,
and towards which the drop moves, be lifted up, the
drop will nevertheless continue its motion, and of
consequence must be attracted; but as the end of
the glasses is raised higher and higher, the drop will
ascend more and more slowly, till at last upon a
certain elevation of the plates the motion ceases, the
gravity of the drop, wherewith it tends downward,
becoming equal to the attractive force which draws
it upward; as appears from this, that upon giving
the plates the least degree of elevation beyond what
is necessary to stop the drop, it straightway begins
to descend, its gravity in that case overcoming the
attraction.

By the help of this phænomenon may the force
be determined, wherewith the drop is attracted at
all distances from the concourse of the glasses. For
that part of a body's gravity whereby it is carried
down an inclined plane, is to its absolute weight,
as the sine of the angle of the plane's elevation, to
the Radius, or as the perpendicular height of the
plane to the length thereof; and therefore may be
denoted by the perpendicular height applied to the
length; and where the length of the plane is given,
that force will be every where as the sines of the
angles of elevation, or the perpendicular altitudes
of the plane; as shall be made appear when I come
to treat of the descent of bodies on inclined planes.
If therefore the sines of such elevations of the plates
as are necessary to stop the motion of the drop, be
taken at two different distances of the drop from

the

the concourfe of the plates; thofe fines will denote the refpective gravities of the drop, and confequently the attractive forces, wherewith the plates
act upon the drop at each of thofe diftances. Thus
for inftance, if the diftances of the drop from
the concourfe of the glaffes be as one and two;
and the fine of the elevation neceffary to ftop the
motion of the drop when at the fmaller diftance be
as four, and when at the greater diftance as one; the
gravity of the drop, wherewith it endeavours to defcend at the forementioned diftances of one and two,
will be as four and one. For the illuftration of

Fig. 2. which, let AB and AC reprefent the plates at different elevations; F and G the places where the
drop ftands upon thofe elevations; then will BD
and CE denote the forces of gravity wherewith
the drop endeavours to defcend along the plates in
the points F and G, which forces are equal to the
attractions of the glaffes in thofe points; and if
BF and CG the diftances of the drop from the
concourfe of the plates be as one and two, and
BD and CE as four and one; then is the attractive power wherewith the glaffes act upon the
drop at F, to the force wherewith they act upon
it at G, as four to one, that is, reciprocally as the
fquares of the diftances of the drop from the concourfe of the glaffes; and this is nearly the cafe, as

Exp. 6. will appear from the experiment.

Tho' the drop be attracted by forces that are in
the reciprocal duplicate ratio of the diftances of
the drop from the concourfe of the glaffes; yet are
the attractions within the fame quantities of attracting furface in the reciprocal fimple ratio only of
thofe diftances: for as the drop moves towards the
concourfe of the glaffes, it muft fpread and touch
each glafs in a larger furface; and this fpreading is
always proportional to the leffening of the interval
between the glaffes; and of confequence from the
nature of fimilar triangles, it is likewife proportional

· to

to the diminution of the diſtance from the concourſe.
So that the force which acts upon the drop is increaſed as the drop approaches the concourſe in the
ſimple reciprocal ratio of the diſtance, on account
of the inlargement of the attracting ſurface in that
proportion; and therefore in a given quantity of
attracting ſurface the force muſt be in the reciprocal
ſimple ratio of the diſtance from the concourſe;
that is to ſay, any given portion of the glaſs ſurfaces taken at the diſtance of one inch from their
concourſe muſt act with twice the force that it does
at the diſtance of two inches, and with thrice the
force that it does at the diſtance of three inches, and
ſo on.   Hence it will be found that the attractive
force of one and the ſame ſlender pipe of a conical
figure is given; or in other words, that the attractive force wherewith a conical pipe is indued
at any one diſtance from the vertex of the cone,
is equal to the attractive force of the ſame at any
other diſtance from the vertex; ſo that the attractive force of a conical pipe is, in every part
equal throughout the whole length of the pipe;
and may be expreſſed by the diameter of a circular
ſection of the pipe taken at any diſtance from the
vertex, applied to that diſtance.   For the attraction in any part of ſuch a pipe, is as the quantity
of attracting ſurface in that part multiplied into
the abſolute force; but the quantity of attracting
ſurface in any part is as the diameter of that part, and
the abſolute force is reciprocally as the diſtance from
the vertex; wherefore if A be put to denote the diſtance
of any part from the vertex and D the diameter,
$\frac{D}{A}$ will expreſs the attraction of that part; but from
the nature of ſimilar triangles the diameters of the
circular ſections of a cone taken at different diſtances from the vertex are to one another as the diſtances, conſequently $\frac{D}{A}$ is a ſtanding quantity.

B 3   Wherefore

L e c t. Wherefore since the attractive force in every part of
I. a conical tube is denoted by a quantity which is invariable, it follows that the force is so too ; so that
in this respect conical pipes do not differ from those
of a cylindrical form ; but herein lies the difference,
that in very slender pipes where the diameters are
equal, the attractions of such as are conical do far
surpass the attractions of those which are cylindrical.   And indeed so exceeding great does this attractive force become with respect to the quantity of
attracting surface in that part of a conical pipe, where
the diameter is but one part of an inch divided into
ten millions (if such minuteness may be supposed)
that if the attraction of a cylindrical tube, whose
diameter is an inch, were as great with respect to
its quantity of attracting surface, it would be able
to support a column of water an inch in diameter
and upwards of three miles in height.   For let us
suppose a conical tube whose base is an inch in diameter to be continued till the diameter is so far diminished as to equal only one part of an inch divided
into ten millions ; it is evident from what was just
now said, that the whole attractive force of such a
pipe, where its diameter is an inch, is equal to the
whole attractive force of the same, where the diameter is but the ten millioneth part of an inch ; consequently if a portion of the larger attracting surface
be taken equal to the smaller attracting surface,
the force of that will be to the force of this, as the
force of the smaller surface divided by the number
of parts in the larger surface, to the force of the
smaller surface, that is, as one divided by ten millions to one.   If therefore a conical, or indeed a
cylindrical tube an inch in diameter (for where the
diameter is so large there is scarcely any difference)
was indued with an attractive force as great in proportion to its quantity of attracting surface, as is a
conical tube of the ten millioneth part of an inch in
diameter, its force would be ten millions of times

greater

greater than it is, and of confequence would raife
the water ten millions of times higher than it doth
at prefent : but it has been found by experience
that in a cylindrical tube of an inch in diameter, the
water will rife to the height of about the fiftieth
part of an inch, and therefore if the force by which
it rifes was augmented in the forementioned proportion, it muft rife to the height of two hundred
thoufand inches, which being divided by fixty three
thoufand three hundred and fixty, the number of
inches in a mile, gives three and a little more in
the quotient.

The quantities of liquor fupported by the attractions of flender conical pipes are to one another, as the
diameters of the little circular furfaces of the elevated liquor, applied to the refpective diftances of the
fame circular furfaces from the vertices of the feveral cones whereof the pipes are portions.   For it
has been proved that the attractive forces of conical
pipes are as thofe quantities ; and therefore the
weights which they fupport muft be fo too. Hence
it follows that the lefs the proportion is, which the
diftance of the elevated liquor's furface from the
vertex of the cone bears to the diameter of the
fame furface, or which amounts to the fame thing,
the fafter the fides of the pipe converge, the ftronger is its attractive force, and the greater the quantity of liquor which is fupported.

The firm union and ftrong cohefion of the particles of folid bodies feems to arife from this force,
wherewith they mutually attract each other ; which
as it appears to be exceeding ftrong in the immediate contact of the particles, fo is it found by experience to reach but a very little way beyond the
fame with any fenfible effect. At very fmall diftances indeed it is fufficient to raife up liquors, as
alfo to produce the many odd and furprizing appearances which are to be met with in chymical operations, and which without the affiftance of this and

some other principles, which I shall hereafter have occasion to mention, are utterly inexplicable. For want of a due knowledge of these powers chymists have fallen into gross mistakes and absurdities in their reasonings. Thus for instance, some who were unacquainted with the principle of attraction, have attempted to give a reason for the floating of the minute particles of solid bodies in menstruums specifically lighter than themselves; by saying that there is an intestine motion in the parts of the menstruums, by virtue whereof the particles of the solid bodies are driven perpetually from place to place, and by that means are kept from falling: not considering that Sir ISAAC NEWTON has demonstrated in the nineteenth proposition of the second book of his principles, that fluids have not naturally any intestine motion; but that setting aside all external causes of motion, the particles of fluids are as perfectly at rest as those of solid bodies. There is indeed during the time of the solution a considerable motion, but as this is occasioned by the mutual attraction between the menstruum and the body, by means of which attraction the parts of the fluid are driven with great force between the parts of the solid, so as to loosen and divide them one from another; as soon as the solution is over the motion ceases, and all the parts are at rest again, and the particles of the dissolved body are kept suspended by their close adhesion to the parts of the menstruum, and not by any imaginary motion, wherewith they are tossed to and fro in the manner of a shuttle-cock; and in truth, could such an intestine motion be allowed, as it must be made in all manner of directions, it would be as apt, nay more apt considering the conspiring gravity of the particles, to precipitate and cast them down, than to raise and keep them up.

Were it not beside my present purpose, I could produce many more instances of false reasonings in the Writings of chymists, occasioned by their ig-

norance

'norance of the true principles of nature; but as chymiſtry is at preſent out of my province, I ſhall reſt contented with the ſingle inſtance which I have given.

LECTURE II.

Of Attraction.

HAVING in my former Lecture proved from experiments, that there is a power in nature whereby the parts of matter, which are brought ſo near as to touch, do in ſome circumſtances mutually attract each other: I ſhall now treat of ſuch kinds of attraction as extend themſelves to conſiderable diſtances beyond the point of contact, and on that account affect the mind more ſtrongly, ſo as to convince it more fully of the reality of ſuch a principle. Of this kind is, Firſt, that attraction which obtains between glaſs and glaſs. Secondly, that of electricity. Thirdly, the attraction of magnetiſm. And laſtly, that of gravity; of all which in their order.

And firſt, if a glaſs bubble be ſet to float on water contained in a glaſs veſſel, at a ſmall diſtance from the ſide of the veſſel, it will from a ſtate of reſt begin to move towards the ſide of the veſſel; and its motion will be continually accelerated, ſo as to make it upon its arrival at the ſide of the veſſel to ſtrike the ſame with ſome force.

Perhaps it may be thought that the motion of the bubble ariſes from ſome declivity in the water towards the ſides of the veſſel: but whoever obſerves the ſurface of the water will find, that it riſes all about the ſides of the glaſs, ſo as to become of a concave figure, and for that reaſon may retard, but can by no means promote the motion of the bubble; and this riſing of the liquor about the ſides of the veſſel is to be attributed to the ſame cauſe with the motion of the bubble, namely, the attraction of the glaſs.

The

The acceleration obfervable in the bubble's motion arifes from two caufes; the firft is, the continued and uninterrupted action of the attractive force
of the glafs; for if we fuppofe the time of the
bubble's motion to be divided into a number of equal
parts, as for inftance ten; and if the attraction of
the glafs be fuppofed to make equal impreffions on
the bubble in each of thofe parts of time, it is plain
that whatever be the motion which is excited in the
bubble by the impreffion of attraction in the firft
portion of time, the fame will be doubled in the
fecond, tripled in the third, and fo on continually
thro' the feveral portions of time; for the motion
produced in the firft portion of time is not loft, and
therefore by the addition of as much more in the
fecond portion of time it becomes double, and in the
third triple, and fo on. Now if inftead of ten parts
we fuppofe the time of the motion to be divided
into numberlefs parts indefinitely fmall, in each of
which the attraction of the glafs makes equal impreffions on the bubble, as before; the motion will
be continually accelerated, tho' the attractive force
of the glafs fhould continue the fame at all diftances
of the bubble; but the attractive force acts more
ftrongly the nearer the bubble approaches, on which
account the motion is more and more accelerated
the nearer the bubble comes to the glafs.

By electrical attraction, I mean that kind of attraction which is excited in bodies when their parts
are heated by friction, and which doth not difcover
itfelf by any fenfible effect when the bodies are
cold. Of this fort are the attractive forces, which
amber, rofin, fealing-wax, and indeed moft ful-
phurous fubftances when heated by rubbing, have
been found to exert towards chaff, feathers, leaf-gold,
lamp-black, and many other light fubftances. But
as the attraction of thefe bodies have fallen within
the notice of vulgar eyes, I think it needlefs to make
any experiment for the proof thereof; but choofe
rather

rather to lay before you some experiments which plainly shew this power to obtain in glass, and that to a very notable degree, tho' it has not till of late been commonly observed. And first,

If a cylindrical tube of flint glass be rubbed briskly with brown paper, or woollen-cloth till it acquires some degree of heat, and be then held near to small pieces of gold or brass leaf; they will begin to move, and some of them will fly towards the tube with great swiftness, and fix themselves upon it so as to adhere thereto, being acted upon by the attractive force of the glass: whilst others during their ascent towards the tube, will before they can reach the same, be driven backward with great violence, as will likewise some of those which touch the glass, being actuated by another force very different from that of attraction, which I shall endeavour to explain to you hereafter. The hotter the tube is made by rubbing, the farther doth its power reach, so as in some cases to act upon the leaf at the distance of a foot or more.

This electrical attraction of glass doth in like manner appear from the following experiments.

If over a globe of glass fixed on an axis, whose Exp. 2. position is horizontal, a parcel of woollen threads be suspended from a semicircular wire, so as that their lower ends may be distant an inch or a little more from the globe, they will suitably to the nature of all heavy bodies, hang down perpendicular to the horizon, and parallel to each other; if then the globe be moved pretty briskly round its axis, the threads will immediately change their position, so as to have their ends bent a little upward, pointing that way towards which the motion tends; the rotatory motion of the globe being communicated to the circumambient air wherein the threads hang, and by means thereof in some measure to the threads themselves. Let then an hand be applied to the lower part of the globe, so as to rub the same, and

as

as foon as it grows warm from the friction, the threads which were before crooked will dart them-felves out into fo many ftrait lines, all pointing to-wards the center of the globe; but as foon as the attrition ceafes, and the globe cools, they quit this direction, and return to their former pofition; whence it evidently appears that they are attracted by the glafs, fince they are made to point towards its center, notwithftanding the contrary directions that were given them by the motion of the air and the force of gravity. In this and the two following experiments there is one remarkable circumftance, which tho' it does not concern the matter in hand, yet becaufe I fhall have occafion to have recourfe to it hereafter, I fhall to prevent the repetition of experiments take notice of it here. And it is this; if while the threads are extended and acted upon by the attraction of the globe, a finger be moved to-wards the extremity of any of them, they will im-mediately recede and fly from the touch, and this they will do upon every approach of the finger.

Exp. 3.   If the axis of the globe inftead of being parallel to the horizon be placed perpendicular thereto, and the femicircular wire which fupports the threads be in the plane of a circle parallel to the horizon, the threads muft by reafon of their gravity hang down in lines parallel to the axis of the globe, yet as foon as the motion and attrition are given to the globe as before, the threads will begin to raife and ex-tend themfelves towards the center of the globe, and appear like fo many rays converging towards that center in a plane parallel to the horizon: fo that in this cafe the attractive force of the glafs does not only draw the threads out of the parallel pofi-tion they have to each other, but likewife raifes them up in a pofition parallel to the horizon, notwith-ftanding the force of gravity which is conftantly acting upon them to carry them down.

If

If the threads inftead of being placed without the globe, be fixed to the axis at the center, and be of fuch a length as to reach within about an inch of the furface; when the globe is turned round, they will bend backward contrary to the direction of the motion; becaufe the included air, tho' it does in fome meafure partake of the rotation of the globe, yet doth it not move with equal fwiftnefs, and for that reafon muft refift the rotation of the threads and bend them backward. When the threads are in this ftate, if the attraction of the glafs be excited by attrition as in the two laft experiments, they will ftraightway extend themfelves towards the concave furface of the globe conftituting as it were fo many rays iffuing from the center, and diverging from one another in a regular manner.

L e c t. II. Exp. 4.

The reafon why the threads in all thefe experiments are ftretched into lines tending either to or from the center of the globe, feems to be this. Whatever be the force wherewith the globe acts on the threads, the direction of it muft be perpendicular to the furface of the globe; confequently in the fame direction muft the threads move; but from the nature of the globe thofe and thofe lines only are perpendicular to its furface, which either iffue from or tend towards the central point.

Exp. 5.

Having faid thus much concerning electrical attraction I now proceed to that of magnetifm. Many and furprifing are the properties both of the loadftone and magnetical needle, which however I fhall not here confider; my intent at prefent being only to fhew from experiment the law of magnetical attraction; or in other words, to fhew in what proportion the attractive power of the loadftone varies according to the different diftances of the iron which it attracts. And in order to this, let a loadftone be fufpended at one end of a balance, and counterpoifed by weights at the other; let a flat piece of iron be placed beneath it at the diftance of four

Exp. 6.

tenth

tenth parts of an inch, the stone will immediately descend, and adhere to the iron : let the stone again be removed to the same distance, and a weight of four grains and four tenth parts of a grain be thrown into the scale at the other end of the balance ; this weight will be an exact counterbalance to the attractive force, and prevent the descent of the stone ; but if any part of the weight be taken out, the attraction will prevail, and carry the stone down. If the stone be placed at half the former distance, that is to say, at the distance of two tenth parts of an inch above the iron, the weight necessary to hinder its descent will be about seventeen grains and an half, that is four times as much as before. Consequently, the attractive force of the stone at the single distance from the iron, is to the same at the double distance as four to one, that is reciprocally as the squares of the distances.

Perhaps it may be objected that Sir Isaac Newton (to whose judgment in natural affairs the utmost regard is due) has said that the power of the loadstone decreases nearly in the triplicate ratio of the increase of the distance. But whoever considers his words in the fifth corollary of the sixth proposition of the third book of his principles, where he mentions this law, will find that he speaks of it with diffidence, as a thing which he rather guessed at from some rude observations, than collected from accurate experiments, for his words are, *Et in recessu a magnete decrescit in ratione distantiæ non duplicatâ, sed fere triplicatâ, quantum ex crassis quibusdam observationibus animadvertere potui.* So that notwithstanding this objection I shall still venture to affirm the law of magnetical attraction to be such as makes it act with forces which are in the reciprocal duplicate ratio of the distance. Because this law is deduced from an experiment made with sufficient exactness, and which does not seem liable to any exception.

Tho'

, Tho' the principle of gravity, which comes next to be treated of, be diffused throughout the solar system, and may probably be extended so far as to reach the other systems of the universe; yet shall I consider it at present with respect only to the globe of earth, which we inhabit; the parts whereof would by reason of the diurnal rotation be apt to fly asunder, were they not kept together by the influence of this principle; whereby likewise all bodies on or near the surface of the earth are made to tend towards its center. This power at equal distances from the center of the earth is always proportional to the quantity of matter in the body whereon it acts; for all bodies, the light as well as heavy, being let fall from the same height descend with equal swiftness, provided they meet with no resistance from the air, as will appear from the following experiment. Let a piece of gold and a feather be let fall from the top of an exhausted receiver at the same instant of time, and they will both arrive at the bottom at the same time very nearly.

Exp. 7.

The reason why the feather doth not reach the bottom quite so soon as the gold, is, that the receiver cannot be perfectly exhausted, and therefore the small portion of air which remains within, though very much rarified, gives some small resistance to the descending bodies, which suitably to the nature of all resistance must retard the lighter body more than the heavier, and consequently cause some little difference in the times of the descent, which otherwise would be exactly equal. This then being the case, it evidently follows, that the forces of gravity, whereby bodies descend, must at equal distances from the center be as the quantities of matter in the descending bodies; for if a certain force of gravity be requisite to carry down a certain quantity of matter with a certain swiftness, then is double the force necessary to carry down a double quantity of matter with the

same

ſame ſwiftneſs; and triple the force to carry down
a triple quantity, and ſo in proportion whatever
be the quantity of matter: ſo that the weights of
bodies at equal diſtances from the center of the
earth are always proportional to the quantities of
matter which they contain; and therefore the
quantity of matter in any body may be meaſured
by its weight.

The gravity of a body at any place beneath the
ſurface of the earth has been proved by Sir Isᴀᴀᴄ
Nᴇᴡᴛᴏɴ, to be directly as the diſtance from the
center; that is, ſuppoſing the earth's radius to be
four thouſand miles, a body which on the ſurface of
the earth weighs a pound, will within the earth at
the diſtance of two thouſand miles from the center
weigh only half a pound, at the diſtance of one
thouſand miles only a quarter, and ſo on till at the
center it loſes all its gravity.

It has been likewiſe proved that the force of gravity on the ſurface of the earth, and at all diſtances
beyond it, is in the reciprocal duplicate ratio of
the diſtance from the center; that is, if a body
weighs a pound at the ſurface of the earth, whoſe
diſtance from the center is four thouſand miles,
it will at double that diſtance weigh only a quarter of a pound, and at triple the diſtance, only
the ninth part of a pound, and ſo on, whatever
be the diſtance the force of gravity will be reciprocally as the ſquare of the diſtance. For is it
not highly rational that the power of gravity, whatever it be, ſhould exert itſelf more vigorouſly in a
ſmall ſphere, and weaker in a greater, in proportion as it is contracted or expanded? and if
ſo, ſeeing that the ſurfaces of ſpheres are as the
ſquares of their *radii*, this power at ſeveral diſtances
muſt be as the ſquares of thoſe diſtances reciprocally. Tho' ſtrictly ſpeaking, this be the law of
gravity, yet where the diſtances from the ſurface
are inconſiderable with reſpect to the earth's radius

the

the force of gravity may be looked upon as equal at all those distances; thus for instance, the gravity of a body at the distance of half a mile from the earth may be looked upon as equal to the gravity thereof at the distance of a quarter of a mile; or at the very surface; because the difference is so small, that if it be rejected it will not occasion any error in calculations. And indeed on this supposition are founded most of the reasonings of GAL-LILÆO, TORRICELLIUS, HUYGENS, and other naturalists concerning the descent of heavy bodies; and by the help of the same supposition have the several theorems been formed relating to the acceleration of falling bodies, the spaces described, the times of the fall, and the velocities thereby acquired; as I shall now shew you.

If the force of gravity whereby a body descends remains unvaried, the motion of a body falling by such a force will be accelerated, and that uniformly; that is, the velocity will increase, and the increments thereof in equal times will be equal. For let us suppose the time of the descent to be divided into a number of equal parts indefinitely small, in each of which by supposition, the force of gravity makes equal impressions on the body to carry it down; whatever velocity therefore the body receives from the impression of gravity in the first portion of time, it must receive as much in every other portion; since therefore setting aside all outward lets and obstacles the effect of every impression remains, the velocity given in the first portion of time, will be doubled in the second, tripled in the third, quadrupled in the fourth, and so on continually thro' the several portions of time. So that the velocity of a body falling by the force of gravity will constantly increase in the same proportion with the time of the descent. Or in other words, the motion of a body carried down by the force of gravity will be uniformly accelerated: and the ve-

C
locities

locities acquired will be as the times of the defcent from the beginning of the fall.

From what has been faid it follows, that if a
Fig. 3.
right line as A B be fuppofed to denote the time of a body's fall, and another right line as BC fet at right angles to the former, to exprefs the velocity acquired by the falling body in the time denoted by A B. The triangle ABC being compleated, and another right line as DE drawn parallel to BC, then will DE denote the velocity acquired by the falling body in a portion of time, which is to the time denoted by A B, as A D to A B. For from the nature of fimilar triangles, A B is to A D as BC to DE; but BC expreffes the velocity acquired where the time is as A B, confequently, fince the velocities are as the times of the defcent, DE will exprefs the velocity acquired in the time denoted by A D.

And what has been thus proved of the line DE, is in like manner true of any other right line, as F G, or H I, drawn within the triangle parallel to the bafe; for F G and H I will exprefs the velocities acquired in the times denoted by A F and A H.

The fpaces defcribed by bodies falling from a ftate of reft by the force of gravity are to one another as the fquares of the times from the beginning
Fig. 4.
of the fall. In the triangle ABC, let AB exprefs the time of a body's fall, and BC the velocity acquired at the end of the fall, let AB be divided into a number of equal parts indefinitely fmall; and from each of thofe divifions fuppofe lines, as DE drawn parallel to BC; it is evident from what has been faid, that thofe lines will exprefs the velocities of the falling body in the feveral refpective points of time; which velocities, inafmuch as the body is given and the portions of time are indefinitely fmall, will be as the refpective fpaces defcribed in thofe times: but the fum of the fpaces defcribed in all the fmall portions of time is equal

5   to

to the fpace defcribed from the beginning of the
fall; and the fum of all the lines, as DE taken indefinitely near each other conftitute the area of the
triangle. And therefore the fpace defcribed by a
falling body in the time expreffed by AB, and where
the velocity acquired at the end of the fall is denoted
by BC, will be as the area of the triangle ABC.
And for the fame reafon the fpace defcribed by a
falling body in the time expreffed by AD will be
as the area of the triangle ADE. But from the
nature of fimilar triangles thefe areas are to one another as the fquares of their homologous fides; that
is, as AB^q to AD^q, or as BC^q to DE^q. But AB
and AD exprefs the times of the fall, and BC and
DE the velocities acquired by the fall; wherefore the fpaces defcribed by a falling body are
as the fquares of the times from the beginning of
the fall, or as the fquares of the velocities at the
end of the fall. And what has been thus demon-
ftrated from the nature of gravity is likewife confirmed by experiments. For if a weight of eleven
hundred grains be let fall from the height of three
inches, fo as to ftrike one end of a balance, its
force will be juft fufficient to raife a pound weight
at the other end of the balance to the height of
about the eighth or tenth part of an inch; whereas
if the fame body be required to raife a weight of
two pounds to the fame height, it muft be let fall
from the height of twelve inches; and if the weight
to be raifed, be three pounds, then muft the moving
body fall from the height of twenty feven inches,
for leffer heights will not fuffice, as will appear from
the experiment.

The forces wherewith the defcending body ftrikes
the end of the balance are meafured by the weights
that are raifed; which in this cafe are as one, two,
and three; but the forces wherewith one and the
fame body ftrikes, are as the velocities of the body;
wherefore in the cafe before us the velocities acquired

by

by the falling body are as one, two, and three; but
the heights from which it defcends in order to acquire thofe velocities are as one, four, and nine;·
that is, as the fquares of the velocities.

Exp. 9.   If this experiment be repeated with a body
double in weight to the former, to wit, with one
of twenty two hundred grains; the weights raifed
by the ftrokes will be two, four, and fix pounds,
to wit, double the former.

From this experiment appears the truth of that·
rule, which collects the quantity of motion in any
body by multiplying the velocity of the body into
its quantity of matter. For the force of a ftroke
is, *cæteris paribus*, always proportional to the quan-·
tity of motion in the ftriking body; confequently
in like circumftances the motions of bodies may be
meafured by the force of their ftrokes; but it has
appeared from the experiment that where the ftriking body is as unity, and the velocities wherewith
it moves at the times of the ftrokes, as one, two,
and three; the forces of the refpective ftrokes are
likewife as one, two and three. But where the
body is as two, the ftrokes are as two, four and fix:
that is, in both cafes the ftrokes are as the products
arifing from the multiplication of the quantities of
matter in each body into the refpective velocities;
wherefore the quantities of motion are as thofe
products. Whence as a corollary it follows, that if
the weight of one body multiplied into its velocity
gives an equal product to what arifes from the
multiplication of the weight of another body by
its velocity, the motions of thofe two bodies are
equal; and this will ever be where the weights of
the bodies are reciprocally proportional to their velocities. Thus when the body whofe weight was as
unity, was let fall from the height of twelve inches,
and thereby acquired a velocity which was as two; it
raifed a two pound weight, which was likewife raifed.
by the body whofe weight was as two, when by fall-·

ing

ing from the height of three inches, it had acquired a velocity which was as unity.

From what has been proved concerning the spaces
described by falling bodies it follows, that if the
time of a body's fall be divided into a number of
equal parts, the spaces thro' which it falls in each
of those parts of time taken separately and in
their order, beginning from the first, are as the
odd numbers taken likewise in their order, beginning from unity. For instance, if the time of the
fall be four seconds, the space described in the first
of those seconds will be as one, in the second as
three, in the third as five, and in the fourth as seven; for where the times of the fall are as one,
two, three and four; the spaces described are as
one, four, nine and sixteen; and therefore if from
the space described in two seconds, to wit, four, be
subducted the space described in the first second, to
wit, one, the remainder, to wit, three, will be the
space described in the next second. And if from
nine, which is the space described in three seconds,
be taken four, which is the space described in two
seconds, the remainder, which is five, will be the
space described in the third second. In like manner
subducting nine, the space described in three seconds,
from sixteen, which is the space described in four
seconds, the remainder, to wit, seven, will be the
space described in the fourth second; and so on according to the number of parts into which the time
of the fall is divided.

From what has been said it likewise follows, that
the velocity acquired by a falling body at the end of
the fall is such as with an equable motion would in
the same time in which the body fell, carry it thro'
a space double that of the fall. That the truth of this
may be made appear, it is necessary that some things
be premised concerning the spaces described by bodies carried with an equable motion. And first, if
the velocity of a body moving uniformly be given,

C 3

the

the space described will be as the time of the motion; for if a body with a given velocity moves
thro' a certain space a foot, for instance, in a second
of time, it will in two seconds, with the same velocity, move thro' two feet, and thro' three feet in
three seconds, and so on, whatever be the time,
the space described will be proportional thereto.
On the other hand, if the time be given, the space
described will be as the velocity; for if a body in
a given time moves thro' the space of a foot with
a certain velocity, with double the velocity it will
pass thro' the space of two feet, and with triple the
velocity thro' the space of three feet, and so on,
whatever be the velocity, the space described will
be in the same proportion.   But if neither the time
of a body's motion, nor the velocity wherewith it
moves be given, the space described will be as the
time and velocity conjointly; for if a body moving
with a certain velocity runs thro' a certain space in a
certain time, it follows from what has been said, that
if the time be increased or diminished in any proportion, in the same also will the space be increased
or diminished, supposing the velocity to remain the
same, but if that likewise be changed, it is plain
that the space will be changed in the same proportion; and therefore universally the space described
by a body moving equably is as the time and velocity conjointly.   For which reason, if in the rec-
Fig. 5. tangle, one side, as A B, be supposed to denote the
time wherein a body moves equably, and B C the
velocity wherewith it moves, the rectangle A B C D
will be as the space described; but the triangle
A B C of the same figure, is as the space described
by a falling body in the time denoted by A B, and
B C is as the velocity acquired at the end of the
fall; and the rectangle A B C D is double the triangle A B C, consequently the velocity acquired by
a falling body is such as will carry the body with an
equable motion in the time of the fall thro' double
the space of the fall.
                                             As

As the motion of bodies falling from a state of
reſt is uniformly accelerated; ſo likewiſe the motion
of bodies thrown upward is uniformly retarded;
for the ſame force of gravity which conſpires with
the motion of deſcending bodies, acts in direct oppoſition to the motion of ſuch as aſcend; and
therefore in whatever manner it accelerates the one,
in the very ſame manner muſt it retard the other.
Whence it follows, that if a body be thrown directly upward, the time of its riſe will be equal to
that wherein a body falling freely from a ſtate of
reſt, acquires the ſame velocity wherewith the body
is thrown up. For ſince the action of gravity is
conſtant and uniform in whatever time it generates
any velocity in a falling body, in the ſame time muſt
it deſtroy that velocity in a riſing body; and therefore the time of the riſe muſt be equal to that of
the fall. It likewiſe follows that the height to which
a body thrown upward riſes is equal to that from
which a body falling freely does at the end of the
fall acquire a velocity equal to that wherewith the
body is thrown up. For ſince the times in which
the velocity of the falling body is generated, and
that of the riſing body is deſtroyed, are equal;
and ſince of the two equal velocities one is generated and the other deſtroyed by the conſtant uniform action of one and the ſame power; it is manifeſt, that whatever be the ſpace thro' which the
falling body moves in order to acquire its velocity,
the riſing body muſt aſcend thro' an equal ſpace in
order to loſe its velocity; that is, it muſt riſe to the
ſame height from which the other falls.

The force of gravity at the ſurface of the earth is
ſuch as, ſetting aſide the reſiſtance of the air, makes
a body falling from a ſtate of reſt to deſcend thro'
a ſpace of ſixteen feet and an inch in a ſecond of
time. For the time wherein a pendulum performs
its ſmalleſt vibrations is to the time in which a body
falls thro' half the length of the pendulum as the

circum-

circumference of a circle to its diameter (as shall be shewn when I come to treat of. the pendulum) wherefore since the spaces described by falling bodies are as the squares of the times, and since the diameter of a circle expresses the time which a body takes to fall thro' half the length of a pendulum vibrating seconds, when the circumference expresses a second ; it follows, that as the square of the diameter is to the square of the circumference, so is half the length of the pendulum to the space thro' which a body falls in a sec^{ond} of time. So that putting D to denote the diameter of a circle, which is as unity, P the periphery which is as 3,1416, L the length of the pendulum vibrating seconds, which is 39¼ inches, and S to denote the space sought; we shall have this analogy, $D^2 : P^2 : : \frac{L}{2} : S$.

Consequently $S = \frac{P^2 \frac{1}{2} L}{D^2}$, or rejecting the divisor as being equal to unity, $S = P^2 \frac{1}{2} L = 193$ inches, or sixteen feet and an inch.

Before I quit this subject I must observe to you, that bodies do not every where descend at the rate of sixteen feet and an inch in a second of time, but in such places only as are in or near the latitude of forty nine degrees ; in places more distant from the line the descent is quicker, and more slow in those less distant. For the force of gravity is less towards the æquator than towards the poles, as has been collected from observations made on pendulums ; for they have been found to vibrate more slowly near the line than in places farther removed ; insomuch that a pendulum which in the latitude of Paris vibrates seconds, must be shortened one sixth of an inch French measure in order to its vibrating seconds under the line. And the length of a pendulum which in the latitude of Paris performs its vibrations in a second, is to the length of a pendulum whose vibrations are performed in the same

time

time under the line as 220 to 219. Since therefore the force of gravity which actuate pendulums that vibrate in equal times are to one another as the lengths of the pendulums (as shall be shewn when I come to treat of pendulums) it is evident that the force of gravity in the latitude of Paris is to the same force under the line as 220 to 219. And indeed it has appeared from a great number of observations, that the force of gravity is least at the æquator, and that it continually increases as we recede from thence and approach the poles, under which it is greatest of all. And the chief cause of this difference is the rotation of the earth about its axis, whereby all bodies on or near the surface of the earth are indued with a centrifugal force, which acts in opposition to that of gravity, and of course must lessen the same; and the diminution of gravity arising from this cause must be greatest under the æquator, and grow less and less in the approach to the poles: and that for two reasons, first, because the centrifugal force is greatest at the æqua-tor, and from thence towards the poles is continually diminished so as at last to vanish in the polar points. For all parts of the earth's surface with the bodies thereto adjacent revolve in the same time either in the æquator or in circles parallel thereto; but the æquator is the largest of all those circles, and the others grow less and less as they are more and more distant from the æquator. Now the centrifugal forces of bodies revolving in the same time in different circles being to one another as the *radii* of the circles (as shall be shewn when I come to treat of those forces) it follows that the centrifugal force must be greatest at the æquator, and thence be continually diminished towards the poles. To illustrate this, let AB be the axis of the earth, Fig. 6. CK the radius of the æquator, DI, EH and FG the radii of so many circles parallel to the æquator, the centrifugal forces in the points K, I, H, G, are

as

as thofe radii; fo that the centrifugal force is greateft in the point K, that is at the æquator, and at I'
it is lefs than at K, and at H lefs than at I, and lefs
again at G, and fo on till at length it vanifhes at
the polar point where there is no rotation. Whence
it is evident that the force of gravity muft be
fmalleft under the line, and muft increafe towards
the poles, inafmuch as the force which acts in oppofition to it is greateft under the line, and leffens
in the approach to the poles. The force of gravity
muft likewife be lefs under the æquator than in
any other place, becaufe under the line the centrifugal force acts in direct oppofition to the force of
gravity, whereas in other places it acts in an oblique
direction to that of gravity, and of confequence
muft act lefs powerfully againft it. Thus in the
point K the force of gravity pulleth from K towards C, whilft the centrifugal force pulleth directly contrary from C towards K; whereas in the
point L gravity pulleth from L towards C, whilft
the direction of the centrifugal force is from O towards L. Let the centrifugal force in the point L
be expreffed by the line LM, and to CL continued to N let fall the perpendicular MN. The force
LM, according to the known method of refolving
forces, of which I fhall fpeak hereafter, may be
refolved into two forces denoted by the lines NM,
and LN; whereof the latter only acts in oppofition to gravity, as pulling directly againft it; the
other no way affecting the fame: confequently, fuppofing the centrifugal force at L to be the fame as
at K, yet will the force of gravity be lefs diminifhed by it at L than at K, becaufe at L part only of
the centrifugal force refifts that of gravity, whereas at K the whole centrifugal force acts in oppofition thereto.

From what has been faid it follows, that the force
whereby gravity is leffened in the æquator is to the
force whereby it is leffened in any other part of the
earth's

earth's furface as the fquare of radius to the fquare of the fine of the compliment of latitude. For the centrifugal force in the point K, the whole of which acts in oppofition to gravity, is to the centrifugal force in the point L, as CK or CL to OL; but the whole centrifugal force in L is to that part of it which oppofes gravity, as LM to LN, that is, becaufe the triangles LNM and COL are fimilar, as CL to OL; wherefore the centrifugal force, or the force which oppofes gravity in the point K, is to that part of the centrifugal force which oppofes gravity in the point L in the duplicate ratio of CL to OL, that is, as the fquare of radius to the fquare of the fine of the compliment of latitude.

L E C T U R E III.

Of Repulsion and Central Forces.

AS experience has convinced us that there are Powers in nature, whereby not only the larger fyftems and collections, but likewife the fmaller parcels and particles of matter are in fome cafes made to tend to one another; the fame experience will inform us of other powers in nature, whereby the parts of matter do in fome circumftances recede and fly from each other. For if the difagreeing pole of a loadftone be moved towards a magnetical needle floating on water, the needle will recede; and the nearer the ftone is brought to it, with the greater violence and precipitation will it fly off; the repelling power, like the attractive, exerting itfelf with greater vigor at fmaller diftances.

This repelling power is likewife evident from the experiments which were made relating to electrical attraction: for it was obfervable that upon holding the glafs tube, when heated by friction, nigh fmall pieces of brafs-leaf; fome of thofe pieces which by
the

Lect. III.

Exp. 1.

LECT.
III.
the attraction had been raised towards the tube, were, before they could reach it, driven back again with great precipitation: and of those which adhered to the tube some were thrown off with a velocity much greater than could possibly arise from the force of gravity in such light bodies, and consequently must have been driven down by some repelling power in the glass. And in the experiments of the glass-globe and woollen threads; when the threads were, by the attractive force of the globe, made to extend themselves towards its surface, upon moving one's finger towards them, they were observed to recede and fly off, and that at considerable distances from the finger; which plainly argues a repelling power interceding the finger and the threads, when under the circumstances of those experiments. From this power it is, that the leaves of the sensitive plant shrink and retire from the touch of an approaching hand. And to the same power we are to attribute the elasticity of the air; as also the shaking off of the particles of light from the sun and other luminous bodies.

Besides the forementioned principles of attraction and repulsion, whereby nature seems to perform most of her operations, and which for that reason are very properly stiled active principles; there is another of a passive nature, commonly called the *vis insita* and *vis inertiæ* of matter, a force arising from the inertness or inactivity of matter; which force in any body is proportional to its quantity of matter. From this force result three passive laws of motion, usually called by modern naturalists the three LAWS OF NATURE*.

The

---

* By virtue of the *vis inertiæ* it is, that the motion of a body produced by a force impressed upon it, is measured by the quantity of matter in the body and its velocity, taken together. For the body by its *vis inertiæ*, resists the force impressed upon it which causes its motion, in proportion to its quantity of matter; and consequently, to produce a given tendency in the body
forward,

The firſt of theſe laws is, That every body, in proportion to its quantity of matter, perſeveres in its preſent ſtate, whether it be of reſt or uniform motion ſtraight forward in a right line. For as every particle of matter is with reſpect to itſelf perfectly unactive, it is utterly impoſſible it ſhould produce any alteration in its own ſtate; for which reaſon (ſetting aſide all impreſſions from external cauſes) if it be at reſt, it muſt continue ſo for ever; or if in motion, it muſt for ever continue its motion, without any change either as to direction or velocity: ſo then the continuation of motion in bodies projected, (the cauſe whereof very much perplexed the naturaliſts of old) is to be attributed to the paſſive nature of matter, which makes it as impoſſible for a body of itſelf to ſtop its own motion when once begun, as it is for it to move itſelf originally, or of itſelf to change its figure.

As a conſequence of this law it follows, that all motion is of itſelf equable and rectilineal. For firſt whatever be the velocity wherewith a body begins to move, the ſame velocity muſt continue during the motion, unleſs a change be made therein by ſome cauſe from without; wherefore the body

forward, by which it moves at a given rate or with a given velocity, the force impreſſed muſt be proportional to the reſiſtance ariſing from its *vis inertiæ*, that is, to its quantity of matter; and if the quantity of matter in the body, and conſequently the reſiſtance ariſing from its *vis inertiæ*, be given, the force impreſſed will be proportional to the tendency forward which it communicates to the body, that is to its velocity; and if neither the quantity of matter in the body nor its velocity be given, the force impreſſed will be in a ratio compounded of the quantity of matter and velocity; that is, putting F for the force impreſſed, Q for the quantity of matter in the body, and V for its velocity, F will be as $Q \times V$. But the motion of the body is the effect produced by the force F, and is proportional to it, that is, putting M for the motion of the body, M is as F. And therefore, by proportion of equality, M will be as $Q \times V$; or the motion of the body will be meaſured by its quantity of matter and velocity taken together.

muſt

muſt in equal times move thro' equal ſpaces with an uniform velocity; that is, the motion muſt be equable. And as motion is by virtue of this law in itſelf equable; ſo is it likewiſe rectilineal: for motion cannot otherwiſe be conceived than as directed and determined towards ſome place or other; and it muſt by the foregoing law keep the direction which it had at firſt, until it be hindered or put out of its way by ſome extrinſic cauſe, that is, it muſt move on in a right line. If therefore a body moves in a curve, that curvature muſt of neceſſity proceed from ſome external force continually acting on the body; and whenever that force ceaſes to act, the body will move forward in a right line touching the curve in that point wherein the body is at the inſtant of time when the force ceaſes to act. Thus for inſtance, if a ſtone, moved about in a ſling, be ſet at liberty by ſlipping one end of the ſling; it will not continue its circular motion, but go on in a right line touching the circle made by the circumvolution of the ſling in that point where the
Fig. 8.　ſtone is let go. If the circle BCDE be the curve deſcribed by the revolution of the ſling AB about the center A; and if the ſtone be let off at the point B, it will move on in the right line BG, which touches the circle in B. For by the law, the natural tendency of the ſtone in the point B is along the line BG, tho' by the force of the ſling it be made to revolve in the curve: And what has been ſaid of the ſtone in the point B, is in like manner true of the ſame at any other point, as C, D, or E; for in thoſe points its tendency is along the lines CF, DH, and EK.

　　Another conſequence of the foregoing law is that all bodies, which revolve about a center, muſt endeavour to recede from the center: for ſince bodies, that are moved round in a curve, do of themſelves in every point of the curve tend to move in the tangents to each point; and ſince all the parts of

the

the tangents are more diftant from the center of motion than are the parts of the curve, as is evident from the figure; it is manifeft that bodies fo moved muft perpetually endeavour to fly off from the center of motion, which endeavour of receding is commonly called the centrifugal force; and it is oppofed to the centripetal force, or that force which by drawing the bodies towards the center makes them to revolve in a curve.

Thefe two forces are by one common name called the central forces: and they are in all cafes equal the one to the other. For let us fuppofe a body to revolve in the orbit EAC, and that being in the point A, the centripetal force ceafes to act; it will then move forward in the direction of the tangent AB, and BC will be the fpace thro' which the body recedes from the orbit by means of the centrifugal force; and if AB be in its nafcent ftate, the centrifugal force will be as BC; but if the centripetal force acts at A, it will make the body defcribe the arc AC in the fame time that it would defcribe the tangent AB, in cafe it were not acted upon by the centripetal force; confequently, the fpace BC is defcribed by means of the centripetal force; and the arc AC being in its nafcent ftate, the centripetal force will be as BC, and of confequence equal to the centrifugal. Fig. 9.

In treating of thefe central forces I fhall proceed in the following manner. Firft, I fhall confider two equal bodies moving uniformly in two different circles; and thence deduce one general expreffion for the central forces in the terms of the circle. Secondly, By fubftituting other proportional quantities in the place of thofe which conftitute the general expreffion, I fhall form other general expreffions for the fame forces. Thirdly, by a proper application of thofe expreffions I fhall determine the laws of central forces in particular cafes,

and

and at the fame time confirm each law by an experiment.

Fig. 10.
11.
As to the firft, if two equal bodies movihg uniformly in the circles marked 1, and 2, do in the fame
portion of time taken indefinitely fmall defcribe the
nafcent arches AC; and if from the points C be
drawn the lines CB, perpendicular to the tangents
AB, thofe lines will exprefs the proportion of the
central forces. For fince the time in which the
arches AC are defcribed is indefinitely fmall, the
bodies will be carried thro' the fpaces BC, by one
fingle impulfe of each central force; for which reafon the motions of the bodies thro' thofe fpaces
will be uniform; confequently, fince the time of
the motion is the fame, and the bodies equal, the
motions will be as the fpaces defcribed, that is, as
the lines BC; but forces which generate equable
motions are to one another as the motions generated; that is, in this cafe, as the lines BC; which
lines being equal to the verfed fines AD of the
arches AC, muft be equal to the fquares of the
arches AC, divided by their refpective diameters
AE. For from the nature of the circle, the verfed
fine of any arch is equal to the fquare of the chord
divided by the diameter; but as in this cafe the
arches AC are fuppofed to be nafcent, they do not
differ from their chords; and therefore in each
circle the verfed fine of the arch AC, (which
verfed fine expreffes the central force) is equal to the
fquare of the arch divided by the diameter: confequently, the central forces are as the fquares of the
nafcent arches applied to their refpective diameters;
and forafmuch as thofe nafcent arches, are to one
another as any other two arches which are defcribed
by the revolving bodies in a given time, the central
forces of two equal bodies revolving uniformly in
different circles, are to one another as the fquares
of the arches defcribed in a given time applied to
their

their refpective diameters; or becaufe the diameters are as the radii, as the fquares of the arches applied to their refpective radii.   Wherefore putting A to denote the arch of a circle defcribed in a given time, D for the radius, and F for the central force. F is as $\frac{A^2}{D}$, as it ftands in the firft place of the firft rank of fymbols.

F is as $\frac{A^2}{D}$.

F is as $\frac{V^2}{D}$,

F is as $\frac{D}{P^2}$.

F is as DN^2.

F is as $\frac{QA^2}{D}$.

F is as $\frac{QV^2}{D}$.

F is as $\frac{QD}{P^2}$.

F is as QDN^2.

As the bodies are fuppofed to move uniformly in the circles, it is evident that the arches defcribed in a given time are as the velocities of the revolving bodies; and therefore in the general expreffion for the central force, the velocity of the body may be fubftituted in the place of the circular arch; whence puttng V for the velocity of the body, F is as $\frac{V^2}{D}$, as in the fecond place of the firft rank of fymbols, which is a fecond general expreffion for the central force.

Again, the velocity of a body moving uniformly in a circle, is as the radius applied to the periodic time, or the time of one intire revolution.   For if the velocity of the body be given, the periodic time muft be proportional to the circumference of the circle, inafmuch as a body, which with a given ve-locity defcribes a certain fpace in a certain time, will require a double or triple time to defcribe a double or triple fpace; and univerfally whatever be the magnitude of the fpace, the time in which it is

defcribed

deſcribed will be proportional to it. If the circum-ference of the circle be given, the periodic time will be inverſly as the velocity with which the body moves; for if a body moves thro' a given ſpace with a certain velocity in a certain time, it will with double the velocity move through the ſame ſpace in half the time, and with a triple velocity in one third of the time; and in general, in whatever proportion the velocity is increaſed, in the ſame proportion will the time be leſſened; that is, the periodic time will be inverſly as the velocity. If therefore neither the circumference of the circle, nor the velocity of the body be given, the periodic time will be directly as the circumference, and in-verſly as the velocity; that is, as the circumference applied to the velocity; or (becauſe the circumfe-rence is as the radius) as the radius applied to the velocity. Wherefore putting P for the periodic time of a body revolving in a circle, P is as $\frac{D}{V}$, and conſequently V is as $\frac{D}{P}$. If therefore in the ſecond general expreſſion $\frac{D}{P}$ be ſubſtituted in the place of V, we ſhall have a third general expreſſion for the central force, wherein F is as $\frac{D}{P^2}$, as in the third place of the firſt rank of ſymbols.

Again the periodic time of a body revolving uniformly is inverſly as the number of revolutions performed in a given time. For if the periodic time of a body be ſuch, as that in a given time it can perform a certain number of revolutions; if the periodic time thereof be doubled, it will perform but half the number of revolutions in the ſame time; and if the periodic time becomes thrice as great, it will perform but one third of the number of re-volutions in the given time; and ſo on, as the pe-riodic time is inlarged the number of revolutions
will

Plate 1.

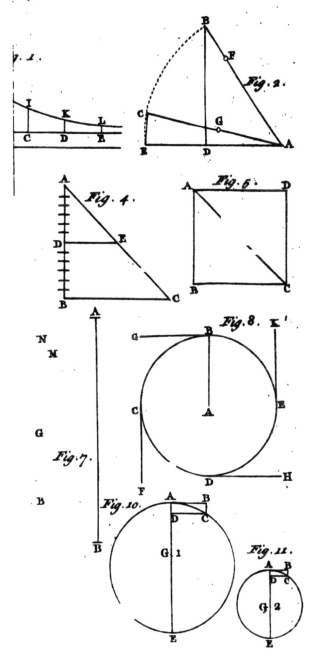

Fig. 1.

Fig. 2.

Fig. 4.

Fig. 6.

Fig. 7.

Fig. 8.

Fig. 10.

Fig. 11.

will be diminished in the same proportion, so that putting N for the number of revolutions in a given time, P will be as $\frac{1}{N}$. Consequently, if in the third general expression $\frac{1}{N}$ be substituted in the room of P, we shall have a fourth general expression for the central force, wherein F is as DN^2, as it stands in the last place of the first rank of symbols.

In collecting these general expressions, I have all along supposed the quantity of matter in the revolving body to be given; and for that reason have not made it a part of those expressions, inasmuch as it may be denoted by unity; and as such, whether it be taken in, or left out, it will not vary the expressions. But the case will be different, if the quantity of matter in the revolving body varies; because the central forces, and consequently the expressions for those forces will likewise vary; so as to be greater *cæteris paribus* in larger quantities of matter than in smaller. For the whole central force of any body, is made up of the forces of each particle whereof the body consists; and therefore the more numerous the particles of matter are in any body, the greater will its central force be; so as to be double in a double quantity of matter, triple in a triple quantity; and so on in proportion to the quantity of matter. In order therefore to render the expressions yet more general, let Q be put for the quantity of matter in the revolving body, and let it be multiplied into each of the four expressions, as in the second rank of symbols.

Before I apply these expressions to the several particular cases, I shall offer an experiment in confirmation of what I just now proved, *viz.* that the greater the quantity of matter in any body is, the greater is the central force.

Let

Exp. 1.

Let three glafs tubes half full, one with mercury and water, and another with water and fmall leaden bullets, the third with water and a piece of cork, be ftopped clofe, and made faft to an inclined plane; and let the plane be fo fixed to a table moveable about its center by means of a wheel and axle, as that the lowermoft part of the plane may reft upon the center of the table. As long as the table continues at reft, the liquors and folids contained in the tubes, will by reafon of their gravity poffefs themfelves of thofe parts of the tubes which lie next the center of the table, leaving the remoter parts empty: and of the two bodies included in each tube, that which is heavieft will be neareft the center; but upon turning the table about, the feveral bodies will by reafon of their centrifugal forces, whereby they are carried from the center of motion, fly to the uppermoft parts of the tubes; and in each tube, the heavier body will poffefs the uppermoft place as being indued with the ftronger centrifugal force.

If bodies moving in equal circles perform their revolutions in equal times, or in other words, if the velocities of bodies revolving in circles be equal, and their diftances from the center likewife equal, their centrifugal forces are as their quantities of matter. For in the fecond general expreffion, fince V and D are given, F is as Q; that is, the central force is as the quantity of matter; which is con-

Exp. 2. firmed by the following experiment. Let two fmall troughs be fo fixed to two moveable tables, as that the centers of the troughs may lie upon the centers of the tables, and let the centers of the tables be fixed to two axles, on each of which is a grooved wheel, with equal diameters; let the two wheels be turned by means of one and the fame chord going round them: it is manifeft, that as the wheels are equal, they, and confequently the tables with their affixed troughs, muft perform their revolutions in the fame time; and the parts of the

tables

tables and troughs, whose diſtances from their reſpective centers are equal, will revolve equally ſwift; and ſo likewiſe muſt all bodies that are placed in the troughs at equal diſtances from the centers; ſo that by this contrivance, if two bodies be placed one in each trough at equal diſtances from the centers, they will revolve equally ſwift. Let then two balls, whereof one is double the other, be laid one in each trough, and let each ball be faſtened to one end of a chord, whoſe other end paſſing thro' an hole in the center of the table is made faſt to a weight, which reſts upon the floor; and let the lengths of the chords be ſuch, as that being ſtretched, and the weights not raiſed, the balls in the troughs may be equally diſtant from the centers. This being done if the weights be to one another as the balls, and if the tables be turned about with ſuch a velocity as that the centrifugal forces of the balls may be ſufficient to raiſe the weights, they will be lifted up preciſely at the ſame time. Whence it appears, that in this caſe the centrifugal forces are as the quantities of matter, inaſmuch as they overcome reſiſtances which are in that proportion.

If equal bodies moving in unequal circles perform their revolutions in equal times; or in other words, if the quantity of matter in the revolving bodies be given, as alſo the number of revolutions performed in a given time, their centrifugal forces are as their diſtances from the center. For in the fourth general expreſſion ſince Q and N are given, F is as D; that is, the force is as the diſtance. For the confirmation whereof, let two equal balls be placed in the troughs at diſtances from the centers, which are as one and two, and when the tables are turned about, that ball, whoſe diſtance from the center is double, will raiſe a double weight. Exp. 3.

If equal bodies move in unequal circles with equal velocities; or more generally, if the quantity of matter in the revolving bodies be given, as alſo the

velocity

velocity wherewith they revolve; their central forces are inverfly as their diftances from the center. For in the fecond general expreffion fince Q and V are given, F is as $\frac{1}{D}$, that is, the force is inverfly as the diftance. Before I mention the experiment whereby this law is confirmed, I muft obferve to you, that to the axle of one of the tables is fixed a fecond wheel, whofe diameter is but one half of the diameter of the other wheel; and therefore when the chord goes round the fmaller wheel, the table muft turn as faft again as when it goes round the larger wheel; fo that the table which is moved by means of the fmaller wheel, will revolve twice in the fame time, that the other table which is turned by means of the larger wheel, performs one revolution.

Exp. 4.   This being premifed, let two equal balls be fo placed in the troughs, as that the diftance of that ball which is to revolve by means of the fmaller wheel, may be but one half of the other's diftance from the center; in which cafe their velocities will be equal: for tho' the peripheries of the circles which the two balls defcribe, are as one and two; yet will the leffer periphery be defcribed twice in the fame time that the larger is defcribed once; and therefore the fpaces thro' which the bodies move in a given time will be equal, and of confequence their velocities will be fo too. If then two weights be made faft to the chords of the balls in the manner of the former experiments; the tables being turned about, the ball whofe diftance from the center is as one, will raife twice the weight, that is raifed by the ball whofe diftance is as two; fo that the weights raifed, and confequently the forces which raife them, will be inverfly as the diftances of the balls from the center.

If equal bodies revolve in equal circles with unequal velocities, their central forces are as the fquares of the velocities, or becaufe the velocities are as the
number

number of revolutions in a given time, the forces are as the squares of the numbers of revolutions performed in a given time. For by the fourth general expression since Q and D are given, F is as N^2, that is, the force is as the square of the number of revolutions in a given time. To confirm this law, Exp. 5. let two equal balls be placed in the troughs at equal distances from the centers; and let that table, whose axle has two wheels, be turned about by means of the smaller, so that it may perform two revolutions in the same time that the other table performs one: in this case the numbers of revolutions performed by the two balls in a given time being as one and two, their squares will be as one and four; in which proportion the weights raised will likewise be.

If unequal bodies revolve in equal circles with unequal velocities, their central forces are as the products of their quantities of matter into the squares of their respective velocities; or, which is the same thing, as the products of their quantities of matter into the squares of the numbers of revolutions in a given time. For by the fourth general expression, D being given, F is as QN^2. Let therefore two balls, whereof one is double the other, be placed at equal distances from the centers; and let the larger revolve twice in the same time that the smaller revolves once. In this case, the quantity of matter in Exp. 6. the lesser ball, which is as unity, being multiplied into the square of its number of revolutions in a given time, which is likewise as unity, gives one for the product. And the quantity of matter in the larger ball, which is as two, being multiplied into the square of its number of revolutions in the given time, which square is as four, gives eight for the product: so that the weights raised by the two balls will be as one and eight.

If unequal bodies revolve in unequal circles with unequal velocities, their forces are as their quantities of matter multiplied into the squares of their re-

spective

LECT.
III. spective velocities, and that product divided by their respective distances from the centers ; or what amounts to the same thing, their forces are as the products arising from the continued multiplication of their quantities of matter into their respective distances from the centers, into the squares of their numbers of revolutions in a given time ; or to use the mathematical phrase, their forces are in a ratio compounded of their quantities of matter, of their distances from the center, and of the squares of their numbers of revolutions in a given time. For by

Exp. 7. the fourth general expression, F is as QDN^2. To confirm this law by an experiment, let two balls, whereof one is double the other, be placed in the troughs, so as that the distance of the smaller from the center may be to the distance of the larger as two to one ; and let the larger revolve twice in the same time that the smaller revolves once. In this case the quantity of matter in the smaller body, which is as one, being multiplied into the distance from the center, which is as two, and the product being multiplied into the square of the number of revolutions performed by the smaller body in a given time, which is as one, gives two for the product. In like manner the quantity of matter in the larger body, which is as two, being multiplied into the distance from the center, which is as one, and the product of that multiplication being again multiplied into the square of the number of revolutions performed by the larger body in the given time; which square is as four, gives eight for the product ; consequently, the weights which are raised, as also the forces which raise them, are as two and eight, or one and four.

If equal bodies revolve in unequal circles in such a manner as that the squares of their periodical times are as the cubes of their distances from the center, their central forces are inversly as the squares of their distances from the center. For since the

quantity

quantity of matter in the revolving bodies is given, and the cubes of the diſtances are as the ſquares of the. times ; if in the third general expreſſion the cube of D be ſubſtituted in the room of the ſquare of P, F will be as D divided by the cube of D, or as one divided by the ſquare of D; that is, the force will be inverſly as the ſquare of the body's diſtance from the center. To confirm this law, let two equal balls be placed in the troughs, ſo as that the diſtance of one from the center may be as two, and the diſtance of the other as three and one ſixth; and let·that which is at the ſmalleſt diſtance revolve twice in the ſame time that the other revolves once; ſo that their periodical times may be as one and two, the ſquares of which being one and four, are very nearly proportional to the cubes of the diſtances; for the cube of the ſmaller·diſtance is eight, and that of the larger thirty two very nearly ; conſequently, the balls muſt raiſe weights which are to one another inverſly as the ſquares of the diſtances from the centers ; that is, the weight raiſed by the ball, whoſe diſtance is as two, muſt be to the weight raiſed by the ball whoſe diſtance. is as three and a ſixth, as the ſquare of the laſt diſtance to the ſquare of the former, that is, as ten to four, or five to two very nearly.

If the ſquares of the periodical times·be proportional to the cubes of the diſtances; and the revolving bodies unequal, ·the central forces are. directly as the quantities of matter in the bodies, and reciprocally as the ſquares of their diſtances·from the center. For in the third general expreſſion, if the cube of D be ſubſtituted in the room of the ſquare of P; F will be as $\frac{Q}{D^2}$. If therefore all things remain as in the laſt experiment, excepting that the body which is at the greater diſtance·from the center is to·the body leſs diſtant, as two to one; the weight which·is raiſed by the former, will be to the weight raiſed raifed by the latter, as two, to two and a half; that is, the weights raifed, will be as the products arifing from the multiplication of the quantity of matter in one body into the fquare of the other body's diftance.

Among the feveral laws of central forces, that which obtains in nature, and by virtue whereof the heavenly bodies are made to revolve in their feveral orbits, is, where the forces are to one another inverfly as the fquares of the diftances of the revolving bodies from the center. For it has been found by obfervation, that all the planets as well primary as fecondary revolve either in circular orbits or fuch as are nearly fo. And that the fix primary planets move about the fun as their center in fuch a manner, as that the cubes of their mean diftances from the fun are very nearly proportional to the fquares of their periodical times. And the fame thing has been difcovered with regard to the four fecondary planets or fatellites that move about JUPITER, as alfo with refpect to the other five that revolve about SATURN. And therefore the forces whereby they are retained in their orbits muft be in the inverfe ratio of the fquares of their diftances from the central bodies about which they revolve.

Exp. 9. If two bodies are by means of their mutual attraction made to revolve about each other, and alfo about a fixed point; and if their diftances from that fixed point, be reciprocally proportional to their quantities of matter, that is to fay, if as much as one body exceeds the other in quantity of matter, fo much is its diftance from the fixed point exceeded by the other's diftance from the fame point; or what amounts to the fame thing, if the product arifing from the multiplication of one body into its diftance from the fixed point, be equal to the product arifing from the like multiplication of the other body into its diftance from the fixed point, their central forces are equal. For as the two bodies muft of necefity necessity perform their revolutions in the same time; the number of their revolutions in a given time is given: and therefore by the fourth general expression, F is as QD, that is, the central force is as the product arising from the multiplication of the quantity of matter into the distance from the center, or fixed point; but by supposition the product of one of the bodies into its distance from the fixed point, is equal to the product of the other into its distance, consequently, their central forces are equal; for which reason neither of them can fly off from the fixed point so as to draw the other after it; for however strongly either of them endeavours to recede by virtue of its own centrifugal force, it is with equal strength drawn the contrary way by the centrifugal force of the other. But if the distances of the bodies from the fixed point be not reciprocally proportional to their quantities of matter; that body, whose distance with regard to the distance of the other is greater than in the forementioned proportion, will fly off and draw the other after it; for in this case, the product of the former body into its distance from the fixed point is greater than the product of the latter into its distance; which products being as the centrifugal forces of the bodies, the former body will have a greater centrifugal force than the latter, and of course must recede from the fixed point, and drag the other after it; all which is fully confirmed by the following experiments. Let two equal balls be tyed together by a small chord; and let them be laid in one and the same trough, one at each end, so as that the chord being stretched may have its middle point just over the center of the table; let then the table be turned about, and the balls will revolve about the center without flying off either way; and continue so to do as long as the motion of the table lasteth. And the same thing will likewise happen tho' one ball be double the other, provided its distance from the

Exp. 10.

center

center of the table be but one half of the diftance of
the fmaller.   But when equal balls are made ufe
of, if one of them be placed at a greater diftance
from the center than the other, upon turning the
table it will fly off and draw the other after it.   So
likewife when unequal balls are made ufe of, fhould
that which is double the other be placed at a diftance from the center greater than one half of the
diftance of the fmaller, it will fly off and draw the
fmaller after it.   And on the other hand, if the
diftance of the larger be lefs than half the diftance
of the fmaller, the fmaller will in that cafe fly off
and draw the larger after it.

LECTURE IV.

Of the Composition and Resolution of Motion.

THE fecond Law of nature, refulting from
the inertnefs of matter, is, that whatever
motion, or change of motion is produced in any
body, it muft be proportional to, and in the direction of the force imprefled. For fince a body cannot
by reafon of its inactivity contribute to the production of its own motion, or of any change therein, it is plain, that whatever motion or change of
motion is generated in any body, it muft intirely
proceed from the force imprefled on the body ; and
of confequence, fince effects are ever proportionate
to their adequate caufes, muft be proportional thereto.   And it muft likewife be directed and determined towards the fame part with the generating
force. Wherefore if the body whereon the impreffion is made, was in motion before the impulfe,
that motion will be retarded or accelerated according as the force imprefled oppofes it, or confpires
therewith, or if it acts obliquely to the fame, the
direction thereof will be changed, and the body
will

will move in a direction situated between the direction of its former motion, and that of the impressed force. For instance, if a body moving from
A towards B, be impelled at the point A by a
force acting in the direction AC, it will move along
a line as AD placed between AB and AC, the
situation of which may be thus determined. Let
the lines AB denote the velocity wherewith the
body moves in the direction AB; and let AC denote the velocity wherewith the body would move
by virtue of the impulse along the line AC, supposing it had no other motion : that is, let AB be
to AC as the space described by the body in a
given time in the direction AB, to the space described by it in the direction AC, each of the motions being considered singly and apart ; then compleating the parallelogram ABDC, and drawing
the diagonal AD, that diagonal is the line in which
the body moves ; for the proof of which, let us
suppose a small inflexible wire equal in length to the
line AB, to pass thro' the center of a ball, and
that whilst the ball moves uniformly on the wire
from A towards B, with a velocity which is as AB,
the wire is also moved uniformly from AB towards CD, with a velocity which is as AC, and
in such a manner as to be always parallel to AB,
and with its extremities to describe the lines AC
and BD. Then, forasmuch as the spaces described
in a given time where the motions are uniform, are
to one another as the velocities of the motions ; it
is evident, that in whatever time the ball moves
the length of the wire, in the same time will the wire
move the length of AC, to wit, from AB to CD;
consequently, at the end of that time the ball will
be found in D at the extream point of the diagonal AD. From any point in the diagonal taken
at pleasure as E, let the line EF be drawn parallel
to DB, and from the nature of similar triangles,
AF will be to FE, as AB to BD, that is, as the
velocity

velocity of the ball to the velocity of the wire; consequently, in the same time that the ball moves the length of A F along the wire, the wire will move the length of F E, from A B to K L; and the point F, which is the place of the ball on the wire, will be found in E. And what has been thus proved in relation to the two points D and E of the diagonal, may in the very same manner be demonstrated of any other point in the same line; wherefore the ball will by virtue of its own motion, and that of the wire, whereof it partakes, be carried in such a manner as to be always found in the diagonal A D; that is, it will by virtue of its compound motion describe the diagonal line. This being so, it plainly follows, that if the wire be taken away, and the ball at A have two motions impressed upon it at once; one in the direction A B, the other in the direction A C; and if the motions impressed, or, which is the same thing, if the forces impressing those motions be to one another in the proportion of A B to A C, the ball will by virtue of the double impression, move along the diagonal A D. For as to the effect it matters not whether the motion which the ball has in the direction A C arises from a force impressed on it at the point A, or whether it be communicated by a wire supporting the ball, and carrying it along with it in that direction.

Exp. 11. To confirm this by an experiment, let three ivory balls of equal size, be suspended from three pins by strings of equal lengths, and let the middle ball rest over one angle of a wooden square; then let each of the extream balls be let fall separately from the same height, in such manner as to strike the middle ball in the direction of one side of the square, and the middle ball will by each of the strokes made separately, be moved along over that side of the square, which correspondeth to the direction of the stroke; but if the two balls be at the same instant

of

of time let fall from equal heights, so as that they may strike the middle ball at once, and in the directions of the two sides of the square, the middle ball will by the double stroke, be driven over the diagonal of the square.

As a COROLLARY it follows, that a body will in the same time describe the diagonal AD of a parallelogram with two forces conjoined, that are to one another as the sides AB and AC, that it would the respective sides with each of those forces separately. As also, that the velocity wherewith a body moves along the diagonal, is to the velocity wherewith it is carried along the sides when acted upon by each force singly, as the diagonal to each side respectively: consequently, if the two forces be given, the velocity along the diagonal, which arises from the conjunction of both forces, will be so much the greater, by how much the angle BAC is less; for as that angle is diminished, the diagonal which in this case denotes the velocity, is lengthened, till at last the angle vanishing by the coincidence of the sides, the diagonal becomes equal to both the sides taken together; and the velocity of the body equal to the sum of the velocities wherewith the body would move, were each of those forces impressed upon it in the same direction. Thus the lines AB and AC being placed at three different angles, so as to constitute the sides of three different parallelograms, (the diagonals whereof are represented by the pricked lines) it is evident to sight, that as the angle BAC grows less, the diagonal grows longer; and that when the angle vanishes by the coincidence of AB with AC, the diagonal AD becomes equal to AC and CD, that is, to AC and AB; and the velocity denoted by AD, is in that case a *maximum*, or the greatest that can arise from the conjunction of those two forces. On the other hand, as the angle inlarges, the velocity along the diagonal must decrease, till at length the angle vanishing

Pl. 2.
Fig. 2.

Pl. 2.
Fig. 3.

nifhing by the two fides becoming one right line,
the velocity becomes equal to the difference of the
velocities, arifing from the impreffion of each force
Pl. 2. Fig. 4. when made fingly and feparately. Thus the lines
AB and AC being as before placed at three different angles, BAC, it is evident that the diagonals
AD reprefented by the pricked lines, grow fhorter
as the angle BAC inlarges; till at laft the angle,
and with it the diagonal vanifhing the two fides
Pl. 2. Fig. 5. BA and AC conftitute one right line as BAC,
wherein the body is, as it were, carried two contrary ways, to wit, from A towards B by the force
which acts in the direction AB, and from A towards C by the force acting in the direction AC;
and the difference of the velocities, which arife
from the impreffions of the two forces when they
act feparately, is the velocity wherewith the body
actually moves in the direction of the ftronger
force, which velocity is a *minimum*, or the leaft velocity that can arife from the joint action of thofe
two forces.

As a fecond COROLLARY it follows, that a body
may be moved thro' one and the fame line by numberlefs pairs of forces acting upon it. For if in-
Pl. 2. Fig. 6. ftead of the force, whofe direction is AB, we fuppofe another, the direction whereof is AE; and if
inftead of the force acting in the direction AC, we
fuppofe one to act in the direction AF, and that
thofe forces are to one another as AE to AF:
then compleating the parallelogram AEDF, the
line AD will be the diagonal of this parallelogram,
as well as of the former; and therefore the body
will from the joint action of thefe two forces defcribe the fame line AD which it did before: and
as AD may be made the diagonal of numberlefs
parallelograms, it is evident that it may be defcribed by a body acted upon by numberlefs pairs
of forces in different directions. And not only fo,
but it may likewife be defcribed by a body, where-

on

on a great number of forces act at the same time; for as the forces acting upon the body in the directions AB and AC make it to move along the diagonal AD, so may the direction along AB arise from the directions of two other forces, and each of those from the directions of two others, and so on without number. Hence we see, that all forces and motions whatever may be resolved into innumerable forces and motions; and any simple direct force or motion may be looked upon as compounded of innumerable oblique forces or motions. For the line and direction of the motion is the same, whether that motion be compounded of two motions arising from forces impressed in the directions AB, AC, or in the directions AE, AF, or arise from the impression of a single force in the direction AD; and therefore the motion along the line AD, tho' it be simple arising from one single force acting in that direction; yet may it be considered as compounded of two or more motions in other directions, such as AB and AC, or AE and AF, since the very same motion would arise from such a composition.

Pl. 2. Fig. 6.

This composition and resolution of motions and forces is of singular use in mechanicks; for by the help thereof, the effects of powers acting in oblique directions are readily determined, as will appear hereafter.

The third LAW OF NATURE arising from the inertness of matter is, that reaction is always equal to action, and contrary thereto; or in other words, that the actions of two bodies, one upon another, are constantly equal, and in directions contrary to each other; so that whatever change is made in the state of one body, whether at rest or in motion, by the action of another; the same change is produced in the state of the other by the reaction of the former; but the tendencies or directions of those changes are contrary ways. Thus, when one presses

E a stone

a ftone with his finger directly downward, the finger is equally preffed by the ftone, and that directly upward. And when a horfe draws a load, he
is equally drawn back by the load; for as much as
he promotes the progrefs of the load, fo much is
he retarded in his own motion; that is, he is in effect drawn back; for the fame force of mufcles and
finews, which he exerts in order to drag on the
load, would, if he was freed from the incumbrance,
carry him forward to a diftance much greater than
what he reaches in the fame time whilft tied to the
load; and confequently, as far as his progrefs fall-
eth fhort of that diftance, fo much is he in effect
drawn back; and whatever motion he communicates to the load, fo much does he lofe of his own,
the load reacting upon him with the fame force
that he acts upon it; for which reafon, if by addition of weight the load be fo far increafed as to require the whole ftrength of the horfe to move it,
no motion will enfue, the whole power of the
horfe, wherewith he endeavours to go forward, being but juft equal to the reaction of the load whereby he is drawn back. This equality of action and
reaction obtains in all kinds of attractions whatever.
When a loadftone attracts a piece of iron, it is
equally attracted by it; as will appear from the
Exp. 12.   following experiment. Let a piece of iron and a
loadftone equal in weight, be fufpended by two cords
of an equal length, and let the diftance between
them be fo fmall, as that they may not be out of
the reach of each other's attraction; then will they,
from a ftate of reft, begin to move towards each
other, and that with equal velocities, fo as to meet
at the middle point of their firft diftance: if they
be again feparated, and the loadftone fixed, the
iron being fufpended at the fame diftance from it
as before, will move towards it, fo as at length to
touch it, and adhere thereto. And on the other
hand, if the iron be fixed, and the ftone moveable,
                                                the

the ftone will approach the iron in the fame manner, as the iron did the ftone; all which plainly
fhews that the attraction between the loadftone and
iron is mutual, the one drawing the other as much
as it is drawn by it: fo that the reaction of the iron
upon the ftone is exactly equal to the action of the
ftone upon the iron.

The equality of action and reaction with refpect
to attractions is likewife manifeft from hence,
that if a man placed in a boat, draws another boat
by means of a rope faftened thereto; the boat
wherein the man is placed will be equally drawn
with the other, and the two boats will approach
one another with equal quantities of motion; fo
that if they be equal in weight, and of the fame fize
and fhape, they will approach with equal velocities,
and meet at the middle point: but if one be heavier
than the other, then by how much it exceeds the
other in weight, by fo much will it be exceeded by
the other in the velocity of its motion; for inftance,
if the weight of one be to the weight of the other
as one to two, then will the velocity of the former
be to the velocity of the latter as two to one; that
is, their velocities will be reciprocally proportional to their weights. To confirm this by an ex- Exp. 13.
periment, let a cord be made faft to one end of a
fmall boat, and let it pafs over a pulley fixed to
the end of another fmall boat of the fame fhape
and fize, and let a weight be tied to the end of the
cord, and hang in the water; this being done,
let the boats be placed at fuch a diftance as that the
cord may be ftretched, then letting go the boats
the weight will defcend, and in defcending draw
the boat to whofe end the cord is faftened towards
the other, and at the fame time the other will move
towards it; and when they come together, the fpace
defcribed by the boat whofe weight is as one, will
be to the fpace defcribed by the boat, the weight
whereof is as two, as two to one; that is, if the

E 2   diftance

diftance between the two boats be divided into **three**
equal parts, that boat which is double in **weight to**
the other, will move thro' one of thofe parts in **the**
fame time that the lighter moves thro' the **other**
two.

As action and reaction are equal with regard to
attractions, fo are they likewife in refpect of ftrokes
or impulfes made by bodies one upon another ; the
force of two bodies, ftriking each other equally,
affecting the motions of both, and producing equal
changes therein towards contrary parts. On this
equality of action and reaction do the feveral laws
which have been collected concerning the collifion
of folid bodies in a great meafure depend ; which
laws, as they relate to bodies void of elafticity, I
fhall now explain ; in doing of which, I fhall lay
down one general PROPOSITION concerning the
collifion of fuch bodies, whence I fhall deduce the
laws of particular cafes, and at the fame time confirm each law by an experiment.

The PROPOSITION is as follows : *If two bodies*
void of elafticity move in one right line, either the fame
or contrary ways, fo as that one body may ftrike di-
rectly againft the other ; let the fum of their motions
before the ftroke when they move the fame way, and
the difference of their motions when they move contrary
ways, be divided into two fuch parts as are propor-
tional to the quantities of matter in the bodies ; and
each of thofe parts will refpectively exhibit the motion
of each body after the ftroke. For inftance, if the
quantities of matter in the bodies be as two and one,
and their motions before the ftroke as five and four,
then the fum of their motions is nine, and the difference is one ; and therefore when they move the
fame way, the motion of that body, which is as
two, will after the ftroke be fix, and the motion of
the other three : but if they move contrary ways, the
motion of the greater body after the ftroke will be
two thirds of one, and of the leffer one third of one.

For

. For since the bodies are supposed to be void of elasticity, they will not separate after the stroke, but move together with one and the same velocity; and of consequence, their motions will be proportional to their quantities of matter; and from the equality of action and reaction it follows, that no motion is either lost or acquired by the stroke when the bodies move the same way, because whatever motion one body imparts to the other, so much must it lose of its own; consequently the sum of their motions before the stroke is neither increased nor diminished by the stroke, but is so divided between the bodies, as that they may move together with one common velocity, that is, it is divided between the bodies in proportion to their quantities of matter; but it is otherwise, where the bodies move contrary ways; for then the smaller motion will be destroyed by the stroke, as also an equal quantity of the greater motion, because action and reaction are equal; and the bodies after the stroke will move together equally swift, with the difference only of their motions before the stroke; consequently, that difference is by means of the stroke divided between them in proportion to their quantities of matter.

The several particular cases concerning the collision of bodies may be reduced to four general ones. For, 1ft, it may be that one body only is in motion at the time of the stroke. Or, 2dly, they may both move one and the same way. Or, 3dly, they may move in direct opposition to each other, and that with equal quantities of motion. Or, lastly, they may be carried with unequal motions in directions contrary to each other. As the bodies may be either equal or unequal, each of these four general cases may be looked upon as consisting of two branches; and as such I shall consider them, and treat of them in the order, wherein I have laid them down.

As

As to the firft, if a body in motion ftrikes another equal body at reft, they will by the propofition move together, each of them with one half of the motion that the body had which was in motion before the ftroke ; and fince the quantity of motion in any body, is as the product arifing from the multiplication of its quantity of matter into its velocity ; the common velocity of the two bodies, will be but one half of the velocity of the moving

body before the ftroke. For the confirmation whereof, let two equal balls of clay be fufpended from two pins of an equal height, by threads of an equal length, and in fuch a manner, as that when they hang freely they may juft touch one another, and that their centers and point of contact may lie in a right line parallel to the horizon. This being done, and one of the balls being at reft, let the other be removed to any diftance from it, and then let fall ; it will in its defcent defcribe the arch of a circle, and by the time it arrives at the loweft point of the arch, that is, when it comes to touch the quiefcent ball, it will have acquired fuch a velocity as would carry it to the fame height from which it fell, as fhall be fhewn when I come to treat of pendulums ; and confequently, if the other ball was removed, would actually afcend to that height ; but upon ftriking the other ball, which is of equal fize, it will communicate one half of its motion to it, and they will move together with half the velocity that the moving body had at the time of the ftroke, fo as to afcend to one half only of the height from which the ftriking body fell.

That the nature of this and the other experiments relating to the collifion of bodies may be more readily comprehended ; I fhall lay down fome things concerning the motion of bodies thro' the arches of circles, the truth whereof fhall be demonftrated in my lecture upon pendulums. And firft, all the arches of a circle, provided they be

not

not large, are defcribed in equal times by bodies defcending along them; and therefore if two bodies be let fall at the fame time, one from C and the other from E, or from D and F, they will both arrive at the loweft point B at one and the fame time; and the ftroke of the fubfequent body upon the preceding will be made at B: and for the fame reafon if one be let fall from C, and the other from D or F; or one from E, and the other from D or F, they will meet and ftrike one another at B.

2dly, The velocity which a body acquires in falling thro' the arch of a circle, is as the chord of the arch; that is, the velocity of a body which has fallen from C to B, is to the velocity of a body that has fallen from E to B, as the chord CB to the chord EB. And here I muft obferve to you, that when in the following experiments I fpeak of a body falling from, or rifing to any height, as four, fix, or ten inches, I would be underftood to mean it of a body's falling thro' or moving up an arch, whofe chord is of fuch a length.

3dly, The velocity wherewith a body begins to rife up thro' the arch of a circle, is as the chord of the arch which the body defcribes in its afcent. Thus the velocity wherewith a body begins to move from the point B towards D, if it afcends as high as D, is as the chord BD; but if it rifes only to F, the velocity is as the chord BF. So that in the experiments the chords of the arches thro' which the bodies defcend, exprefs the velocities of the bodies in the point B at the time of the ftroke; and the chords of the arches thro' which the bodies afcend after the ftroke exprefs the velocities of the bodies immediately after the ftroke.

Thefe things being laid down, I fhall now proceed to determine the laws of the four general cafes. As to the firft, it has been already fhewn, that

where

where the moving body is equal to the quiescent, the common velocity of the two bodies after the stroke, is but one half of the velocity of the moving body before the stroke; and of consequence, the motion of each body after the stroke, is equal to one half of what the moving body had before the stroke. But if the quiescent body differs in size from the moving body, then the common velocity after the stroke will be so much less than the velocity of the moving body before the stroke, by how much the sum of the two bodies exceeds the body which was first in motion. Thus, if the moving body be to the quiescent as two to one, the common velocity after the stroke will be to the

Exp. 15. velocity of the moving body before the stroke, as two to three; wherefore if a ball of clay, falling from the height of nine inches, strikes another at rest, and of one half the magnitude, they will ascend together to the height of six inches only: and on the other hand, if the larger be quiescent, and the smaller falls from the height of nine inches, they will ascend to the height of three inches only; and the quantity of motion in each body immediately after the stroke will be had, by multiplying each of them into the common velocity.

As to this and all other experiments of this nature, it must be observed, that they do in some measure vary from the theory, and that for two reasons. First, because clay or any other body, wherewith these experiments can be made, is not perfectly void of elasticity. Secondly, because the air resists the motions of the balls, and by so doing diminishes their velocities.

As to the second general case, where both the bodies are in motion before the stroke, and move one and the same way: In order to find their common velocity after the stroke, let the sum of their motions before the stroke be divided by the sum of the bodies,

bodies, and the quotient will express the common velocity. Wherefore, if two equal balls of clay be let fall at the same time, one from the height of three inches, and the other from the height of six, after the stroke they will ascend to the height of four inches and an half; for as in this case the bodies are equal, their motions are as their velocities, that is, as six and three, the sum of which being divided by two, the sum of the bodies, gives four and an half for the common velocity after the stroke.

Lect. IV.

Exp. 16.

Where the bodies are unequal, let us suppose the preceding body to be as one, and to fall from the height of three inches as before, so that its quantity of motion will be as three; and let the subsequent body be as two, and fall from the height of six inches, so that its quantity of motion will be twelve; and the sum of the two motions will be fifteen, which being divided by three, the sum of the two bodies, gives five in the quotient; so that in this case, after the stroke, the balls will ascend to the height of five inches, and the motion of the greater will be as ten, and that of the smaller as five.

Exp. 17.

As to the third general case, where the bodies move in direct opposition to each other, if they have equal quantities of motion, they will upon the stroke lose all their motion, and continue at rest; for by the proposition, the bodies after the stroke will be carried with the difference of their motions before the stroke; which difference is supposed to be nothing. Wherefore, if two equal balls of clay be let fall at once from equal heights, upon the stroke they will cease to move; and the same thing will happen where the balls are unequal, provided the heights from which they fall are reciprocally proportional to their quantities of matter; for instance, if the balls be as one and two, let the former fall from the height of six inches, and the latter from the height of three, and upon their

Exp. 18.

Exp. 19.

meeting

LECT. meeting they will ſtand ſtill, for in this caſe, the
IV.   quantities of motion, wherewith they oppoſe each
other, will be equal.

When two bodies meet with unequal quantities
of motion, if the difference of their motions be divided by the ſum of the bodies, the quotient will
expreſs their common velocity after the ſtroke;
for by the propoſition, the difference of their motions before the ſtroke, is equal to the ſum of their
motions after the ſtroke; conſequently, that difference divided by the ſum of the bodies muſt give
Exp. 20. the velocity.   Wherefore, if two equal balls of
clay be let fall at the ſame time, one from the
height of three inches, and the other from the
height of ſix, after the ſtroke they will aſcend together to the height of an inch and an half; for
ſince the balls are equal, their motions will be as
their velocities, that is, as ſix and three, the difference whereof is three, which being divided by two,
the ſum of the bodies gives one and an half in the
Exp. 21. quotient.   If the balls be unequal in the proportion, for inſtance, of two to one; and if that which
is as two falls from the height of ſix inches, and the
other from the height of three; after the ſtroke
they will aſcend together to the height of three
inches; for the greater ball being as two, and its
velocity as ſix, its motion is as twelve; whereas the
ſmaller being as one, and its velocity as three, its
motion is likewiſe as three, which being ſubducted
from the greater motion leaves a remainder of nine;
and this being divided by three, the ſum of the bodies, gives three for the common velocity, or the
height to which the bodies will riſe.

In order to diſcover the quantity of motion communicated by one body to the other, I ſhall lay
down four rules adapted to the four general caſes.
And Firſt, if one of the bodies be quieſcent at the
time of the ſtroke, let that body be multiplied into
the common velocity after the ſtroke, and the product

duct will exprefs the communicated motion. For
fince that body had no motion before the ftroke, it
is manifeft, that whatever motion it has after the
ftroke muft be communicated to it by the ftriking
body; but that motion is as the product arifing
from the multiplication of the quantity of matter in
the body into the common velocity, confequently,
that product expreffes the communicated motion.

Since the body which is at reft before the ftroke
has no motion, but what is imparted to it by the
ftriking body; and fince the motion of the ftriking
body is by the propofition to be divided between
the two bodies in proportion to their quantities of
matter; it follows, that where the ftriking body is
greater than the quiefcent, it will communicate lefs
than half its motion, and where it is equal to it, it
will impart one half; and where it is lefs, more
than one half: and if the quiefcent body be infinitely great with refpect to the ftriking body, which
is in effect the cafe where the quiefcent body is fixed,
fo as not to give way to the ftroke, the ftriking
body will impart all its motion to the other; for
as the quiefcent body is fuppofed to be infinitely
greater than the ftriking body, the motion, which
it receives from the ftriking body, muft bear an infinite proportion to the motion remaining in the
ftriking body; but as the motion communicated is
a finite quantity, it cannot bear an infinite proportion to the remaining motion, unlefs that remaining
motion be in its evanefcent ftate, and reduced to
nothing.

When both the bodies are in motion before the
ftroke, and their motions are directed the fame
way, which was the fecond general cafe; the rule
for determining the quantity of motion communicated is as follows. Let the preceding body be
multiplied into the common velocity after the ftroke,
and from the product let the motion which it had
before the ftroke be fubducted, and the remainder
will

LECT.
IV.
will be the motion communicated. For the product arising from the multiplication of the preceding body into the common velocity, gives the whole motion of that body after the stroke, and therefore, if from thence be taken the motion which it had before and independent of the stroke, the remainder must be the motion acquired by the stroke.

When the bodies move towards one another with equal quantities of motion, as in the third general case; the motion communicated is equal to the motion of either before the stroke. For as in this case, both their motions are destroyed by the stroke; it is plain, that whichever of the bodies is considered as giving the stroke (and either of them may) it must communicate just as much motion to the other, as the other has at the time of the stroke; for by this means the motion communicated, as it is directly opposed to the former motion of the body, will be just sufficient to destroy the same, and by so doing cause the body to rest.

When the quantities of motion in two bodies moving directly towards each other are unequal, which is the fourth general case; the motion communicated is determined by the following rule. Let the body which had the lesser motion before the stroke be multiplied in the common velocity after the stroke, and to the product let the motion which it had before the stroke, be added, and the sum will be the motion communicated. For as the body, to which the motion is communicated, does after the stroke move in a direction contrary to what it did before, it is evident, that besides the motion, wherewith it is carried in that contrary direction, it must have received as much more in the same direction, as was sufficient to withstand the motion it had before the stroke in an opposite direction; for till that motion was destroyed by an equal motion opposed thereto, the body could not change its direction, and move backward.

LECTURE

LECTURE V.

OF THE COLLISION OF ELASTICK BODIES.

HAVING given you an account of the collision of bodies void of elasticity, I come now
to consider the effects thereof in such as are elastick;
by which I mean bodies that consist of such parts
as yield and give way when pressed, and which restore themselves upon the removal of the pressure:
if the force wherewith they restore themselves be
exactly equal to the pressure whereby they are bent
inward, then are the bodies said to be perfectly
elastick; and such are all those bodies supposed to
be, wherewith experiments are usually made for
confirming the theory relating to the collision of
elastick bodies; but as there is not perhaps in nature
any body perfectly elastick, if among the experiments that are now to be made, any shall be found
to vary a little from the theory, such variation must
be looked upon as rising rather from the want of
perfect elasticity in the bodies, than from any error
in the theory itself, or in the calculations grounded
thereon.

The method which I shall observe in treating of
the percussion of elastick bodies is this; First, to
lay down one general proposition concerning such
percussion, and then, Secondly, to deduce the laws
relating to the four general cases mentioned in my
last lecture, and to confirm each of those laws by
experiments.

Before I lay down the proposition, I must observe
to you, that wherever I mention the striking body,
I thereby mean that body which is in motion where
one of the two is quiescent, as also that body which
moves swiftest when they both move the same way;
and lastly, that body which has the greatest quan-

4 tity

tity of motion, when they move in oppofition to one another, or in this cafe, if their motions be equal, then either of them may be taken indifferently for the ftriking body.

This being premifed, the PROPOSITION is as follows.

If of two bodies perfectly elaftick, one be at reft, and the other in motion; or if they both move either the fame or contrary ways, fo as that one fhall ftrike the other; let them be confidered as void of elafticity, and by the propofition laid down in my laft lecture, let the motion of each body after the ftroke be found, and by one of the four rules laid down in the fame lecture, let the motion communicated by the ftriking body to the other be likewife found; and let this motion be fubducted from the motion of the ftriking body after the ftroke, and added to that of the body which received the ftroke, and the refidue will be the motion of the ftriking body, and the fum the motion of the other body after reflexion. For, fince the bodies are fuppofed to be perfectly elaftick, their parts which are bent in by the ftroke will reftore themfelves with a force equal to that which bends them in; but the force which bends them in, is meafured by the quantity of motion communicated by the ftriking body to the other, and therefore the parts of each body which are bent inward will reftore themfelves with fuch a force, as is fufficient to generate a motion equal to that which is communicated; confequently, the bodies will by virtue of their elafticity throw one another contrary ways, each with a quantity of motion equal to that which the ftriking body communicates to the other; for which reafon, if that motion be fubducted from the motion remaining in the ftriking body after the ftroke, as being contrary thereto, and added to the motion of the other body after the ftroke, as confpiring therewith, the refidue and fum will give the true motions of the bodies after reflexion.

To

. To apply what has been said to the four general L ᴇ ᴄ ᴛ. cafes, the firſt whereof is where one of the bodies V. is at reſt at the time of the ſtroke. If a body perfectly elaſtick ſtrikes another of the fame kind and of equal magnitude at reſt, the ſtriking body will communicate all its motion to the other and remain at reſt, for by the firſt of the four rules laid down in my laſt lecture, the ſtriking body will upon the ſtroke, communicate half its motion, and by the propoſition now laid down, a quantity of motion equal to that which is communicated, muſt be fubducted from the motion remaining in the ſtriking body, and be added to the motion of the body which receives the ſtroke, by which means the ſtriking body will have no motion left ; but the other body will have a quantity of motion equal to what the ſtriking body had before the ſhock. For the confirmation of which, let two equal ivory balls Exp. 1. be fufpended as were thoſe of clay ; and let one of the balls fall from any height, and ſo as to ſtrike the other at reſt, the ball which receives the ſtroke will aſcend to the fame height from which the other fell, and will leave the other at reſt.

If inſtead of one there be two, three, or more Exp. 2. quieſcent balls contiguous to one another, that which is fartheſt removed from the ſtriking ball will fly off with the velocity of the ſtriking ball, and all the intermediate balls together with the ſtriking ball will quieſce ; for as the ſtriking ball imparts all its motion to the firſt of the quieſcent balls, ſo does that in like manner to the ball which lies next beyond it, and that again to a third, and ſo on ; till at length the laſt ball meeting with none other to refiſt it, flies off with all the motion of the ſtriking ball, leaving that and the intermediate ones at reſt.

If two balls be let fall together contiguous to one Exp. 3 another, upon the ſtroke the two fartheſt will fly off, leaving the others at reſt; for as the foremoſt
of

of the two moving balls is carried equally swift with the subsequent, it cannot during its motion receive any impression from the subsequent ball; consequently, when it makes the stroke, it will produce the same effect in the quiescent balls as if the subsequent ball was away; that is, it will by means of the intermediate balls communicate all its motion to the last, and make that fly off; but no sooner has it made the stroke, and thereby parted with its own motion, but the subsequent ball impels it, and imparts to it all its motion, and this motion being propagated thro' the several intermediate balls as before, makes the last but one to fly off, and that in such a manner as to keep pace with, and closely pursue the other; because in the same instant of time that the foremost of the two moving balls makes its stroke, it likewise receives the stroke from the hindmost ball, and of consequence, the flying off of the two last balls, which is the effect of the double stroke, must happen at one and the same time.

Exp. 4.　For the same reason that two balls fly off where the number of striking balls is two, three will fly off when there are three striking balls, and four, where there are four, and so on, whatever be the number of striking balls, an equal number will constantly go off.

Exp. 5.　If two elastick balls be unequal, for instance, if one be double the other, and if the greater be let fall from the height of nine inches, and strike the smaller at rest; they will both move forward after the stroke, the striking body with one third of the motion which it had before the stroke, and the other with two thirds; and the striking body will ascend to the height of three inches, and the other to the height of twelve. For since the striking body is to the quiescent as two to one, it will by the first of the four rules laid down in my last lecture, communicate one third of its motion to it, and on account

of

of the elasticity a quantity of motion equal to what is communicated, must be taken from the motion remaining in the striking body, and added to the motion of the other; consequently, the striking body will retain one third only of its motion, the other two thirds being communicated to the body which receives the stroke; wherefore since the striking body is as two, and the height from which it falls as nine, its motion must be as eighteen, one third of which, to wit, six, it will retain after the reflexion; and the other two thirds, to wit, twelve, will be the motion of the other body, and these motions being divided by the bodies, will give three and twelve for the quotients; which quotients are as the velocities of the bodies after reflexion, or as the heights to which they ascend.

On the other hand, if the larger ball be quiescent, and the smaller be let fall from the height of nine inches, its motion will be as nine, whereof two thirds will by the first of the four rules be communicated by the stroke to the greater, and one third only will remain in the striking ball, from which on account of the elasticity must be taken as much as was communicated to the larger ball, that is, two thirds, but upon subducting two thirds from one third, there will remain one third negative; which shews, that the striking ball will be reflected with one third of the motion it had at the time of the stroke, so as to ascend backward to the height of three inches; and the quiescent ball, to which two thirds of the striking ball's motion was communicated by the stroke, will likewise on account of the elasticity receive two thirds more, so as to be carried forward with a motion equal to what the striking ball had at the time of the stroke, and one third more; that is to say, with a motion which is as twelve, which being divided by two, the quantity of matter in the ball gives six for the velocity, or the height to which that ball must ascend.

Exp. 6.

From

From what has been said it follows, that when the quiescent ball is smaller than the striking ball, there can be no reflexion, because in that case the striking ball will by virtue of the stroke communicate less than half its motion, and the motion which is to be taken from the striking ball on account of the elasticity, being equal to the motion communicated, will upon the subduction always leave some motion in the striking ball to carry it forward, consequently, it cannot be reflected. Where the two balls are equal there will likewise be no reflexion, but the ball which was quiescent will go forward with all the motion of the striking ball, and the striking ball will become quiescent; as is evident from what was said concerning that case. But where the striking ball is less than the quiescent, it will be reflected, and there will likewise be an augmentation of motion in the greater ball; for the smaller ball must upon the stroke communicate more than half its motion to the greater ball, and there must likewise, on account of the elasticity, as much motion be subducted from the smaller ball, and added to the larger, as is communicated; wherefore, since two equal quantities of motion, each of which exceeds half of the smaller ball's motion, are to be subducted from the smaller ball, and given to the larger; it is plain, that the smaller must lose all its motion and something more, that is, it must be carried backward or reflected; and the greater ball must go forward with more motion than was in the smaller at the time of the stroke, that is, there will be an augmentation of motion; and the excess of motion in the greater ball, above the motion which the smaller ball has at the time of the stroke, is ever equal to the motion wherewith the smaller ball is reflected after the stroke, as is evident from what has been said. If therefore motion be communicated from a smaller elastick body to a larger, by means of several intermediate bodies each

larger

larger than the other, the motion will be augmented in each of them, and the motion of the laſt will
greatly exceed that of the firſt ; and this augmentation of motion is greateſt when the bodies are in
a geometrical progreſſion ; for inſtance, if there be
two bodies which are as one and four, and if the
ſmaller communicates motion to the larger by means
of one intermediate body, the motion will be greater in the larger body, if the middle body be as two,
that is, a geometrical mean between the two, than
if it be as one and an half, or two and an half, or
three, or in ſhort in any other proportion whatever but that of the geometrical mean. For the
proof of which, let the leſſer body be expreſſed by
unity, and the larger by the ſquare of a, and the
geometrical mean will be expreſſed by a ; ſo that
the three bodies taken in their order from the leaſt,
will be expreſſed by the ſymbols in the firſt ſtep,
and the motion produced in the ſecond body by the
ſtroke of the firſt, will be expreſſed by the ſecond
ſtep ; and the motion produced in the third by the
ſtroke of the ſecond, will be expreſſed by the third
ſtep. Again, let another body greater or leſs than a
be ſubſtituted in the room thereof, and let the difference between that body and a be called x, in
this caſe, the bodies will be expreſſed by the ſymbols in the fourth ſtep, and the motion produced
in the ſecond by the ſtroke of the firſt, will be expreſſed by the fifth ſtep ; and the motion produced
in the third by the ſtroke of the ſecond, will be expreſſed by the ſixth ſtep ; but this fraction of the
ſixth ſtep is leſs than that of the third ſtep, for if
from the product ariſing from the multiplication of
the denominator of this fraction into the numerator
of that be ſubſtracted, the product which ariſes from
the multiplication of the numerator of this into the
denominator of that ; that is, if from the ſeventh
ſtep the eighth be ſubducted, there will remain

the

Lect. the quantity which is expressed in the ninth step.
V.
    Whence it appears, that the former product is greater than the latter; and therefore, by the 2d Corol. of the 19th Prop. of the 7th book of the elements, the numerator of the former fraction bears a greater proportion to its denominator, than that of the latter fraction does to its denominator; that is, the fraction in the third step which expresses the motion of the greatest body when the intermediate one is a geometrical mean, is greater than the fraction in the sixth step, which expresses the motion of the greatest body when the middle body is not a geometrical mean; consequently, the motion is more augmented when the intermediate body is a geometrical mean, than when it is greater or less in any proportion.

1st. $1, a, a^2.$

2d. $\dfrac{2a}{1+a}.$

3d. $\dfrac{a^3}{a^3+2a^2+a} \times 4.$

4th. $1, a \pm x, a^2.$

5th. $\dfrac{2 \times \overline{a \pm x}}{1+a \pm x}.$

6th. $\dfrac{a^3+a^2x}{a^3+2a^2+a+a^2x+2ax \pm x+x^2} \times 4.$

7th. $a^6+2a^5+a^4+a^5x+2a^4x+a^3x+a^3x.$

8th. $a^6+2a^5+a^4+a^5x+2a^4+a^3x.$

9th. $- - - - - - - - - - a^3x^2.$

    To give you an instance, how prodigiously motion may be augmented, by being successively communicated to several bodies in a geometrical pro-
gression;

greffion; if twenty elaftick bodies be placed one after L E C T. another, each fucceeding body exceeding the fore- IV. going in the proportion of twenty to one; and if motion be communicated thro' the feveral interme- diate bodies from the firft to the laft, it will be fo far augmented, as to be two-hundred thoufand times greater in the laft body than in the firft; fo that if we fuppofe the firft to be a cannon-ball, moving with the fame velocity wherewith it flies from the mouth of a cannon, which, from the obfervations of Mr. DERHAM, I fhall fuppofe to be at the rate of 612 feet in a fecond, tho' there are pieces of cannon which difcharge their balls with double that veloci- ty, the motion of the laft body will be fo great as if applied to the ball would carry it at the rate of above twenty three thoufand miles in one fecond of time; which velocity is five thoufand times as great as the velocity of a body revolving about the earth, by the force of gravity at a fmall diftance from its furface; for a body fo revolving will not come round in lefs than an hour and twenty four mi- nutes.

From the increafe of motion in elaftick bodies, a reafon may be drawn for the augmentation of found in fpeaking trumpets; for as the fpeaking trumpet is narroweft at the mouth-piece, and thence widens and inlarges continually to the extremity, the air within it, which is an elaftick fluid, as fhall be fhewn hereafter, may be confidered as divided into a great number of cylindrical bodies, of very fmall but equal altitudes, the bafis of the firft being equal to the mouth of the trumpet, and the bafis of the reft increafing one above another as they are more and more removed from the mouth; upon which ac- count the motion that is impreffed by the force of the voice on the firft cylindrical body of air, grows larger in the fecond, and larger ftill in the third, and fo on, till at length at the exit of the tube it becomes fo large as to magnify the found to a great

degree;

degree; and of the several kinds of trumpets, those
magnify the sound most, that are of such a figure as
arises from the revolution of the logarithmick curve

Pl. 2.
Fig. 8.
about its axis; that is, let AG be the logarithmick
curve, and HO its axis, the figure arising from the
revolution of AG about HO, is such as a speaking
trumpet ought to have in order to give it the greatest advantage possible. For from the nature of the
curve, if HI, IK, KL, LM, and so on, be taken
equal; the ordinates HA, IB, KC, LD, and
so on, are in geometrical proportion; wherefore
if HI, IK, and so on, be taken very small, they
will represent the equal altitudes of the cylindrical bodies of air in the trumpet; and the ordinates HA, IB, and so on, will be the radii of their
bases, and the bodies of air being of equal heights
will be to one another as their bases, that is, as the
squares of their radii; but the radii being to one
another in a geometrical proportion, their squares
will be so too; consequently, the little cylindrical
bodies of air will be in a geometrical progression,
the smallest whereof lies next the mouth, and the
largest at the exit of the tube; for which reason
the augmentation of sound will be greater, *cæteris
paribus*, in a trumpet of such a form than of any
other form whatever.

But to proceed to the second general case, wherein both the bodies move one and the same way, but
the subsequent more swiftly than the preceding.

Exp. 7.   If two equal elastick bodies move in the same direction, and in such a manner as that one may over-
take and strike the other, upon the stroke they will
change their quantities of motion with each other;
for instance, if the motion of the subsequent body
before the stroke be double the motion of the preceding body, then will the preceding body after the
stroke have double the motion of the subsequent
body after the stroke; and the preceding body after the stroke, will move with the same velocity
where-

wherewith the subsequent body moved before the
stroke; and the subsequent body will after the
stroke be carried with the velocity of the preceding
body before the stroke; so that upon the stroke the
bodies will change their motions and velocities. For
since by supposition the sum of the motions is three,
and since the bodies are equal, the motion of each
after the stroke, setting aside the elasticity, must be
one and an half; and by the second rule for determining the quantity of motion communicated by
the striking body to the other; the motion communicated in this case will be as one half, and so
likewise will the motion arising from the elasticity,
which being deducted from the motion which remains in the striking body after the stroke, and
added to that of the preceding body, leaves the
motion of the former as one, and of the latter as
two; so that upon the stroke the motions will be
changed. Wherefore if two ivory balls of an equal
size be let fall at the same time, one from the height
of six inches, and the other from the height of
three, after the stroke, the preceding ball will rise
to the height of six inches, and the subsequent to
the height of three only.

If the bodies be unequal and move the same way,
their motions and velocities after the stroke may in
like manner be discovered by the help of the proposition. For instance, if the subsequent body be as
two, and have twelve parts of motion, and the preceding body as one, and its motion as three; the
motion of the subsequent body after the stroke will
be as eight, and that of the preceding body as seven, and the velocity of the former will be as four,
and that of the latter as seven; for the sum of the
two motions before the stroke being fifteen, and
the bodies being as one and two, the motion of the
lesser body after the stroke, setting aside the elasticity, will be as five, and that of the greater as ten,
but the motion of the lesser body before the stroke

was

Lect. was at three, consequently, the communication
V. tion is as two; wherefore adding so much on account
of the elasticity to the motion of the lesser body,
and subducting as much from that of the greater
body, which in this case is the striking body, we
shall have eight for the motion of the greater, which
being divided by two, the quantity of matter in the
greater, gives four for its velocity; and we shall
have seven for the motion of the lesser body, which
because the quantity of matter in the lesser is as
one, will likewise express the velocity. Wherefore

Exp. 8. if two ivory balls, one double of the other, be let fall
at the same time, the larger from the height of six
inches, and the smaller from the height of three,
after the stroke the lesser will ascend to the height
of seven inches, and the greater to the height of

Exp. 9. four.   On the other hand, if the smaller ball be let
fall from the height of six inches, and the greater
from the height of three; after the stroke the lesser
will ascend to the height of two inches, and the
greater to the height of five, and the motion of
the former will be two, and that of the latter ten;
for since the smaller ball is as unity, and falls from
the height of six inches, its motion at the time of
the stroke is six; and since the larger ball is as two,
and falls from the height of three inches, the motion thereof at the time of the stroke is likewise
six; and the sum of those two motions, which is
twelve, being divided between the bodies in proportion to their quantities of matter gives eight for
the motion of the greater, and four for the motion
of the lesser, which motions they would have after
the stroke, supposing they were not elastick; and
since the motion of the greater body before the
stroke was six, the motion communicated to it by
the stroke is two, which by reason of the elasticity
being subducted from four, the motion of the
striking body, and added to eight, the motion of
the other body, gives two and ten for the motions
of

of the two bodies, which motions being divided by the refpective bodies, give two and five for the velocities.

Lscr.
V.

If two equal bodies meet one another with equal quantities of motion, which is one branch of the third general cafe, they will rebound with the fame motions and fame velocities wherewith they approached; for were they void of elafticity they would upon the ftroke ftand ftill, becaufe they communicate to one another a quantity of motion equal to that which each of them has at the time of the ftroke, and that in a contrary direction; but by the propofition, each of them muft on account of the elafticity receive as much motion as was communicated by the ftroke; and the motions which are thus received by the bodies being equal, and contrary to the motions wherewith the bodies met, and which were deftroyed by the ftroke, muft carry the bodies backward with the fame velocities wherewith they approached. Wherefore, if two equal ivory balls be let fall at the fame time from equal heights, fo as to meet one another, upon the ftroke they will be reflected back to the heights from which they fell.

Exp. 10.

If the balls be unequal, for inftance, if one be double the other; let the larger fall from one half only of the height from which the fmaller defcends, by which means when they meet their motions will be equal, and upon the ftroke they will be reflected each to the height from which it fell.

Exp. 11.

Where the bodies meet one another with unequal motions, which is the fourth general cafe, if the bodies be equal, they will both be reflected, and each of them will recede with the motion and velocity wherewith the other approached; that is, they will change their motions and velocities; for let us fuppofe the motions of the two bodies to be as fix and three; if they were void of elafticity the body which has the fmalleft quantity of motion would

would upon the ſtroke be turned back, and the two. bodies would be carried with the difference of their motions divided equally between them, that is, the motion of each would be as one and an half, and the motion communicated would by the fourth rule be as four and an half; but a quantity of motion equal to what is communicated, muſt be ſubducted from the motion remaining in the ſtriking body, and added to the motion of the other, that is, four and an half muſt be ſubducted from one and an half, and likewiſe added thereto; whereby there will be three negative for the motion of the ſtriking body, which ſhews that it will be carried back with a motion which is as three; and there will be ſix poſitive for the motion of the other body, which ſhews that it will be carried with a motion which is as ſix, in the direction of the ſtriking body before the ſtroke; that is, it will be reflected; ſo that each of them will be carried back with the motion wherewith the other approached. Where-

Exp. 12.
fore, if two equal balls of ivory be let fall at the ſame time, one from the height of ſix inches, and the other from the height of three, upon the ſtroke they will return back; but that which fell from the height of ſix inches will riſe only to the height of three, whereas that which fell from three inches will riſe to ſix.

If the balls be unequal, and meet one another with unequal quantities of motion, their motions after the ſtroke may in like manner be determined by the help of the rule laid down in the propoſition;

Exp. 13.
for inſtance, if two ivory balls which are as one and two be let fall at the ſame time, the greater from the height of ſix inches, and the ſmaller from the height of three; in this particular caſe, the greater ball will upon the ſtroke loſe all its motion, and the ſmaller will be reflected with the difference of their motions, ſo as to riſe to the height of nine inches; for ſince the larger ball which deſcends from the

height

height of fix inches is as two, its motion is as twelve, whilft the motion of the fmaller ball, which is as unity, and defcends only from the height of three inches, is as three, the difference of which motions is nine; and this being divided between the bodies in proportion to their quantities of matter, gives fix for the motion of the larger, and three for that of the fmaller; and with thefe motions the bodies would be carried after the ftroke, fuppofing they were void of elafticity; but becaufe of the elafticity, a quantity of motion equal to what is communicated by the ftriking body to the other, which in this cafe is fix, muft be taken from the motion of the greater body, and added to that of the fmaller, which two motions being fix and three, the remainder after fubduction, which expreffes the motion of the greater body, will be nothing; and the fum arifing from the addition, which expreffes the motion of the fmaller ball, will be nine.

LECTURE VI.

OF THE CENTER OF GRAVITY, BALANCE, AND LEVER.

MY defign in this lecture is to give you an account of the firft and fecond of the mechanick powers, commonly called the balance and the lever; but I fhall firft take notice of fome things relating to heavy bodies, the knowledge of which is in a great meafure neceffary to the right underftanding of what fhall be faid concerning the mechanick powers in general. And Firft, in every body there is a certain point, commonly called by the writers of mechanicks, the center of gravity; the nature of which will beft appear from its chief properties, which are thefe.

1ft, If a body be fufpended by its center of gravity, it will continue in any pofition whatever

wherein

Lect. wherein it is placed; whereas if it be fufpended by
VI. any other point, it will not reft in any other pofition
but where the center of gravity is either directly
above, or directly beneath the point of fufpenfion;
Exp. 1. thus, if two beams be fupported, the one by an
axle paffing thro' its center of gravity; the other
by an axle which doth not pafs thro' the center of
gravity, but thro' fuch a point, as when the beam
is parallel to the plane of the horizon, lies directly
above the center of gravity; the former will reft
in any pofition, whether it be perpendicular, parallel, or inclined to the horizontal plane; but the
latter will reft in the parallel pofition only; and
fhould it by any force be removed from that pofition, it will, upon the removal of the force, begin
to move in order to recover the parallel pofition, and
after feveral vibrations will at length fettle therein.

A fecond property of the center of gravity is, that
where that is fupported the whole body is likewife
fuftained; for which reafon the whole weight of a
body may be looked upon as applied to that fingle
point, and as centered therein.

A third property of this center is, that it continually endeavours to move downward towards the center of the earth, and where all lets and impediments
are removed does actually defcend; and therefore if
in any cafe a body feems to move upward by the force
of gravity, it will be found that the center of gravity
defcends notwithftanding any appearance to the con-
Exp. 2. trary. Thus, if two rulers be fo placed as to meet
in an angle at one of their ends, and there to reft
upon an horizontal plane, whilft at their other end
they are raifed a little above the plane; and if a
body confifting of two equal fimilar cones united
at their bafes, be laid upon the rulers in fuch a manner, that the edge of their bafes may lie between
the rulers, it will when left to itfelf begin to roll
towards the elevated extremities of the rulers, and
upon that account appear to afcend, whereas in reality

ality it moves downward; for if a string be stretched horizontally beneath the rulers so as that it may
touch the edge of the bases of the cones at the
concourse of the rulers, it will be found that the
edge of the bases descends below the string, and
that more and more as the body moves nearer to
the higher end of the rulers.

Whilst the body rolls upon the rulers, the parts
of the cones which rest thereon, do by reason of the
widening of the rulers grow continually smaller;
upon which account, at the same time that the body ascends along the plane of the rulers, it is as it
were carried down another plane equal in length to
the side of the cone, and whose perpendicular altitude is equal to the semidiameter of the bases of the
cones; and therefore, if the perpendicular altitude
of the rulers in that part where their distance is
equal to the length of the double cone, be less than
the semidiameter of the bases, the body will move
up along the rulers, because, by so doing, it will in
reality descend, and the descent thereof will be
equal to the difference between the semidiameter of
their bases, and the perpendicular altitude of the
rulers in that part where their distance is equal to the
length of the cones; but if that perpendicular altitude be equal to the semidiameter, the body will
rest on any part of the rulers, being carried as
much upward on one account, as it is downward
on the other; and if the altitude of the rulers be a
little increased, so as to exceed the semidiameter of
the bases of the cones, the body will roll down
the rulers, and thereby descend thro' a space equal
to that excess.

If a cylinder be so contrived as to have its center
of gravity near one of its sides, which may be done
by making a wooden cylinder hollow towards one
side, and then filling it with lead; when it is placed
on an inclined plane in such a manner as that the

side

LECT.
VI.
side which is neareft to the center of gravity may lean
towards the upper part of the plane, it will afcend,
provided the inclination of the plane be not too
fmall, but the center of gravity will at the fame
time defcend; for it will fuitably to its nature endeavour to move downward, and thereby caufe the
cylinder to revolve about its axis; and this revolution will make the cylinder, and confequently its
center of gravity, to move up the plane; fo that
the center of gravity will have as it were two motions, one upward arifing from the progreffion of
the cylinder along the plane, the other downward
occafioned by the rotation of the cylinder about its
axis; but the defcent occafioned by the latter motion, will be greater than the afcent arifing from
the former; as will appear by ftretching a line horizontally at the fame height with the center of gravity before the cylinder begins to roll, for after the
rotation ceafes the center of gravity will be beneath
the line; fo that upon the whole, that center will
be found to defcend notwithftanding the afcent of
the cylinder on the plane.

When the elevation of the plane becomes fo great
that the afcent arifing from the progreffion becomes
equal to, or greater than the defcent arifing from
the rotation, the cylinder will in the former cafe
continue at reft, and in the latter roll down the
plane.

A line drawn from the center of gravity of any
body, perpendicular to the plane of the horizon, is
called the line of direction of the center of gravity,
becaufe when the body is carried downward by the
force of gravity, if it meets with no let or obftacle, its center of gravity will defcribe that line.
The chief property of this line is, that as long as
it falls within the bafe of the body, fo long the body ftands, whereas no fooner does it fall beyond the
bafe, but the body tumbles; as will appear from
the

the following experiment; let a piece of wood be
set on a moveable plane with a plummet hanging
from its center of gravity, and let the plane be gradually elevated, till at length the plum-line (which,
as it is always perpendicular to the horizon, will
reprefent the line of direction) falls beyond the bafe;
the wood will not tumble as long as the plummet
line falls within the bafe, whatever be the elevation
of the plane whereon it ftands, but the moment
that line gets beyond the bafe the body falls.

The reafon why a body ftands during the continuance of the line of direction within its bafe is,
that no motion can arife in any body from the force
of gravity, unlefs the center of gravity can by fuch
motion be carried downward; but as long as the
line of direction of any body falls within the bafe,
its center of gravity is fupported, and therefore
cannot defcend; and confequently, the body will
remain unmoved; whereas upon the removal of the
line of direction beyond the bafe, the center of
gravity ceafes to be fupported, and is therefore at
liberty to defcend.

From what has been faid it appears, why among
bodies defcending on inclined planes, fome, for inftance cubes, only flide, whilft others as globes or
cylinders roll; the lines of direction falling beneath the bafes of the former, but not the latter.

The center of motion in any body is a fixed
point or axis about which the feveral parts of a
body do move, and in moving defcribe circular
arches.

The direction of any power or weight is, that
ftrait line wherein it moves or endeavours to
move. And the moment of any power or weight
is, that force wherewith it either moves or endeavours to move, and it is always proportional to the
product arifing from the multiplication of the power or weight into the velocity wherewith it moves
or would move if it were not hindred by fome oppofite

posite power or weight; and therefore if the product arising from the multiplication of one weight or power into its velocity, be equal to the product arising from the like multiplication of any other weight or power into its velocity, the momenta of those two weights or powers must be equal; and this will always be where the weights or powers are to one another reciprocally as their velocities; consequently, two weights or powers may balance, if as much as one exceeds the other in magnitude, so much must it be exceeded by the other in velocity; and herein consists the whole force and efficacy of all mechanical engines; for they are so contrived as to diminish the velocity of one weight or power and to increase that of the other, by which means a very small weight or power may become a balance to one exceedingly great, as will appear from what shall be said concerning the mechanick powers, which are commonly reduced to six, namely, the *balance*, the *lever*, the *pulley*, the *axle in the wheel*, the *wedge*, and the *screw*, of each of which in their order.

The BALANCE, strictly speaking, is a beam supported by an axle whereon it turns; which axle therefore is the center of motion; the parts of the beam which lie on each side of the axle are called its arms, and those parts of the arms to which the weights are applied are called the points of suspension; concerning which it must be observed, that the appending weight, whatever be the length of the cord by which it hangs, acts with the same force and in the same manner as if its center of gravity was applied to the point of suspension; so that it matters not what the distance is between the weight and point of suspension, as will appear from the following experiment.

Let a weight appended at one arm of a balance be counterpoised by a weight at the other, and let it by means of a cord be hung at different distances below the point of suspension; the position

of

of the balance will remain unvaried, and the weights will continue to counterpoise each other at all thofe diftances.

The moment of any weight appended at the arm of a balance, is proportional to the product arifing from the multiplication of the weight into the diftance of the point of fufpenfion from the axis of the balance; for as was before faid, the moment of a weight is proportional to the product of the weight into its velocity, and in this cafe the velocity of the weight is as the diftance of the point of fufpenfion from the axis; for fince the weight acts in the fame manner as if its center of gravity was applied to the point of fufpenfion, whatever be the velocity wherewith that point moves round the axis, the fame will the velocity of the weight be; but the velocities wherewith the feveral points in the arm of a balance move round the axis, are as the fpaces, that is, as the circular arches, which they defcribe in the fame time, which arches from the nature of the circle are to one another as their refpective radii, that is, as the diftances of the points from the axis. Thus, if A B reprefents the arm of the balance moving round the axis at A, the velocities of the points B and D, which defcribe the arches B C and D E, will be as thofe arches, becaufe they are defcribed in the fame time; but from the nature of the circle, thofe arches are to one another as their radii A B, and A D, that is, as the diftances of thofe points from the axis; confequently, the moment of a weight appended at the arm of a balance, is as the product of the weight into the diftance of the point of fufpenfion from the axis. Whence it follows, that if two weights be appended at the arms of a balance in fuch a manner, as that the diftances of the points of fufpenfion from the axis fhall be reciprocally proportional to the weights, thofe weights will counterpoife each other, and the balance will be in *æquilibrio*; for inftance, if two equal weights

Pl. 2.
Fig. 9.
Exp. 6.

be

be applied at equal distances from the axis, the balance will not incline to either side, but remain parallel to the horizon, the weights in this case counterpoising one another. Again, if one weight be larger than the other in any proportion, for instance, in the proportion of three to one, if the point at which the smaller is applied be thrice as far distant from the axis as the point at which the larger is applied, the balance will be in *æquilibrio*.

Exp. 7.

    On this *æquilibrium* arising from the suspension of weights at distances reciprocally proportional to the weights, is founded the *Statera Romana*, otherwise called the steel-yard, which consists of two arms very unequal in length, but equally poised by means of a weight annexed to the shorter, from which likewise hangs a scale in order to receive such things as are to be weighed; the longer arm is divided into a number of equal parts beginning from the axis, and sustains a weight with slides from one end to the other; which weight being applied to the second division, will counterpoise double the weight in the scale of the shorter arm, that it will when applied to the first division; and triple, when applied to the third division; and so on, whatever be the division to which it is applied, the weight in the scale of the shorter arm must be proportional thereto; otherwise the products arising from the multiplication of the weights into their respective distances from the axis would not be equal, and consequently would not balance each other.

Exp. 8.

Exp. 9.

    On the same *æquilibrium* is likewise founded the deceitful balance, which is so contrived, as that one arm be longer than the other, yet is the shorter made so much thicker than the longer, as thereby exactly to poise the same; upon which account the balance appears to be just, and consequently such weights as counterpoise are judged equal, whereas in truth that which is appended at the longest arm is less than the other, and that in the proportion of
the

ie length of the shorter arm to that of the longer; for instance, if the longer arm be to the shorter as ten to nine; a weight of nine ounces applied at the longer arm, will counterbalance ten ounces appended at the shorter.

Several weights appended at several distances from the axis in one side of a balance, will counterpoise several others appended likewise at several distances on the other side; provided the sum of the products which arise from the multiplication of the weights on one side into their respective distances from the axis, be equal to the sum of the products arising from the like multiplication of the weights on the other side into their respective distances. Thus, if on one side a weight of one ounce be appended at the distance of two inches from the axis, and another of two ounces at the distance of three inches, and a third of three ounces at the distance of four inches; and if on the other side be appended one weight of five ounces at the distance of an inch from the axis, and another of three ounces at the distance of five inches; the two latter will balance the three former; for the product of five into one, being added to the product of three into five, gives the sum of twenty; as does likewise the addition of the three products of one into two, two into three, and three into four.

The chief use of the balance, commonly called a pair of scales, is to compare the weights of different bodies together; and that this machine may be as exact and perfect as possible, it is requisite, 1st, that the center of gravity of the beam be placed a little below the axis, because in this case, when there is an equilibrium, the beam will not rest in any position but the parallel; consequently, the weights which are compared together will appear to be equal, as they really are; whereas if the axis be placed beneath the center of gravity, should the center of gravity be moved out of the perpendicu-

G 2   lar

lar line, which can scarcely be avoided
return, but from its tendency down
carried lower, so as to give the beam an
sition; for which reason the weights wi
be unequal, tho' in reality they are not
same inconvenience will arise if the axis
the center of gravity, for in that case
already shewn, that the beam, notwithstand
equilibrium, will rest in any position.

Secondly, the arms of the beam ought
actly equal both as to weight and length
son of which is evident, from what wa
cerning the deceitful balance.

Thirdly, the points from which the
suspended, ought to be in one right li
thro' the beam's center of gravity; for
trivance the weights will act directly
other, so that no part of either will be lo
count of any oblique direction.

Fourthly, the friction of the beam
axis ought to be as little as possible; becau
the friction be great, it will require a co
force to overcome it; upon which accou
one weight should a little exceed the oth
not preponderate, the excess not being, ju
overcome the friction, and bear down th

That the friction may be as little as po
parts of the beam which play upon the a
the axis itself, should be well polished, and
should be made as small as the uses of the
will admit; but as friction cannot be enti
vented, to remedy the inconveniences arisin
it as much as possible, the arms of the beam
to be made as long as they conveniently
cause the longer the arms are, the le
weight be that is requisite to overcome the
the moments of weights increasing in pro
their distances from the center of motion
been already shewn.

I shall close what I had to say concerning the balance, by laying before you one property of it, which is somewhat singular and surprising; tho' it has not that I can find been taken notice of by any of the mechanick writers *, namely, that if a man standing in one scale and counterpoised by a weight in the other, lays his hand to any part of the beam, and presses it upward, he will thereby destroy the balance, and make the scale wherein he stands to preponderate.

In order to account for this property, let A B Pl. 2.
Fig. 10. represent the beam of a pair of scales playing on the axis at C, and let a man standing in the scale D, and counterpoised by a weight in the scale E, lay his hand to some part of the beam, either on the same side of the axis with himself as at H, or on the other side as at K, and press the same upward; inasmuch as action and reaction are always equal, it is manifest that with whatever force the hand presses upward against the point H or K, with the same the hand, and consequently the man's whole body, is pressed downward; and therefore the scale D wherein he stands bears the same pressure from his feet that the point H or K does from his hand; but the pressure upon the scale D may be looked upon as applied to the beam at the point A from which the scale hangs; consequently, the same force which presses up the point H or K, presses down the point A; wherefore putting F to denote that force, $F \times HC$ will express the moment wherewith the arm AC is pressed upward when the hand is applied at H, and $F \times KC$ the moment wherewith the arm

---

* The property here mentioned, had not been taken notice of by any of the Mechanick Writers, when the Author composed this Lecture; but has been published since, both in the Philosophical Transactions for the year 1729, and in a course of experimental Philosophy, by Dr. Desaguliers, to whom the Author communicated it, as he told me and many others, about thirteen or fourteen years ago when he was in *London*.

BC

BC is pressed upward, the hand being applied at K;
and in both cases $F \times AC$ will express the moment
wherewith the arm AC is pressed downward by
means of the reaction; if therefore the hand be
applied at H, it is manifest that as the arm AC is at
one and the same time pressed upward by a force
which is as $F \times HC$, and downward by a force
which is as the same $F \times AC$, and as HC is ever
less than AC, the arm AC must descend with the
difference of those forces, that is, with a force equal
to $F \times AH$, which is the distance of the hand from
the point A; if the hand be applied at K, the arm
CB is pressed upward, and consequently AC downward, with a force equal to $F \times KC$, and upon account of the reaction AC is likewise pressed downward with a force equal to $F \times AC$; and therefore
it must descend with a force equal to the sum of
those two forces, that is, with a force equal to
$F \times AK$ the distance of the hand from the point A;
so that the scale D must preponderate, whether the
hand be applied to that part of the beam which
lies on the same side of the axis with the man, or to
that which lies on the other side; and if D be put
to denote the distance of that point to which the
hand is applied from the point A, the force wherewith the preponderating scale descends will be universally as $F \times D$, that is, as the force which the
hand exercises against the beam, multiplied into
the distance of the hand from the point A. And
if the force wherewith the hand presses the beam
be required, it may be discovered by throwing in
as much weight into the scale E as is sufficient to
balance the force of the hand, and to prevent the
descent of the scale D; for putting W to denote that
weight, its moment is as $W \times BC$ or AC, which being equal to $F \times D$ the moment of F, F will be
found equal to $W \times \dfrac{AC}{D}$, that is, to the weight
multiplied into half the length of the beam, and
divided

divided by the diftance of the hand from A. For
inftance, if the balancing weight be twenty pounds,
and the diftance of the hand from A be to half the
length of the beam as one to two, the force wherewith the hand preffes the beam is equal to twenty
pounds multiplied by two and divided by unity,
that is, it is equal to forty pounds ; from what has
been faid it follows, that when the hand is applied
to that part of the beam which lies on the fame fide
of the axis with the man, the force of the hand
upon the beam is greater than the weight which bal-
lances it in the fcale E, and lefs than the fame when
the hand is applied to that part of the beam which
lies on the other fide of the axis with refpect to the
man; for in the firft cafe, $W \times \frac{AC}{D}$ is greater than
W, and in the latter lefs, inafmuch as AC is in the
former cafe always greater, and in the latter lefs
than D.

The fecond, and indeed the moft fimple of all the
mechanick powers is the LEVER ; an engine chiefly
made ufe of to raife large weights to fmall heights.
By the writers of mechanicks, it is fuppofed to be
an inflexible line void of all gravity ; tho' fuch as
are in common ufe are both flexible and weighty.
In every lever there is one immoveable point, about
which as a center all the parts of the lever turn ;
and whatever fupports that point is called the prop ;
and with regard to the different fituations of the
moving power, and the weight to be moved in refpect to the prop, the lever is divided into three
kinds ; the firft of which is where the prop is placed
between the moving power and the weight to be
raifed ; which kind of lever is reprefented, wherein C denotes the prop, B the weight, and A the
power. In this lever there will be a balance between the power and the weight, provided they be
to one another reciprocally as their diftances from
the prop ; that is to fay, if the power at A be to

Pl. 2.
Fig. 11.

G 4   the

L E C T.
VI.

the weight at B, as CB to C A; for upon the motion of the lever round its fixed point C, the power
at A will deſcribe the arch AD in the ſame time
that the weight at B deſcribes the arch BE; conſequently, the velocity of the power will be to the
velocity of the weight, as the arch AD to the arch
BE; that is, becauſe the arches are ſimilar, as is evident from the manner wherein they are generated,
as AC to CB. That therefore the product ariſing
from the multiplication of the power into its velocity, may be equal to the product of the weight into its velocity; or in other words, that their moments may be equal, the power muſt bear the ſame
proportion to the weight, that BC the diſtance of
the weight from the prop bears to AC the diſtance
of the power from the prop. For inſtance, if BC be
to AC as one to two, and if a man's ſtrength be

Exp. 12.

ſuch as that without the help of a machine he can
ſupport an hundred weight, he will by the help of
this lever be enabled to ſupport two hundred; becauſe as BC is to AC, which by ſuppoſition is as
one to two, ſo muſt the power at A be to the
weight at B; but the power at A is ſuppoſed to be
equal to one hundred, conſequently the weight muſt
be equal to two.

As in this lever the prop may be placed either
at the middle diſtance between the moving power,
and the weight, or nearer to one than the other, it
is evident that there may be a balance between the
power and the weight, either when they are equal,
or when the one exceeds or is exceeded by the
other according to the different ſituations of the
prop.

To this kind of lever may be reduced ſeveral
ſorts of inſtruments, ſuch as ſciſſars, pincers, ſnuf-
fers, each of which may be conſidered as made up
of two levers, whoſe prop is the ſame with the pin
which rivets them together. Quarry crows are
likewiſe levers of this kind, concerning which it

muſt

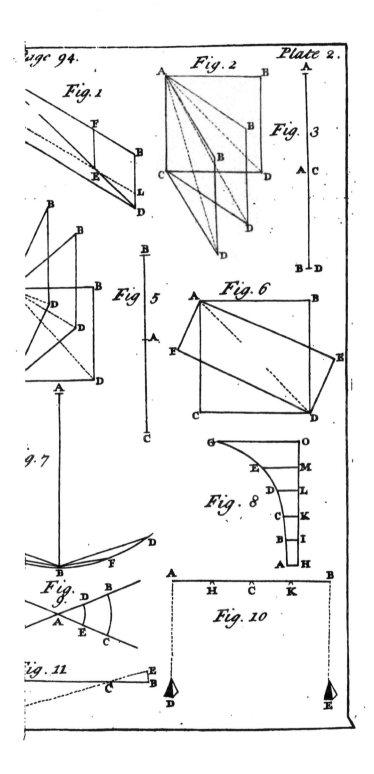

Fig. 1

Fig. 2

Plate 2.

Fig. 3

Fig. 5

Fig. 6

9.7

Fig. 8

Fig. 9.

Fig. 10

Fig. 11

muſt be obſerved, that the larger and more ponderous they are, provided they are not ſo big as to
become unmanageable, the more uſeful they muſt
be, becauſe the weight of that part of a crow which
lies on the ſame ſide of the prop with the power,
and which uſually far exceeds the other part in
length, acts in conjunction with the power, and
thereby facilitates the raiſing of the ſtones.

If the arms of this lever, inſtead of lying in a
right line, meet each other at the prop in a right
angle, where A C and B C repreſent the arms of a
lever united at the prop C, in ſuch a manner as
to conſtitute a right angle A C B; if to one arm as
C B placed horizontally, a weight be appended at B,
and to the other as A C ſtanding perpendicularly a
power be applied at A acting in the direction A D.
In order to a balance the power muſt be to the
weight as B C to A C, that is, the power and weight
muſt be in the inverſe *ratio* of the lengths of the
arms to which they are applied. For as the arms
turn together upon the prop C, in the ſame time
that the point B deſcribes any arch as B K, the point
A muſt deſcribe a ſimilar arch as A H; conſequently, the velocity of A will be to the velocity of
B as A C to B C; but as the moment of the power
at A is ſuppoſed equal to the moment of the
weight at B, the power muſt be to the weight, as
the velocity of the latter to the velocity of the former, that is, as B C to A C.

To confirm this by experiment, let B C be one·
fourth of A C, and a weight of twelve ounces be
appended at B; to the cord A D F made faſt to the
point A and paſſing over a pulley at D, let a weight
of three ounces be hung at F ſo as to pull the arm
A C in the direction A D, and there will be a balance. And if B C be one third or one half of A C,
then a weight at F, which in the former caſe is one
third, and in the latter one half of P, will balance,

the

the fame; and if AC and BC be equal, the balancing weights muft be fo too.

From the experiments, and what has been faid
concerning them, it is evident, that the greater the
proportion is which AC bears to BC, the greater is
the force of the lever, or the lefs the power at A
requifite to balance a given weight at B. And forafmuch as the hammer, when made ufe of in drawing nails is a lever of this kind; it is manifeft,
that the longer the handle is in proportion to
that part of the hammer which lies between the
handle and that portion of it which gripes the nail;
the lefs will the force be that is requifite to draw the
nail.

The fecond kind of lever has its prop at one end,
the power at the other, and the weight between; as
Pl. 3. where C is the prop, A the power, and B the
Fig. 2. weight; in this lever, in the fame time that the
Exp. 14. power at A moves thro' the arch of a circle whofe
radius is AC, the weight at B moves thro' a fimilar arch of a lefter circle whofe radius is BC; confequently, the velocity of the power is to the velocity of the weight as AC to BC; in order therefore
to a balance, the power muft be to the weight as
BC to AC; that is, as much as AC, the diftance
of the power from the prop, exceeds BC, the diftance of the weight from the prop, fo much muft
the weight exceed the power.

As in this lever the diftance of the weight from
the prop is always lefs than the diftance of the
power from the prop, it is evident that there cannot be a balance in any cafe but where the weight
exceeds the power.

To this kind of lever may be reduced the oars
and rudders of fhips; cutting-knives fixed at one
Exp. 15. end, and doors moving upon hinges.

If in this lever we fuppofe the power and the
weight to change their places; fo as that the power
may

may be applied at B between the weight at A and the prop at C, it will become a lever of the third kind; wherein in order to a balance, the power at B must so far exceed the weight at A, as BC the distance of the power from the prop, is less than AC the distance of the weight from the prop.

It is evident, that the moving power receives no advantage from this kind of lever, and, therefore is never made use of but in cases of necessity, and where the weights to be raised cannot be managed in a more convenient manner; as is the case of ladders, which being fixed at one end are by the force of a man's arms reared against a wall.

As levers are of service in raising weights, so are they likewise in carrying and supporting the same; concerning which it is to be observed, that when two powers support a weight by help of a lever, the sum of the powers must equal the weight; and the weight, being placed between them, their respective distances therefrom must be reciprocally as the powers; thus, if a weight resting on the lever at B be supported by two powers, one at A and the other at C, the distance of A from B must be to the distance of C from B, as the power at C is to the power at A. For in this case the lever is of the second kind, where each of the powers is in its turn to be looked upon as the prop, and then the other power must be to the weight as the distance of the weight from the prop to the distance of the power from the prop; that is, when A is considered as the prop, the power at C must be to the weight at B, as AB to AC; and when C is considered as the prop, the power at A must be to the same weight at B, as CB to CA. Consequently, since the power at A is to the weight, as BC to AC; and since the same weight is to the power at C, as AC to AB, the power at A must be to the power at C, as BC to BA, that is, the powers must be to one another inversely as their distance from the weight; and thus

Pl. 3.
Fig. 5.

LECT.
VI.

Exp. 16.

Pl. 3.
Fig. 4.

Exp. 17.

Pl. 3.
Fig. 5.
Exp. 18.

it will appear to be from experiments. For if from the point B of the lever A C a weight as D be suspended, and if two other weights as E and F be suspended from the extream points A and C by cords passing over pullies, so as that they may draw the lever directly upward; they will support the weight D provided the sum of those two weights be equal to the weight D, and the weight E be to the weight F as BC to BA.

The same thing will happen, if the three weights be made to pull the lever horizontally, which may be done by passing the cords over small wheels or pins placed on a level with the lever.

In shewing what the proportion ought to be between two powers which support a weight placed upon a lever, I have supposed the position of the lever to be parallel to the plane of the horizon; what the proportion ought to be, and in what manner such proportion is determined in inclined positions of the lever, shall be shewn, when I come to treat of powers acting in oblique directions.

If instead of a single lever, several be combined together in such a manner, as that a weight being appended to the first lever, may be supported by a power applied to the last, as in the machine, which consists of three levers of the first kind, and is so contrived as that a power applied at the point L of the lever C, may sustain a weight at the point S of the lever A. The power must be to the weight, in a *ratio* compounded of the several *ratios*, which those powers that can sustain the weight by the help of each lever when used singly and apart from the rest, have to the weight; for instance, if the power which can sustain the weight P by help of the lever A alone, be to the weight as one to five; and if the power whereby the same weight can be sustained by the help of the lever B alone, be to the weight, as one to four; again, if the power which can support the same weight by the help of the lever C alone,

alone, be to the weight as one to five; the power L
which supports the weight by means of those three
levers joined together will be to the weight in a
ratio compounded of one to five, one to four, and
one to five, that is, it will be as one to an hundred.
For since in the lever A, a power equal to one fifth
of the weight P pressing down the lever at L, is
sufficient to balance the weight; and since it is the
same thing whether that power be applied to the
lever A at L, or the lever B at S, the point S bear-
ing on the point L, a power equal to one fifth of
the weight P being applied to the point S of the
lever B, and pressing the same downward, will support the weight; but one fourth of the same power being applied to the point L of the lever B, and
pushing the same upward, will as effectually depress
the point S of the same lever, as if the whole power
was applied at S; consequently, a power equal to
one fourth of one fifth, that is, to one twentieth
part of the weight P; being applied to the point L
of the lever B, and pushing up the same, will support the weight; but it matters not whether that force
be applied to the point L of the lever B, or to the
point S of the lever C; since if S be raised, L
which rests thereon must be so too; but one fifth
of the power applied at the point L of the lever C,
and pressing it downward, will as effectually raise
the point S of the same lever, as if the whole power
was applied at S and pushed up the same; consequently, a power equal to one fifth of one twentieth,
that is, to one hundredth part of the weight P, being applied to the point L of the lever C, will balance the weight at the point S of the lever A; that
is, a power which is to the weight, in a *ratio* compounded of the three *ratios*, which the powers have
to the weight in each lever taken separately, will be
a balance to the weight, when the three levers are
used jointly. And by the same way of reasoning it
will

will be found, that in all machines of this kind, the power requisite to sustain the weight, is to the weight, in a *ratio* made up of the several *ratios* of the power to the weight in each lever taken separately, whatever be the number of levers.

In all that has been hitherto said concerning the lever, the power and the weight are supposed to act in direct opposition to each other; and on this supposition, the power must be to the weight in each of the three kinds of levers, in the reciprocal ratio of their distances from the prop, as has been fully proved with regard to each kind; but where the directions of the power and weight are inclined to each other, the proportion will vary from what has been here determined, as shall be shewn, when I come to treat of powers acting in oblique directions.

L E C T U R E VII.

Of the Pulley.

IN this lecture I shall give you an account of the *Pulley*, the *Axle in the Wheel*, the *Wedge*, and the *Screw*. The PULLEY is a small wheel that turns about its axis; and which has a drawing rope passing over it. It is made use of in raising large weights to considerable heights; and is of two kinds, fixed and moveable; the sole use of the fixed pulley, is to change the direction of the moving power; which in all cases where weights are to be raised to great heights, is exceedingly convenient,

and very often of absolute necessity; for instance, if the weight P is to be raised by the force of a man's hand to any height as A above the reach of the hand, the man must quit his place and ascend in order to carry up the weight, which for the most

part

part is found to be inconvenient, and sometimes impracticable; whereas if to a rope as P A F passing over the fixed pulley at A, the weight be made fast at one end as P, and the hand applied to the other end at F, the man by drawing the rope A F downward, will without moving from his place raise the weight as effectually, as if his hand was applied to it and moved upward from P to A; so that in raising weights to great heights the fixed pulley is of singular service, in as much as by changing the direction of the power, it takes off the necessity that a man would otherwise lie under of ascending along with the weight, and by so doing lessens his labour; besides, it has this farther convenience attending it, that by means thereof the joint strength of several persons may be made use of to raise one and the same weight, which in many cases cannot be done, at least not so conveniently, where the weight is raised by the immediate application of the hands; but this pulley does not in the least assist the power, by increasing its moment; because it neither lessens the velocity wherewith the weight rises, nor augments that of the power; for whatever be the space thro' which the power moves by drawing the rope A F, the weight must in the same time be drawn up thro' an equal space; the rope A P constantly shortening in the same proportion, that the rope A F is lengthened; and therefore, wherever any power supports a weight by means of a fixed pulley, that power must be equal to the weight.

When a pulley rises and falls along with the weight, as does this pulley, it is said to be moveable, and with regard to its use, it is just the reverse of the fixed pulley; for it adds to the moment of the power, but causes no change in its direction: for if the hand be applied at F to the rope D, in order to raise the weight P appended to the moveable pulley E, it must move directly upward

Pl. 3.
Fig. 7.

ward in the very fame manner, as if it was applied
immediately to the weight; confequently, the direction of the hand which raifes the weight is no
way altered by this pulley, but the moment thereof
is doubled, becaufe it is made to rife twice as faft
as the weight; for in the fame time that the hand
moves upward from F to G, thro' the fpace FG
equal in length to the two equal ropes D and C;
the pulley, and confequently the weight annexed,
will be drawn up thro' the fpace EH, whofe length
is equal to one of the ropes only.

In machines confifting of feveral pullies, whereof fome are fixed and fome moveable, and which
have one common rope that goes round them all; if
one end of the rope be fixed, as is the cafe in the
Pl. 3. machines reprefented by thefe figures, in order to a
Fig. 8, 9,
10. balance, the moving power muft be to the weight,
as one to twice the number of moveable pullies;
becaufe the velocity wherewith the power moves in
raifing weights by the help of fuch engines, is to
the velocity of the rifing weight, as twice the number of moveable pullies to unity; as I fhall now
Pl. 3.
Fig. 8. fhew you in the machine, which confifts of one
fixed pulley as A, and another moveable as E.
Since it is one and the fame rope that is continued
from G to F, the part AF which lies beyond the
fixed pulley, cannot be drawn down and thereby
lengthened, unlefs the two parts D and C, which
lie on each fide of the moveable pulley, be at the
fame time drawn up and fhortened, and that equally; whence it is evident, that the part AF will be
lengthened as faft again as either D or C is fhortened, inafmuch as what each of thofe parts lofe of
their length is added to the length of AF; but the
point F to which the power is applied, defcends
as faft as AF is lengthened, and the point E to
which the weight is faftened, afcends as faft as
D or C is fhortened; confequently, the velocity of
the power is to the velocity of the weight, as two

to one, that is, as twice the number of moveable pullies to unity; if therefore a weight appended at F, be to a weight appended at E, as one to two, they will balance each other, as being to one another in the reciprocal *ratio* of their velocities.

Pl. 3.
Fig. 9, 10.
Exp. 3.

In the machines, each of which confifts of two fixed and as many moveable pullies, and which differ only in this, that in one the pullies of the fame kind move upon one and the fame axis, and in the other upon different axes; I fay, in thefe machines, the velocity wherewith the power moves is to the velocity wherewith the weight rifes, as four to one, that is, as twice the number of moveable pullies to one; for as the part of the rope A F is drawn down and lengthened, the four parts B, C, D, H, which lie on each fide of the two moveable pullies are drawn up and fhortened, and that equally; and what each of them lofes of its length is added to the length of A F; confequently, A F is lengthened four times as faft as each of the other parts fhortens; but the power moves as faft as A F lengthens, and the weight rifes as faft as the other four fhorten; and therefore, the velocity of the power at F is to the velocity of the weight at E, as four to one, or as twice the number of moveable pullies to unity: for which reafon, if a weight be appended at F, which is to the weight at E as one to four, that is, in the reciprocal *ratio* of their velocities, there will be a balance.

What has been thus proved with regard to the three laft machines, namely, that the velocity wherewith a power moves in raifing a weight is to the velocity wherewith the weight rifes, as twice the number of moveable pullies to unity, is in the fame manner demonftrable with regard to any other machine of the fame kind, whatever be the number of pullies whereof it confifts; and therefore, in all machines confifting of feveral pullies, whereof fome are fixed and others moveable, and round

H   which

LECT.
VII.
which goes one common rope, fixed at one end, it may be laid down as a general rule, that in order to a balance between the moving power and the weight, the former muſt be to the latter, as one to twice the number of moveable pullies.

Exp. 4.   If the rope which goes round the pullies, inſtead of being fixed at one end, be faſtened to the weight or to the block which ſupports the moveable pul-

Pl. 3.
Fig. 11.
lies, ſo as to riſe therewith, as in this machine, which conſiſts of five pullies, whereof three are fixed and two moveable, and in which the end of the rope is joined at G to the block which ſup-ports the two moveable pullies ; the velocity of the power is to the velocity of the weight, as the ſum of twice the number of moveable pullies increaſed by unity to one ; for in this caſe, the parts of the rope which are equally ſhortened in order to lengthen the part A F, are more in number by one than the ſum of the moveable pullies when doubled ; conſequently, ſince the powei at F moves as faſt as A F is lengthened, whilſt the weight at E riſes in proportion only to the ſhortening of the ropes B, C, D, H, K, the velocity of the power bears the ſame proportion to the velocity of the weight, as the ſum of twice the number of moveable pullies increaſed by unity does to one ; and therefore, if the power be to the weight in the inverſe *ratio*, that is, as one to twice the number of moveable pullies added to unity, there will be a balance. Thus, if

Pl. 3.
Fig. 11.
in the machine a weight appended at F be to an-other at E, as one to five, they will balance, and remain unmoved.

Exp. 5.   If to any of the forementioned machines be added a runner, that is, a ſingle moveable pulley, which has its own rope diſtinct from that which is common to the other pullies, one end whereof is

Pl. 3.
Fig. 12.
fixed as at L, the other being faſtened to the block at E, and the weight appended at M, the force of the former machines will be doubled by this
addi-

additional pulley; for since the point E moves with twice the velocity of the point M, as I shewed when speaking of the single moveable pulley, whatever be the proportion which the velocity of the power at F bears to the velocity of the weight when appended at E, it will be doubled if the weight be appended at M; consequently, the power will by the help of the runner be able to sustain twice the weight that it did before.

If a machine be combined of one fixed and several moveable pullies, put together in such a manner as that each of the moveable pullies has a separate rope, one end whereof being fixed, the other either passes over the fixed pulley, as does that of the first-moveable pulley E, or is joined to the moveable pulley which lies next above it, as is the case of the ropes B, C, D, which belong to G, H, and I, the second, third and fourth moveable pullies; B being joined at N to the first moveable pulley, C at K to the second, and D at L to the third; the weight being appended to the last moveable pulley at H. The velocity wherewith the weight rises in such a machine is to the velocity of the power, as one to the last term of a duple progression, whereof the first term is unity, and the number of terms more by one, than the number of moveable pullies.

Pl. 3.
Fig. 13.
Exp. 6.

For as I proved when speaking of the single moveable pulley, the velocity of the power at F is to the velocity wherewith the pulley E rises, as two to one; and so likewise is the velocity of E, to that of G, and that of G, to that of I, and so on, whatever be the number of moveable pullies, the velocity of each succeeding pulley is but one half of the velocity of the preceding; wherefore, if the velocity of the last pulley, which is the same with the velocity of the weight, be put equal to unity, the velocity of that which immediately precedes it, to wit H, will be as two, and the velo-

H 2   city

city of G, as four, and of E, as eight, and so on; if there be more moveable pullies, the velocity will be continually doubled, and since the velocity of the last pulley is expressed by unity, that of the first will be expressed by the last term of a duple progression whose first term is unity, and the number of terms equal to the number of moveable pullies; and consequently, since the velocity of the power is double that of the first moveable pulley, if the duple progression be continued to one term more, that term will express the velocity of the power, the velocity of the weight being as unity; thus, in this machine, the number of moveable pullies being four, the velocity of the weight at M is to that of the power at F, as one to sixteen; if therefore a weight appended at F be to the weight at M, as one to sixteen, there will be a balance.

Tho' this engine be of greater force than any other wherein there is the same number of moveable pullies, yet inasmuch as it does for that very reason raise weights more slowly; men for the sake of dispatch choose rather to make use of such combinations of pullies as are represented in the 9th and 10th Figures, and where they have occasion to raise very large weights, they double the force of those machines by the addition of a runner.

The fourth mechanick Power is called the AXLE IN THE WHEEL; which is a simple engine consisting of one wheel fixed to the end of an axle that turns along with the wheel; the manner of raising weights by the help of this machine is thus; the power being applied to some part of the wheel's circumference, turns the wheel and together with it the axle, by which means a rope that is tied to the weight at one end, and made fast to the axle at the other, is wound about the axle, and thereby the weight drawn up; and for as much as the wheel and its axle revolve together, in whatever time the power moves thro' a space equal to the circumference

of

of the wheel, the weight muft in the fame time be
raifed up thro' a fpace equal to the circumference
of the axle, confequently, the velocity of the power
is to the velocity of the weight, as the circumference of the wheel to the circumference of the axle;
that is, from the nature of the circle, as the diameter of one to the diameter of the other; if therefore the power be to the weight in the inverfe *ratio*
of thofe diameters; that is to fay, if the power be
to the weight, as the diameter of the axle to the
diameter of the wheel, there will be a balance; the
power in that cafe being juft fufficient to fupport the   Exp. 7.
weight. For inftance, if the diameter of the wheel
be five inches, and that of the axle one, a weight
of one ounce hanging from any point in the circumference of the wheel, will fupport a weight of
five ounces hanging at the axle; and if the diameter of the axle be but half an inch, then will ten
ounces at the axle be fupported by one at the wheel.

Where the parts of the axle differ in thicknefs,
if weights be hung at the feveral parts, they may
be fuftained by one and the fame power applied to
the circumference of the wheel, provided the product arifing from the multiplication of the power
into the diameter of the wheel be equal to the
fum of the products arifing from the multiplication of the feveral weights into the diameters of
thofe parts of the axle from which they are fufpended. Thus a weight of five ounces hanging   Exp. 8.
from the part of an axle whofe diameter is one
inch, and another of ten ounces from a part whofe
diameter is half an inch, will be balanced by a
weight of two ounces hanging from the circumference of the wheel whofe diameter is five inches;
for the fum of the products of five into one, and
of ten into one half, which exprefs the moments
of the weights, is equal to ten, as is alfo the product of two into five, which expreffes the moment
of the power.

                         If

If to the axle in the wheel be added one or three
wheels with teeth, so that motion may be communicated from the first wheel to the last; the weight
being hung from the axle of the last wheel, whilst
the moving power is applied to the circumference
of the first wheel; in order to a balance, the power must be to the weight in a *ratio* compounded of
the inverse *ratio* of the diameter of the first wheel
to the diameter of the last axle, and of the inverse
ratio of the number of revolutions made by the first
wheel, to the number of revolutions made by the
last axle in a given time; for if the first wheel and
the last axle revolved in the same time, the ratio of
the diameter of the wheel to that of the axle, would
express the *ratio* of the velocity of the power, to
the velocity of the weight; but if the wheel revolves oftener than the last axle in a given time, it
is evident, that the *ratio* of the velocity of the
power to that of the weight, will be greater in that
proportion; consequently, the velocity of the power
must be to the velocity of the weight in a *ratio*
compounded of the *ratio* of the diameter of the
first wheel to the diameter of the last axle, and of
the revolutions of the first to those of the last axle in
a given time; and therefore, that there may be a
balance between the power and the weight, the former must be to the latter inversly in the same compounded *ratio*. For instance, in a machine con-

Exp. 9. sisting of two wheels with their axles, wherein the
diameter of the first wheel is four inches, and that
of the second axle a quarter of an inch, and wherein the cogs or teeth of the first axle, by applying
themselves successively to the teeth of the second
wheel, turn it about, and therewith its axle; but
the teeth of the first axle being in number but
one fourth of the teeth of the second wheel, that
axle, and consequently the first wheel, must revolve four times in order to turn the second wheel
and its axle once; so that, the revolutions of the
first

firſt wheel in a given time are to the revolutions of
the ſecond axle, as four to one : in this machine,
in order to a balance, the power muſt be to the
weight inverſly in a *ratio* compounded of ſixteen to
one, and of four to one ; that is, it muſt be to the
weight inverſly as ſixty-four to one ; ſo that a
weight of one ounce at the circumference of the
firſt wheel, will ſupport a weight of ſixty-four
ounces faſtened to the ſecond axle.

Again, in a machine compoſed of three axles, the two laſt having wheels with teeth, and the firſt a perpetual ſcrew, which in each revolution of the firſt axle moves one tooth only of the wheel of the ſecond axle ; which wheel having twenty-eight teeth, moves round once in the ſame time that the firſt axle turns twenty-eight times ; and there being a ſmall wheel with fourteen teeth at the other end of the ſecond axle, and theſe teeth applying them-ſelves continually to the teeth of a wheel fixed on the third axle, which are twenty-eight in number ; the wheel of the third axle muſt revolve but once in the ſame time that the wheel of the ſecond axle revolves twice, and of conſequence the third wheel and its axle move round but once whilſt the firſt axle performs fifty-ſix revolutions ; and the diame-ter of the firſt axle is to that of the laſt as two to one ; in order therefore to a balance between the power which is applied to the firſt axle, and the weight which is applied to the laſt ; the power muſt be to the weight inverſly in a *ratio* compounded of two to one, and of fifty-ſix to one ; that is, the power muſt be to the weight as one to an hundred and twelve ; ſo that one ounce hanging from the firſt axle will ſupport an hundred and twelve ounces hanging from the laſt axle.

In order to exhibit the force of the WEDGE, which is the fifth mechanick power, let AD repre-ſent the baſe of a wedge, from whoſe middle point B let the line BE be drawn perpendicular to the

Exp. 10.

Pl. 3.
Fig. 14.

H 4   ſide

repeated ftrokes of the mallet from B to C, (for I
fuppofe the edge of the wedge to be placed on the
top of a piece of timber at B in order to rend it)
the fpace defcribed by the wood as it yields on each
fide of the wedge in lines perpendicular to thofe
fides, is equal to B E. Confequently, that the moment of the mallet may be equal to the refiftance of
the wood, the abfolute force of the mallet muft be
to the force wherewith the parts of the wood cohere,
as BE to BC, that is, as the fine of the angle BCD
to radius; whence it follows, that all fimilar wedges
are of equal force, for in fuch the angle BCD is
given; it likewife follows, that the powers of dif-
fimilar wedges are inverfly as the fines of the angles
BCD, or in other words, that the forces requifite
to rend timber with fuch wedges, are directly as the
fines BE, which is confirmed by the following experiment,

Exp. 11. Let a machine be fo contrived, as to confift of
two equal cylinders, rolling upon their axles in an
horizontal pofition along the edges of two rulers,
and let them be drawn and kept together by a
weight of 2000 grains, hanging freely by a rope,
faftened at each end to the cylinders, and let the
edge of a wedge be placed between the cylinders,
fo as that when a fufficient weight is hung to it, it
may be drawn down between the cylinders; in this
machine the force wherewith the cylinders are drawn
together, added to the attrition of their axles in
rolling upon the rulers, may be looked upon as
the refiftance of the timber, and the weight of the
wedge

wedge together with the appending weight whereby it is pulled down between the cylinders, as the
force of the mallet upon the wedge; now, if three
wedges be made ufe of, each three inches long, in
which the fines B E are as one, two, and three, their
weights likewife being in the fame proportion, the
firft will be drawn down by a weight of 300 grains,
the fecond by one of 600 grains, and the third by
one of 900 grains.

To the wedge may be reduced the axe or hatchet,
the teeth of faws, the chizel, the augur, the fpade
and fhovel, knives and fwords of all kinds, as alfo
the bodkin and needle, and in a word, all forts of
inftruments which beginning from edges or points
grow gradually thicker as they lengthen; and the
manner wherein the power is applied to fuch inftruments, is different according to their different fhapes
and figures, and the various ufes for which they
were contrived.

The next and laft mechanick power is the
screw, which confifts of two parts, whereof the
firft is called the male or outfide fcrew, being a
cylinder cut in, in fuch fort as to have a prominent
part going round it in a fpiral manner, which prominent part is commonly called the thread of the
fcrew; the other part which is called the female or
infide fcrew, and by common workmen the nut, is
a folid body that contains an hollow cylinder, whofe
concave furface is cut in the fame manner as the
convex furface of the male fcrew, fo that the prominent parts of the one may fit the cavities of the
other. The chief defign of this machine is to
prefs the parts of bodies clofely together, and in
fome cafes to break and divide them; when it is
made ufe of one part is commonly fixed, whilft
the other is turned round, and in each revolution
the moveable part is carried in the direction of the
axis of the cylinder thro' a fpace equal in length to
the interval between two contiguous threads, where-

by

by the parts of the body whereon the preffure is made are forced to move towards one another thro' a fpace equal to that interval; which interval there- fore does exprefs the velocity wherewith the feveral parts of the body give way to the preffure, whilft the circular periphery, which is defcribed by the power whereby the moveable part of the fcrew is turned round, expreffes the velocity of the power; for the moveable part of the fcrew is ufually turned by means of an handle or handfpike, to fome part of which the power is applied, and by moving round with that part defcribes the circum- ference of a circle; if therefore the moving power be to the refiftance of the body which is preffed, as the diftance between two contiguous threads of the fcrew to the circular periphery defcribed by the power, there will be a balance; and if the power be ever fo little increafed beyond that proportion, it muft overcome the refiftance, and move the fcrew; and thus it would conftantly be, provided there was no refiftance from the attrition of the parts of the fcrew one againft another; but as that is very con- fiderable, there is an addition of power requifite to overcome it, over and above what is neceffary to overcome the refiftance of the body whereon the preffure is made: for which reafon fuch experi- ments as are made to fhew the force of the fcrew, muft vary more from the theory, than thofe which have been made concerning the other mechanick powers, wherein the attrition is far lefs confider- able; however, it will appear from the following experiment, that fmall powers are fufficient by the help of the fcrew to overcome great refiftances in the bodies which are preffed.

Exp. 12.   Let a wheel whofe diameter is four inches, be fixed at its center to the head of a male fcrew in an horizontal pofition, and let the end of a rope, which is wound about the groove of the wheel, pafs over a pulley in fuch a manner as that having

a weight

a weight faftened to it, it may be drawn in a line, that is a tangent to the wheel, by which means the intire gravity of the weight will be employed in turning the wheel; to one end of a lever, fupported by a prop at the middle, let a weight of feven pounds be hung, and let the bottom of the male fcrew reft on the other end of the lever; and let the diftance between the threads of the fcrew be equal to one fifth of an inch, and a weight of three ounces and 250 grains being hung to the end of the rope which paffes over the pulley, will juft turn the wheel, and thereby thruft down the fcrew, and with it the end of the lever whereon it refts, and by fo doing raife up the weight at the other end.

In this cafe the power which moves the fcrew, is to the weight raifed whereby the refiftance that is made to the preffure is meafured, as one to 24 nearly; whereas it ought not to exceed the proportion of one to 63; for the diameter of the wheel being four inches, the circumference is twelve and an half nearly, but 12.5 is to $\frac{1}{5}$, which is the interval between the threads of the fcrew, as $62\frac{1}{2}$ to one; confequently, if the power which turns the fcrew be to the weight that is to be raifed in the inverfe *ratio* of thofe numbers, that is, as one to $62\frac{1}{2}$, it ought to balance the weight, and if it be increafed ever fo little, it fhould overpower and raife the weight: fince therefore the force that is requifite to turn the wheel is nearly three times as great as what is neceffary to overcome the refiftance of the weight to be raifed, it is evident, that almoft two thirds of that force is employed in overcoming the refiftance arifing from the attrition of the parts of the fcrew one againft another; what the nature of this refiftance is, and in what proportion it varies, fhall be fhewn hereafter.

LECTURE

LECTURE VIII.

Of Compound Engines.

THE mechanick powers, which for the moſt part are made uſe of ſeparately, may in many caſes be combined together, and engines thereby formed of ſuch efficacy, as that by the help thereof exceeding great weights may be raiſed by very ſmall powers. In all ſuch compounded machines the proportion which the moving power bears to the weight when they balance each other, is compounded of the ſeveral *ratios* which thoſe powers have to the weight which balance it in each ſimple machine, whereof the compounded engine conſiſts. Thus when a machine is compoſed of an axle in the wheel and a pulley, by faſtening the drawing rope of the one to the axle of the other; the power which balances the weight in ſuch a machine, muſt be to the weight, in a *ratio* compounded of the *ratio* which that power has to the weight which balances it by means of the axle in the wheel alone, and of the *ratio* which that power has to the weight, which balances the weight by means of the pulley alone. For inſtance, if the nature of the pulley be ſuch, as that a power equal to one tenth part of the weight balances it; and if the axle in the wheel be ſuch, as that a power equal to one fifth part of the weight can ſupport it; the power which balances the weight in the compounded machine, will be to the weight in a *ratio* compounded of one to ten, and of one to five, that is, it will be to the weight as one to fifty; for, ſince the weight is in effect faſtened to the axle of the wheel by means of the rope which goes round the pullies, it is evident that the axle will be drawn by a force equal to that,

Exp. 1.

which

which when applied to the drawing rope of the pulley is requisite to sustain the weight by means of the pulley, which force is by supposition equal to one tenth part of the weight; but that force at the axle is balanced by a fifth part thereof applied to the wheel; consequently, the power requisite to balance the weight in this machine, is equal to one fifth of one tenth part of the weight, that is, the power is to the weight, as one to fifty. So that one ounce at the wheel will support fifty ounces at the pulley.

If a machine be composed of the lever, the axle, and the perpetual screw; the lever being thirteen inches long, and fixed at its center to an axle, whereon is a perpetual screw, the tooth whereof adapts itself to the teeth of the wheel of an axle, the teeth of that wheel being twenty-four in number, and the diameter of the axle belonging to that wheel equal to six tenths of an inch; in such a machine the power being applied to one end of the lever, and the weight to the axle of the toothed wheel, the former will balance the latter, if it be in proportion thereto, as one to 520; for if the lever to which the power is applied, moved round in the same time with the axle of the toothed wheel whereunto the weight is fastened, the power would be to the weight, as the diameter of the axle to the length of the lever, that is, as six tenths of an inch to thirteen inches, or in whole numbers, as six to an hundred and thirty; but as there are 24 teeth in the wheel of that axle which sustains the weight, and as the endless screw moves but one of those teeth in each revolution of the lever, the lever must go round 24 times in order to turn the axle, which sustains the weight, once; upon which account the power must be to the weight, as one to 24, which *ratio* of one to 24 being combined with the former of six to 130, gives a *ratio* of six to 3120, or of one to 520; so that an ounce weight being made to act with all its gravity at one end of

Exp. 2.

the

the lever in order to turn it round, which may be
done by fixing a wheel to the lever, will balance a
weight of 520 ounces at the axle of the toothed
wheel.

Exp. 3.   If to the laft machine one moveable pulley be
added, it will conftitute a machine of double the
force; for the *ratio* of the power to the weight in
the foregoing machine, being as one to 520, and in
a fingle moveable pulley, as one to two; the *ratio*
compounded of both, will be as one to 1040; fo
that in this machine an ounce will balance 86
pounds 8 ounces; and if the ftrength of a man's
hand be fuch, as that it can without the affiftance of
an engine fupport an hundred pounds, it will by
the help of this machine fuftain 104000 pounds.

In all that has been hitherto faid concerning the
mechanick powers, the moving force and the
weight or refiftance have been fuppofed to act in
direct oppofition to one another. I fhall now confider the effects of powers acting obliquely, and
fhew in what cafes they balance each other.

And firft, if three powers acting in oblique directions, be to one another, as the refpective fides of
a triangle formed by the concourfe of three lines
drawn parallel to the directions of the powers;
thofe powers will balance one another. For inftance, if three powers drawing the point A in the
directions A B, A C, and A E, be to one another, as
the fides of the triangle A D E, or A D C, made by
the concourfe of the lines A D, A E, and E D; or
A D, C D, and A C, which lines are parallel to the
directions of the powers; they will balance one another, and the point A will remain unmoved.

For if the line A D be fuppofed to denote a power
equal to that which acts in the direction A B, but
contrary thereto; the power denoted by A D will
draw the point A as forcibly towards D, as it is
drawn by the oppofite power towards B; confequently, there will be a balance between the two
powers;

Pl. 4.
Fig. 1.
Exp. 4.

powers; but the power denoted by A D may be
resolved into two powers denoted by A E or C D,
and A C or E D; which two powers acting together upon the point A in their proper directions A E and A C, will draw it as strongly towards
D, as it is drawn by the single power denoted by
A D; as is evident from what has been said concerning the resolution and composition of motions and
forces; consequently, two powers which are as A E
or C D, and E D or A C, acting in the directions
A E and A C, will balance the third power which
is as A D acting in the direction A B; that is, two
powers, which are as the two sides of a triangle,
acting in directions parallel to those sides will balance a third power, which is as the third side, and
which acts in a direction parallel thereto; and what
has been thus proved in particular of two of the
powers with regard to the third, is in like manner
demonstrable of any two of the powers with respect
to the other; consequently, any three powers
which are to one another respectively as the sides
of a triangle, and which act in directions parallel to
those sides, will destroy each the other's effect, and
remain in *æquilibrio*. To confirm this by an experiment; let the sides of a triangle A B C drawn on
an horizontal plane be as two, three, and four;
and let C E be parallel to the side A B, and the side
A C continued towards D. Let three small cords
be joined together at C, and stretched over three
pullies in such a manner, as that one of them may
cover the line C D, another the line C E, and the
third the line C B; this being done, if a weight of
four ounces be hung to the cord which passes over
C D, and one of three ounces to that which covers
C B, and one of two ounces to that which covers
C E, there will be a balance; the weights, which
in this case are the moving forces, being to one
another as the sides of the triangle to which the direction of the weights are parallel.

Pl. 4.
Fig. 2.
Exp. 5.

If

If the weight A hangs freely from one end of a balance, so as to have its line of direction D A perpendicular to the arm of the balance; and if another weight as B, be hung at the other end E, in such a manner, as that its line of direction E C by passing over a pulley at C may be oblique to the arm of the balance, the weight B must be to the weight A when it counterbalances it, as E C to C F, that is, as radius to the sine of the angle C E F made by the oblique direction of B with the arm of the balance; for if the whole force of gravity in the weight B acting in the direction E C, be denoted by the line E C, it may be resolved into two forces denoted by E F and F C, acting in the directions of those lines, of which two forces, the latter only which acts in the direction F C perpendicular to the arm of the balance withstands the force of gravity in the weight A, the other force which acts in the direction E F being entirely employed in pressing the balance against the axis of its motion; since therefore, that part of the weight B which acts in opposition to the weight A, is to the whole weight B, as F C to E C; it is manifest, that in order to make the weight B balance the weight A, it must exceed the weight A in the same proportion that the line E C exceeds the line F C; and thus it is found to be from experiments; for if the pulley be so ordered as that E C may be to F G as three to two, then a weight of three ounces appended at E, will balance one of two ounces appended at D.

As a COROLLARY it follows, that the perpendicular distances of the lines of direction from the center of motion, are to one another inversly as the weights; for, if from G the center of motion be let fall G H perpendicular to E C, that line will be the perpendicular distance of the direction E C from G; and E G, equal to D G, is the perpendicular distance of the direction D A; but the triangles

E F C

EFC and EHG are similar, because their angles
at E are equal, and they have each a right angle;
consequently, as EC is to CF, so is EG to HG;
but the weight B is to the weight A, as EC to FC,
that is, as EG or DG to HG; so that wherever two
powers, which act in oblique directions, are to one
another in the inverse *ratio* of the perpendicular or
shortest distances of their lines of direction from the
center of motion, they must balance one another;
whence it follows, that if two weights as A and B,
be suspended from two points as D and E in the
plane of a wheel placed in a vertical position; and
if the line DE which is drawn thro' the two points
of suspension, passes thro' C the center of motion,
the weights will balance, provided they be to one
another inversly as the distances of their points of
suspension from the center of motion, that is, if A
be to B, as CE to CD; for since the weights hang
freely, their lines of direction DA and FB, will
be perpendicular to the horizon, and of consequence,
parallel to each other; wherefore, if the line HCF
be drawn thro' the center of motion perpendicular
to the two lines DA and FB, the triangles DHC
and ECF will be similar, consequently, DC will
be to EC, as HC to FC; but by supposition, the
weight A is to the weight B, as CE to CD; that
is, as CF to CH; so that the weights are to one
another inversly as the perpendicular distances of
their lines of direction from the center of motion;
consequently, they must balance; and tho' the
wheel should be turned upon its axis, and the distances of the lines DA and EB from C be thereby
altered, yet will the similarity of the forementioned
triangles continue, and of consequence the balance
between the weights will be preserved; as will appear from the following experiment. Let a weight
of one ounce be suspended from the point D, and
another of two ounces from the point E; DC be-

Pl. 4.
Fig. 5.

I   ing

ing to EC, as two to one, that is, inverfly as the weights, there will be a balance, and the wheel will continue at reft. And if by the force of the hand it be turned about its axis either to the right from I towards K, or to the left towards M, the balance will ftill continue, and the wheel will remain unmoved when the hand quits it, whatever be its pofition.

Pl. 4.
Fig. 6.

If the points of fufpenfion D and E be fo pofited, as that the right line DE which joins them, does not pafs thro' C the center of motion ; let that line be divided any where as in G by another line as IL paffing thro' the center C, and there will be a balance, if the appending weights be to one another inverfly as the parts of the line DE, that is, if A be to B as EG to DG, provided the pofition of the wheel be fuch, as that the line IL may be perpendicular to the horizon ; for fince the lines EF, GC, and DH are parallel, FC is to HC, as EG to DG ; but by fuppofition, as EG is to DG, fo is A to B ; wherefore A is to B, as FC to HC, that is, the weights are inverfly as the perpendicular diftances of their lines of direction from the center of motion, confequently, their moments are equal ; but if by turning the wheel about its axis the line IL be put out of its perpendicular pofition, the balance will be deftroyed ; becaufe, in that cafe, one of the lines of direction will approach nearer to the center of motion, whilft the other recedes ; and of courfe their perpendicular diftances will not continue in the inverfe *ratio* of the weights ; for if the wheel be moved upon its axis from I towards K, fo as to have the line SCR perpendicular to the plane of the horizon ; the line of direction DA will approach towards the center fo as to become DP, and its perpendicular diftance from the center of motion will be NC, whilft the other line of direction recedes as far as EQ, and its perpendicular diftance

from

from C becomes equal to OC; for which reason
the weight B muſt preponderate, and move the
wheel about its axis in the direction IKL.   And
as the wheel continues to move in that direction,
the direction of the weight A will approach nearer
and nearer to the center of motion, and at length
paſs beyond it, ſo as to be on the ſame ſide with
the direction of the weight B; ſo that the wheel
will then be moved by the joint force of both
weights, and continue ſo to be, till ſuch time as
the direction of the weight B getting on the other
ſide of C, B begins to act in oppoſition to A, and
at length the point I being brought into the place
of L, the weights do again balance each other,
the line BE being divided by the perpendicular
line IL in the reciprocal *ratio* of the weights.   To
confirm what has been ſaid by an experiment, let
the line DE in the plane of a wheel be divided in
G by the line IL in ſuch a manner, as that DG
may be double of EG; then ſetting the line IL
perpendicular, let a weight of one ounce be hung
from D, and another of two ounces from E, and
the wheel will remain unmoved; let then the wheel
be turned a little upon its axis, either to the right
hand or to the left; in the former caſe, the two
ounce weight will prevail, and carry the wheel
downward to the right hand, but in the latter the
ſmaller weight will preponderate, and make the
wheel to revolve towards the left.

Exp. 8.

If the line DE be divided in another point as T,
by the line SR, ſo as that DT may be one third of
ET; and if a weight of three ounces be ſuſpended
at D, and another of one ounce at E, the ſame
things will happen as in the former experiment;
for the line SR being placed vertical there will be
a balance; and upon moving it out of that poſition the balance will be deſtroyed.

Exp. 9.

If the crooked lever FCD be ſo placed on its
prop at C, as that the arm CF may be parallel to

Pl. 4.
Fig. 7.

the plane of the horizon, and the arm CD inclined
thereto; if two weights as B and A, appended at
D and F, be in the reciprocal proportion of the
perpendicular diftances of their lines of direction
from the prop; that is, if B be to A as FC to EC
there will be a balance; for as long as the arm CF
continues parallel to the horizon, the weight B hanging from the point D acts in the fame manner in
oppofition to the weight A, as if it hung from E
the extremity of the ftrait lever FC continued on
to E, in which cafe the weight B that balances the
weight A muft bear the fame proportion to it that
FC does to EC; if therefore the arm DC be bent
in fuch a manner, as that EC may be one half or
one third of FC, in the former cafe a weight of two
ounces, and in the latter one of three ounces hanging from D, will be counterpoifed by one ounce
hanging from F.

If by moving the lever, the arm FC be put out
of its parallel pofition, the balance will be deftroyed; for that cannot be preferved, unlefs the diftance of B's direction from the prop continues to
bear the fame proportion to the diftance of A's direction, that EC does to FC; which in this cafe is
impoffible; for firft, if the point F be moved upward towards H, and of courfe the point D downward towards G, it is manifeft, that the diftances
of both directions will be leffened; but the decreafe
of EC in a given time will bear a greater proportion to the decreafe of FC than EC does to FC;
for by that time the point D has moved from D to
G thro' the arch DG, which meafures the angle of
CD's inclination, EC will vanifh; whereas FC
cannot vanifh till fuch time as the point F has moved
from F to M thro' the quadrantal arch FM; but in
the fame time that the point D moves from D to
G thro' the arch DG, the point F can move only
from F to H thro' the arch FH fimilar to DG;
which arch being always lefs than the quadrant, the

perpen-

perpendicular diſtance of A's direction from the prop, to wit FC, will not vaniſh upon the arrival of the point F at H, that is, it will not vaniſh ſo ſoon as EC; conſequently, the decreaſe of EC in a given time muſt bear a greater proportion to the decreaſe of FC, than EC does to FC: wherefore EC as diminiſhed in any given time, will be to FC as diminiſhed in the ſame time, in a leſs proportion than that of EC to FC; or in other words, the perpendicular diſtance of B's direction from the prop will bear a leſs proportion to the perpendicular diſtance of A's direction, than EC does to FC; and therefore, the weight A will preponderate. If the point F be moved downward, and conſequently D upward, it is manifeſt from the inſpection of the figure, that the diſtance of A's direction from the prop continually diminiſhes, at the ſame time that the diſtance of B's direction increaſes; and therefore the weight B muſt in that caſe overbalance the weight A.

If FCD be a crooked lever placed as the laſt, and if a weight, inſtead of being hung from the arm DC, be laid thereon at D, and by a vertical plane, as HK, ſet cloſe to it, be hindered from falling off; from the point D whereon the weight reſts, let the line DE be drawn perpendicular to the arm FC continued on towards G; the weight at D will be balanced by the weight A hanging freely from F, provided the weight D be to the weight A, in a *ratio* compounded of EC to CD, and of FC to CD; that is, as a rectangle under EC and FC the perpendicular diſtances of the directions of the two weights from the prop, to the ſquare of CD the inclined arm of the lever. For whatever be the moment wherewith the weight A preſſes down the arm FC, the arm DC muſt with an equal moment be preſſed upward, and with it the weight D in the direction DG perpendicular to

Pl. 4.
Fig. 8.

I 3

.CD;

CD; and forasmuch as the same weight presses perpendicularly against HK the vertical plane, it must be pressed backward by the same in an horizontal direction; and at the same time it must have a tendency downward from the force of gravity in the direction ED; so that it is acted upon by three forces in the directions DG, GE and ED; in order therefore to a balance, the forces must be as the sides of the triangle DGE; and the force of gravity which presses it in the direction ED, must be to the force pressing it in the direction DG, as ED to DG, or, because the triangles DGE and CDE are similar, as EC to CD; but as the force which presses it in the direction DG is of equal moment with the weight A, that force must be to the weight A, as FC to CD; consequently, the force of gravity in the weight D must be to the force of gravity in the weight A, that is, the weight D must be to the weight A, in a *ratio* compounded of EC to CD, and of CF to CD, or as the rectangle under CE and CF to the square of CD. To confirm

Exp. 10. this by experiment, let a crooked lever as FCD consist of equal arms, and let it be bent in such a manner, as that EC may be to CD, as one to two; and let a weight of one ounce be laid on at D, and another of two ounces be hung from F, and they will balance each other; for in this case the product of EC which is as one, into CF which is as two, will be two; and CD being as two, the square thereof will be four; so that the rectangle under EC and CF, is to the square of CD, as two to four, or as one to two; in which proportion therefore the balancing weights must be.

Exp. 11. All things remaining as in the last experiment, excepting that the arm CF is as long again as CD, so that EC, CD, and CF are as one, two, and four; a weight of one ounce at D will be balanced by one ounce hanging freely from F; for CD being

ing as two, its square is four; and the product of
EC, which is as one into FC, which is as four, is
likewise four.

In wheels turned by the force of water falling Exp. 12.
upon them from an height, and which on that account are commonly called overshot wheels, the
moving power is partly the percussive force of the
water which falls into the uppermost bucket, and
partly the gravity of the water contained in the
other buckets, which are lodged on the rim of the
uppermost quarter of the descending part of the
wheel; and the effects which these forces have upon the wheel are greater or less in proportion to their
absolute quantities, and the distances of their lines
of direction from the center of the wheel. Thus,
where AIOP represents an overshot wheel, C its Pl. 4.
center, K, L, M, N four buckets fixed on the up- Fig. 9.
permost quarter of the descending part of the wheel;
AB the direction of the water flowing into the uppermost bucket K, CB the perpendicular distance
of that line from the center C; DE, FG, and HI,
the lines of direction of the centers of gravity of
the several portions of water contained in the buckets L, M, N; CE, CG, and CI, the perpendicular distances of those lines from the center C.
The force of the water flowing into K is proportional to the quantity flowing in in a given time, as
also to the velocity wherewith it flows, and the diftance of its line of direction from the center; and
therefore, where the quantity and velocity are given,
the force will be as BC the perpendicular distance
of AB the line of direction, from C the center of
motion; consequently, the nearer AB approaches to
the tangent in the point A, or the more obliquely
the water flows in upon the wheel, the greater will
its force be. The portions of water contained in
the buckets L, M, N, have different forces according to their different quantities, and the different
distances of their lines of direction from the cen-

I 4

ter

ter C, their quantities being greatest, when the diftances of their directions are least, for the buckets
empty as they descend ; so that their force lessens as
they descend, by reason of the diminution of their
quantities, but at the same time it likewise increases
on account of the increase of the distance of their
lines of direction from the center of motion ; so
that upon the whole, the force in each bucket may
be looked upon as invariable ; but whether this be
so or not, certain it is, that if the wheel be truly
centered, and the buckets be equal and alike, and
if the water flows in uniformly, the whole moving
force must continue the same as long as the wheel
continues to move ; and since it acts incessantly,
the motion of the wheel must be continually accelerated, and that uniformly ; and thus it would
be, were it not that when the wheel arrives at a certain degree of velocity, the resistance which is given becomes so great as to destroy the increments
of motion as fast as they are generated by the
moving force ; by which means the wheel is made
to revolve with one uniform velocity, which is the
greatest that can be given it by that moving
power.

Pl. 4.
Fig. 10.
A plane as A B placed obliquely to BC, which
represents an horizontal plane, is called an inclined
plane ; the angle A BC is called the angle of elevation, and its complement B AC the angle of inclination ; the line A C perpendicular to BC is called
the height of the plane, and AB its length. If a
weight as P be laid on an inclined plane as A B, and
be thereon sustained by a power acting in a direction,
as PF, parallel to the inclined plane ; in order to a
balance, the sustaining power must be to the weight
as the height of the plane to the length thereof,
that is, as A C to A B, or, putting B A for the radius,
as the sine of the angle of elevation to radius ; for
the weight P is acted upon by three powers in different directions, the first of which is the force of
gravity,

gravity, which preffes it downward in the direction
PD perpendicular to BC; the fecond is the power
which draws it in the direction PF parallel to BA,
and the third is the plane BA, which does as it were
prefs it upward in the direction PH perpendicular
to BA; for as the weight P preffes the plane in a
direction perpendicular thereto, it is reacted upon
by the plane in a contrary direction. If therefore
the line EG be drawn parallel to PD, the fides of
the triangle PEG will be proportional to the three
powers, and the force which fupports the weight on
the inclined plane, and which acts in the direction
PF, will be to the abfolute weight of the body acting in the direction PD parallel to GE, as PG to
GE; but inafmuch as the triangles PEG and CBA
are fimilar, as PG is to GE, fo is AC to AB;
confequently, the power neceffary to fupport a
weight on an inclined plane muft bear the fame proportion to the weight fuftained, that the height of
the plane does to its length; which is confirmed by
experiments; for if a weight of four ounces be laid   Exp. 13.
on a plane whofe length is to its perpendicular
height, as two to one, it will be counterbalanced
by a weight of two ounces, provided the whole gravity thereof be made to act in drawing the other
weight in a direction parallel to the inclined plane,
which may be done by faftening one end of a cord
to the greater weight, and then ftretching the cord
along the plane, fo as to keep it parallel thereto,
and paffing it over a pulley at the top of the plane;
for the fmaller weight being tied to the end of the
cord which lies beyond the pulley will hang freely,
and for that reafon act with all its gravity in a direction parallel to the plane.

The fame weight of four ounces being laid on   Exp. 14.
an inclined plane, whofe length is to its height as
four to one, will be fuftained by a weight of one
ounce hanging freely as before.

The

The force wherewith a body resting on an inclined plane presses the same, is to the weight of the body, as the sine of the angle of inclination to radius; for in the triangle PEG, PE denotes the force wherewith the body presses the plane, and GE the weight of the body; but from the similarity of the triangles, as PE is to GE, so is BC to BA; and putting BA for the radius, BC is the sine of BAC the angle of inclination; wherefore as BC the sine of the inclination is to the radius AB, so is the force wherewith the body presses the plane to the absolute weight of the body. Hence, if upon an inclined lever

Pl. 4.
Fig. 11.
as AB, resting on the two props A and B, a weight be laid any where as at P, it will be easy to determine what proportion of the weight each prop bears; for drawing the horizontal line AE, equal in length to AB, and from the point P whereon the weight rests letting fall. PD perpendicular to AE, if AE be supposed to denote the whole weight of the body, AD will denote that part of it which is sustained by the uppermost prop, and DE that part which is supported by the lower; for if the lever was horizontal, so as that the body might press it with all its gravity, the whole weight of the body would be to that part of it which presses the prop B, as BA to PA, as is evident from what has been said concerning the second kind of lever; but as in the inclined position of the lever the whole weight of the body does not press upon it, that part of the weight which the prop B sustains in the horizontal position, must be to the part sustained in the inclined position, in the same proportion with the absolute weight of the body to the force wherewith it presses the inclined plane, that is, as PA to AD; for putting PA for the radius, AD is the sine of the inclination of the lever; consequently, the whole weight of the body must be to that part which presses on the prop B in the inclined position of the lever, in a *ratio*
compounded

compounded of BA to PA, and of PA to AD, that is, it muſt be as BA to AD, or becauſe AB and AE are equal, as AE to AD; and of conſequence the part ſupported by the other prop A muſt be as DE.

Hence it follows, that if two perſons carry a load fixed upon a lever, the load being placed between them, which is the caſe of chairmen, upon deſcents the foremoſt man will bear the greateſt burthen, and upon aſcents the hindermoſt. It likewiſe follows, that in coaches and all other four-wheel carriages, which have the foremoſt wheels ſmaller than thoſe behind, the load muſt be thrown more upon the former than the latter; what effect this has upon the draft, ſhall be ſhewn in my next lecture.

LECTURE IX.

Of FRICTION.

IN my laſt lecture I ſhewed you what force is requiſite to ſuſtain a body on an inclined plane. If a body be laid on a plane parallel to the horizon, it does not ſtand in need of any force to ſupport it; for as the direction of gravity is perpendicular to the plane of the horizon, the whole weight of the body muſt be ſuſtained by the horizontal plane whereon it reſts: whence it follows, that if any power endeavours to move a body reſting on an horizontal plane in a direction parallel to the plane, it will meet with no reſiſtance from the weight of the body, that being intirely taken off by the reaction of the plane whereon the body preſſes; but a reſiſtance will ariſe from the attrition of the body againſt the plane; for the ſurfaces of all bodies whatever, even ſuch as are of the fineſt poliſh, being in ſome meaſure rough and unequal, (as is evident from the

2 obſer-

observations that have been made by the help of
microscopes (when a body is moved upon a plane,
the prominent parts both of the body and plane
must of necessity fall into each others cavities, and
thereby create a resistance to the motion of the body, inasmuch as the body cannot be moved unless
the prominent parts thereof be continually raised
above the prominent parts of the surface whereon
it slides; and this cannot be done unless the whole
body be at the same time lifted up, and as it were
raised on an inclined plane equal in height to the
forementioned protuberant parts; upon which account the moving power must sustain some part of
the weight of the body, even in moving it along an
horizontal plane. But as this is occasioned by the
inequalities in the surface, if those were intirely taken off, so as to leave the surface perfectly smooth
and even, the resistance arising from friction would
likewise be removed; and setting aside the resistance of the medium, the smallest force would be
sufficient to move the most ponderous body along
an horizontal plane. But since there are not in nature any bodies, whose surfaces are perfectly equal,
there will ever be some resistance arising from friction; which resistance will remain unvaried, whatever be the magnitude of the surfaces that rub one
against the other, provided the weight which presses
those surfaces together, as also the roughness of
the surfaces, continue the same; for the same weight
will ever require the same force to raise it over prominencies of a given height, whatever be the magnitude of the surface whereon the weight rests;
consequently, the quantity of resistance will not be
varied by varying the magnitude of the surface;
which may be confirmed by the following experiment. Let four pieces of polished box be laid on
a polished horizontal plane, and let each piece be
so loaded as that its own weight, together with that

Exp. 1.

of

of its load, may be 6685 grains, and let the basis of
one be two inches long, and half an inch broad,
and those of the other three be each four inches in
length, but let their breadths be half an inch, an
inch, and an inch and an half, so that the magnitudes of the bases may be as one, two, four, and
six; let then a small cord be fastened to the end of
each piece, and by passing over a pulley, be kept in
a position parallel to the plane, and a weight of
2030 grains hanging from the end of the cord
which lies beyond the pulley, will just suffice to
move each piece along the plane; so that the resistance arising from friction is the same in each piece,
notwithstanding the different magnitudes of the
surfaces whereon they rest.

If the roughness of the surfaces whereon the bo- Exp. 2.
dies move be given, the resistance arising from friction will vary with the weights of the bodies, and
be proportional thereto; for if a certain force be
sufficient to raise a certain weight over prominences
of a given height, it is manifest that a double or
triple weight will require a double or triple force
to raise it to the same height. If therefore the
pieces of box be so loaded, as that each of them
with its load may weigh 13370 grains, that is, as
much again as in the last experiment, a weight of
4060 grains, that is, twice as much as before, will
be necessary to move them along the same plane.

If the roughness of the surface whereon a body
moves be increased, the resistance will likewise increase tho' the weight of the body remains the
same; but as the degree of roughness in any surface cannot otherwise be determined than by experiment, so neither can the resistance arising therefrom: if the plane made use of in the last experi- Exp. 3.
ments be thinly covered with fine sand, the resistance will thereby become greater in the proportion
of about five to four; for the same pieces of box
which were set a going by 2030 grains when the
plane

plane was free from fand, will in this cafe require 2500 grains, that is, about one fourth more.

To avoid as much as poffible the refiftance arifing from friction, which in rough and uneven roads muft needs be very great, WHEEL CAR-RIAGES have been contrived; the advantages whereof I fhall endeavour to explain to you, but I fhall firft fhew you from what caufe it is that wheels turn round during their progreffive motion along a
Pl. 4.
Fig. 12.
plane. If a wheel as ACB playing freely on the axis at A, be lifted off the plane BD by a power applied to the axle, and be carried in any direction whatever, it will not revolve about the axle; for fince in all wheels that are truly made the axle paffes thro' the center of gravity, it is evident, that in this cafe the wheel is fufpended by its center of gravity, and of confequence will not of itfelf change its pofition, but each point thereof will defcribe a line parallel to the direction of the moving power without any rotation about the axle, in the very fame manner as if the wheel was fixed to the axle; but if one point of the wheel as B refts upon the plane BD, and if a power applied to the axle draws the wheel in any direction as AP, fo as to move it along the plane BD; the motion of the point B will be retarded by the refiftance arifing from friction, whilft the point C which meets with no refiftance is carried forward without any retardation of its motion, and confequently muft move forward fafter than the point B; but as all the parts of the wheel cohere, the point C cannot move forward fafter than the point B, unlefs the wheel revolves about its axis from C towards E; and as the feveral points of the wheel's circumference, which are fucceffively applied to the plane, fuffer a retardation in their motion whilft the oppofite points move freely, the wheel during its progreffive motion along the plane, muft continue to revolve about its axle.

By

By this rotation of wheels about their axles, the resiftance arifing from friction is very much diminifhed, and drafts thereby rendered more eafy; for in plain roads, where the height of the prominent part is inconfiderable with refpeft to the diameter of the wheel; the parts of the revolving wheel which apply themfelves fucceffively to the road, may be looked upon in fome meafure as defcending upon the minute prominencies, and of courfe muft pafs over them without any confiderable friction. And fo much is the refiftance arifing from friction diminifhed in wheel carriages, that if upon the fame plane whereon the pieces of box were drawn, a carriage be laid with four equal wheels, each three quarters of an inch in diameter, and loaded in fuch a manner, as that the weight of the carriage and load may amount to 6685 grains, which was the weight of each piece of box with its load; it will be fet a going by a weight of 420 grains drawing it horizontally, whereas 2030 grains were requifite to move the pieces of box along the fame plane.

Exp. 4.

From this experiment it appears, that the friction is very much leffened by means of wheels; which diminution is not to be attributed to the wheel's touching the plane in a few points, as may poffibly be imagined, but to the rotation of the wheels; for if the wheels of a carriage loaded as before be made faft to the axle, fo as not to revolve in their motion; 2030 grains will be neceffary to fet the carriage a going, that is, juft as much as was requifite to move the pieces of box.

Exp. 5.

As wheel carriages in general meet with lefs refiftance in their motion than any other, fo thofe of larger wheels, *cæteris paribus*, are lefs refifted than thofe of fmaller; for the proof whereof, it will be neceffary to premife two LEMMAS; the firft of which is, *that the fecants of angles are to one another inverfly as the fines of their complements, that is,* A D, *which is the fecant of* B A D, *is to* A C, *which is the*

Pl. 4.
Fig. 13.

fecant

fecant of BAC, as AF, which is the fine of the complement of BAC, to AH, which is the fine of the complement of BAD. For from the nature of similar triangles AD is to AC, as AE to AK, that is, as AE to AG; but AE is to AG, as AF to AH; consequently, AD is to AC, as AF to AH.

The fecond LEMMA is, that if two arches of unequal circles have their verfed fines equal, the arch of the leffer circle is greater in proportion to the whole periphery, than the arch of the greater circle; or in other words, the angle meafured by the arch of the leffer circle, is greater than the angle meafured by the arch of the greater circle. Let HF and DB be two Pl. 4.
Fig. 14. arches of unequal circles, whofe verfed fines FG and BC are equal; I fay, the angle HEF is greater than the angle BAD; for fince EF is lefs than AB, and GF and CB are equal, EF is to GF in a lefs proportion than AB to CB; confequently, EG is to EF in a lefs proportion than AC to AB, that is, the fine of the angle GHE is lefs than the fine of the angle CDA, and of courfe the angle GHE is lefs than the angle CDA; confequently, the angle HEG, which is the complement of the leffer angle GHE, is greater than the angle DAC, which is the complement of the greater angle CDA.

Pl. 4.
Fig. 15. Thefe two LEMMAS being premifed, let HM reprefent a plane whereon move the two wheels ABH and KLR, which are of different magnitudes, but equal in weight, and let BC and LM be two obftacles of equal heights, and of fuch a nature as that the wheels cannot otherwife pafs than by furmounting thofe obftacles; the force requifite to draw the larger wheel over the obftacle BC, is lefs than what is requifite to draw the leffer wheel over the obftacle LM equal in height to the former; for fince the wheels revolve in paffing over the points B and L, their centers of gravity A and K may be looked upon as revolving about the fixed points B and L, and defcribing the arches AF and KP; consequently,

consequently, the forces which move the wheels may be looked upon as drawing them upon inclined planes, whose directions coincide with the directions of the curves in the points A and K, that is, they coincide with the tangents AE and KO; which tangents being parallel to the tangents of the wheels in the points B and L, that is to say, to DB and NL; the centers of gravity of the two wheels, and consequently, the wheels themselves may be looked upon as drawn up the inclined planes DB and NL; but since the wheels are supposed to be equal in weight, the forces which support them on the inclined planes DB and NL, the height whereof is given, must be to one another inversly as the lengths of the planes; that is, the force which supports the larger wheel on the plane DB, must be to the force supporting the smaller wheel on the plane NL, as NL to DB; that is, putting BC or LM for the radius, as the secant of the angle NLM to the secant of the angle DBC; or, because KS and LM, as also AI and BC are parallel, as the secant of the angle KSL to the secant of the angle AIB; but, from the nature of similar triangles, the angle KSL is equal to the angle KLQ, as is also the angle AIB to the angle ABG; and therefore the force which sustains the greater wheel on the inclined plane DB, is to the force sustaining the lesser wheel on the inclined plane NL, as the secant of the angle KLQ to the secant of the angle ABG; but, by the first *Lemma*, the secant of KLQ is to the secant of ABG, as the sine of BAG to the sine of LKQ; and, by the second *Lemma*, the sine of BAG is less than the sine of LKQ; consequently, the force which raises the greater wheel over the obstacle BC, is less than the force which raises the lesser wheel over the obstacle LM equal in height to the former; but the forces requisite to make the wheels surmount the obstacles are the measures of the resistances, and therefore,

K　　　　　　　　　*cæteris*

ceteris paribus, the greater wheel muſt meet with leſs reſiſtance from the ſame obſtacle than the ſmaller. To confirm this by an experiment, let an obſtacle one tenth of an inch in height be fixed on an horizontal plane, and cloſe behind it let there be placed the carriage with four equal wheels, each three quarters of an inch in diameter, and if it be loaded in ſuch a manner as that the weight of the carriage and load may amount to 6685 grains, it will not be raiſed above the obſtacle by leſs than 2850 grains drawing it in a direction parallel to the plane ; whereas if four wheels, each an inch and an half in diameter, be fitted to the ſame carriage, the weight of the whole being the ſame as before, it will be raiſed above the obſtacle by 2050 grains, that is, by 800 grains leſs than were requiſite to raiſe it with the ſmaller wheels.

From this experiment it appears, that the reſiſtance which larger wheels meet with in ſurmounting obſtacles, is leſs than the reſiſtance given to ſmaller wheels by the ſame obſtacles ; and from what has been demonſtrated it is evident, that the reſiſtance given to the greater wheel is to the reſiſtance given to the ſmaller, as the ſine of an angle meaſured by an arch of the greater wheel, to the ſine of an angle meaſured by an arch of the ſmaller wheel, the verſed ſine of each angle being equal to the height of the obſtacle ; ſo that putting R and r for the radii of the two wheels, and x for the verſed ſine or the height of the obſtacle, it follows from the nature of the circle, that as $\dfrac{\sqrt{2Rx - xx}}{R}$ is to

$\dfrac{\sqrt{2rx - xx}}{r}$, ſo is the reſiſtance given to the larger wheel to the reſiſtance given to the ſmaller, or dividing by $x^{\frac{1}{2}}$, as $\dfrac{\sqrt{2R - x}}{R}$ to $\dfrac{\sqrt{2r - x}}{r}$; but as the

proportion

proportion of these sines is not fixed, but varies with
the height of the obstacle, so likewise must the proportion, which the resistance given to the greater
wheel bears to the resistance given to the smaller;
and all that can be determined in this case is, that
larger wheels ever meet with less resistance in sur-
mounting obstacles than smaller; and that the disproportion between the resistances suffered by each
wheel, increases with the height of the obstacle.
Indeed where the obstacle vanishes, which is the
case when wheels move upon planes, the expressions
for the resistances, and consequently the resistances

themselves, are as $\dfrac{1}{\sqrt{R}}$ and $\dfrac{1}{\sqrt{r}}$, that is, the resist-

ances are inversly as the square roots of the semidiameters of the wheels; so that where the heights
of the wheels are as one and two, the forces requisite to draw them along the same horizontal
plane, are as fourteen and ten, that is, inversly as
the square roots of one and two, which is confirmed
by experiments; for whereas the carriage whose
wheels are three quarters of an inch in diameter,
required 420 grains to move it along the horizontal
plane, the weight of the carriage and load being
6685 grains; the carriage whose wheels are $1\frac{1}{2}$
inch in diameter, when loaded in the same manner,
will be set a going by 300 grains; but 420 is to
300, as 14 to 10, that is, the forces requisite to
move the two carriages along the same plane, are
inversly as the square roots of the heights of the
wheels.

If the nature of the obstacle be such, as to be
bore down by the pressure of the wheel, the larger
wheel will in this respect likewise have the advantage over the smaller, and depress the obstacle with
greater force. For let L K be continued to T, so
that T L may be equal to A B; and since the wheels
are supposed to be equally weighty, let A B and

Pl. 4.
Fig. 15.

T L ex-

T L expref the abfolute forces of the two wheels acting againft the obftacles in the directions A B and K L; it is evident from what has been faid concerning the refolution of forces, that the force denoted by A B may be refolved into two forces; one whereof may be denoted by A G, and the other by G B, whereof A G alone acts in depreffing the obftacle B C, inafmuch as it bears directly down upon it; whereas the other force denoted by G B, inafmuch as its direction is perpendicular to the obftacle, may thruft it forward, but can contribute nothing towards preffing it downward from B towards C. In like manner the force denoted by T L, is refolvable into two forces, which may be denoted by T V and V L; whereof T V alone acts in depreffing the obftacle L M; confequently, the force wherewith the greater wheel depreffes the obftacle, is to the force wherewith it is depreffed by the leffer, as A G to T V, or as the fine of the angle A B G to the fine of the angle T L V or K L Q; but by the fecond *Lemma*, the angle B A G, which is the complement of G B A, is lefs than the angle L K Q the complement of K L Q; confequently, the angle A B G is greater than T L V, and A G the fine of the former greater than T V the fine of the latter; but as A G is to T V, fo is the depreffing force of the greater wheel to the depreffing force of the leffer; confequently, the fame obftacle is more eafily depreffed by the larger wheel than the fmaller, and of courfe muft give lefs refiftance to the former than to the latter.

If the obftacle be fuch, as that it can neither be furmounted nor depreffed, but muft be driven forward, then indeed the fmaller wheel has the advantage of the larger; for the forces of the wheels being refolved as before, the lines G B and V L will exprefs the forces which act in driving the obftacle forward; but it has been demonftrated, that G B the fine of the angle G A B, is lefs than V L the fine of the
angle

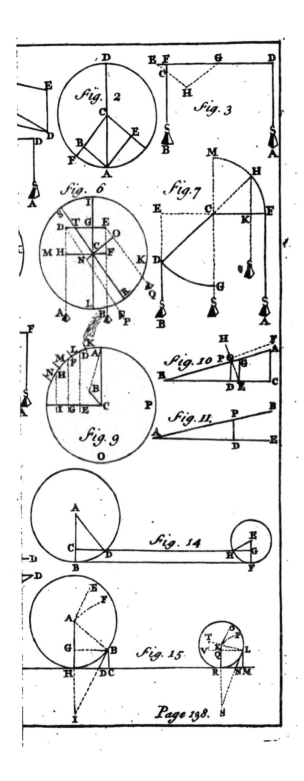

Page 138.

angle VTL equal to QKL; and therefore the
force wherewith the greater wheel propels the obstacle, is less than the force wherewith the smaller
wheel propels the same; besides, as the greater
wheel presses the obstacle directly downward with a
greater force than the smaller, the resistance made
by the same obstacle to the propelling force of the
larger wheel, will be greater than what is made to
the propelling force of the smaller; so that where
the obstacle is to be propelled, the smaller wheel is
preferable to the larger; but as in drafts this is
rarely if at all the case, the obstacles which are
commonly met with in roads being such as must
either be surmounted or depressed by the wheels,
such wheels are to be preferred as best serve both
those purposes, and those I have shewn to be the
larger wheels; which likewise are attended with
other advantages besides what have been already
mentioned; for first, it frequently happens in rough
and uneven roads, that two obstacles are placed so
near each other, that before the wheel has quit-
ed one it meets with the other, and resting upon
each, hangs between them; in which case the
smaller the wheel is, the lower it descends between
the obstacles, and thereby renders the draft more
difficult; inasmuch as it must be raised to a greater
height in order to pass over the foremost obstacle,
than when the wheel is larger: For the illustration Pl. 5.
of which, let F E and H G represent two obstacles Fig. 1.
placed at so small a distance, that the wheel having
surmounted the first but not quitted it, may meet
with the second, so as to hang between them; it is
manifest, that as the arch FDH of the lesser wheel,
which lies between the obstacles, has a greater curvature than FBH the arch of the greater wheel,
which lies between the same obstacles, the point D
must descend lower than the point B; consequently,
the smaller wheel must be raised to a greater height
than the larger, in order to pass over the same ob-
K 3

stacle;

ftacle; and therefore a greater force will be neceſſary to pull up the ſmaller wheel, than what is requiſite to raiſe the larger; which is confirmed by
the following experiment. Let a carriage with four
wheels, each an inch and an half in diameter, and ſo
loaded as that the weight of the carriage and load
may amount to 6685 grains, be ſo placed on the
plane before made uſe of, as that the two foremoſt
wheels may hang between two obſtacles whoſe diſtance is half an inch, and their height likewiſe half
an inch, and a weight of 1150 grains drawing the
carriage horizontally, will move the wheels from
between the obſtacles; whereas if four ſmaller
wheels be made uſe of each three quarters of an
inch in diameter, a weight of 2700 grains will be
requiſite to draw them from between the obſtacles.

As wheels cannot always run upon the nail,
but muſt frequently meet with heavy roads, they
will ſink down, and thereby render the draft more
difficult; but the larger the wheels are, the leſs cæ-
Pl. 5.
Fig. 2. teris paribus will the depth be to which they ſink.
For if ABC denotes the plane of the road, and if it
be of ſuch a nature as to ſuffer the ſmaller wheel to
ſink down as far as E; it is manifeſt that the gravity of the wheel muſt overcome the reſiſtance of
as much of the earth whereon it preſſes, as is equal
to the ſegment HED; for it cannot otherwiſe ſink,
than by forcing ſuch a quantity of the earth out of
its place; and ſhould the larger wheel ſink to the
ſame depth, the gravity thereof muſt overcome
the reſiſtance of as much earth as is equal to the
ſegment AEC, that is, it muſt overcome a greater
reſiſtance in order to ſink to the ſame depth with
the ſmaller; but it cannot poſſibly overcome a
greater reſiſtance, becauſe it is ſuppoſed to have the
ſame gravity with the ſmaller; conſequently, it
will not ſink as deep as the ſmaller, and for that
reaſon will make the draft leſs troubleſome.

As

As large wheels have the advantage of small ones with regard to the resistance arising from the obstacles and impediments in the roads, so have they likewise in relation to the resistance occasioned by the friction of the box against the arm of the axle; not that this resistance is less in greater wheels than in smaller; for since it is not varied by varying the magnitude of the surface, as has been shewn, if the boxes and arms are truly fitted and of an equal smoothness, and the weights whereby the arms and boxes are pressed together be equal, the quantity of resistance will be given, whatever be the magnitude of the wheels, as also of the arms of the axle whereon they play; but where the arms of the axles are of equal diameters, (which is commonly the case in one and the same carriage, tho' the wheels be unequal) a less force is requisite to overcome the given resistance in a larger wheel than in a smaller; for in this case the semidiameter of the wheel may be looked upon as a lever, whose prop or fixed point is at the center of the arm, and the impediment arising from the friction of the box against the arm may be looked upon as a weight placed upon the lever at the distance of the arm's semidiameter from the prop, whilst the moving power is applied to the extremity of the wheel's semidiameter; and therefore in order to a balance, the power must be to the resistance, as the semidiameter of the arm to the semidiameter of the wheel; since then the impediment is given, as also the distance thereof from the prop, it is evident, that the larger the lever is, and consequently, the larger the wheel, the less is the force requisite to overcome the resistance. Thus, if B E F represents the circumference of the arm of an axle, whereon the wheels A G H and D I K revolve, C the center of the arm, B C its semidiameter, D C the semidiameter of the smaller wheel, and A C that of the larger; in the

Pl. 5.
Fig. 3.

K 4   bigger

bigger wheel the length of the lever is AC, and in the smaller DC; since therefore the same impediment is in both levers placed at the same distance from the prop C, to.wit at B, it will be balanced by a less force at A than at D; and the force at A is to the force at D, as DC to AC, that is, inversly as the semidiameters of the wheels; for the force at A is as BC applied to AC, and the force at D is as the same BC applied to DC; that is, the force at A which balances the resistance at B, is to the force at D which balances the same resistance, as BC divided by AC, to BC divided by DC, that is, multiplying crosswise, and throwing out BC, as DC to AC. Whence it follows, that when the semidiameter of the arm is given, the more the wheel is enlarged, the less will the force be that is requisite to overcome the resistance arising from the friction of the wheel against the arm; so that upon this account as well as the former, large wheels are to be preferred to small ones.

In order to lessen the resistance arising from the friction of the box against the arm of the axle, there has been a late contrivance, whereby the axle, contrary to what is usual in most carriages, is made to revolve, and its arms, instead of pressing against the boxes, are made to bear on the circumferences of moveable wheels, which wheels from their use in diminishing the friction, are by the author of this contrivance called *friction wheels*. Now that such wheels, where they can be made use of, do take off much of the resistance occasioned by friction, will

Exp. 8. appear from the following experiments; from the axle of the machine called the axle in the wheel, in which the diameter of the wheel is to the diameter of the axle, as nine to one, let a weight of 23163 grains be hung, and a weight of 2770 grains hanging at the circumference of the wheel, will turn the machine, provided the axle turns on the circumferences

rences of two moveable wheels; whereas, if it turns in the pivets it will be neceffary to add 600 grains more, fo as to make the whole 3370 grains; confequently the refiftance occafioned by friction in the latter cafe, is more than four-fold what it is in the former; for fince the diameter of the wheel is nine times as great as that of the axle, a weight of 2574 grains at the wheel is requifite to balance the weight of 23163 at the axle, which balancing weight being deducted from 2770, and likewife from 3370 grains, leaves 196 grains for overcoming the refiftance in one cafe, and 796 in the other; but 796 is to 196, as four and a little more to one.

Again, let a fmall cart with friction wheels be fo loaded, as that its own weight added to that of the load, may amount to 20000 grains: a weight of 54 grains drawing horizontally, will move it along a fmooth level table: whereas, if the friction wheels be taken off, 322 grains will be neceffary to fet it a going. If the cart be fo loaded, as that the weight of the whole may amount to 40000 grains, then in each cafe, a double force will be requifite to move it, that is to fay, 108 grains with the friction wheels, and 644 without them; fo that in this cart the friction wheels take off five parts in fix of the refiftance; for 54 is but a little more than a fixth part of 322, as is likewife 108 of 644. And from thefe experiments it does again appear, that under like circumftances the refiftance arifing from friction, is proportional to the weight, whereby the furfaces which rub one againft the other are preffed together.

Seeing then that great wheels have in fo many refpects the advantage over fmall ones, it will not be improper in this place to fhew you, on what account it is, that the wheels of common carts, as alfo the foremoft wheels of coaches, chariots, and moft other four-wheel carriages, are commonly made fo fmall as feldom to exceed two feet and an half in diameter; and the firft reafon of this contrivance

is for the convenience of turning; for as in moſt
roads, but more eſpecially ſuch as are narrow, there
are windings of ſuch a nature as to allow but a ſmall
ſpace for carriages to turn in, it is neceſſary to make
uſe of ſuch wheels as can turn in the narroweſt compaſs, and ſuch are ſmall ones; for it is a thing well
known to carters and all others who are uſed to
drive wheel carriages, that the larger the wheels are,
the greater compaſs do they require in order to turn
with eaſe and ſafety; and ſhould they at any time
attempt to turn carriages with large wheels as ſhort
as thoſe which have ſmaller, the wheels will drag,
and thereby render the draft very difficult, and
ſometimes endanger the overſetting of the carriage.

But the ſecond, and indeed the principal reaſon
for the uſe of ſmall wheels is, that upon aſcents, and
in paſſing over obſtacles in rough and hilly roads,
as little of the horſe's force may be loſt as poſſible;
if roads were level and ſmooth without riſings or impediments, the moſt convenient ſize for wheels, ſetting aſide the neceſſity of turning, would be where
the axle is upon a level with the breaſt of the horſe;
for ſince the whole force of the horſe in drawing is
applied to that part of the tackle which lies upon
the breaſt; and to which the traces are joined; and
ſince the traces are faſtened to the carriage in ſuch
a manner, as that being continued they muſt paſs
thro' the axle of the foremoſt wheels, it is manifeſt,
that if that axle be of an equal height with the cheſt
of the horſe, the traces, in whoſe plane the line of
direction lies, will be parallel to the road whereon
the carriage is drawn; conſequently, the whole force
of the horſe will be employed in drawing the carriage directly forward, without any loſs or diminution; whereas if the wheels be of ſuch a ſize as that
the height of the axle is either greater or leſs than
that of the horſe's cheſt, the whole force of the horſe
will not be employed in the direct draft; but in the

- former

former cafe, fome part of the force will be fpent in preffing the carriage directly downward, and in the latter, in lifting the fame directly upward. For the proof and illuftration whereof, let the firft of the three wheels be of fuch a fize, as that its axle A may be of an equal height with the horfe's breaft at B; and let the fecond wheel be fo large as that its axle A may ftand higher than the horfe's cheft at B, and in the third, let the axle be lower than the breaft of the horfe; and in each wheel let the lines of direction of the horfe's draft, to wit, A B be taken equal, and let each of thofe lines exprefs the force of the horfe; it is manifeft, that in the firft wheel, the whole force denoted by AB, is employed without any lofs in drawing the wheel forward, becaufe the line of direction AB, wherein the force draws, is parallel to EF, the road whereon the wheel moves; whereas, in the fecond and third wheels the lines AB, wherein the forces draw, being inclined to EF, whereon the wheels move, fome part of each force muft be loft; for if each force denoted by AB be refolved into two, to wit CB and AC, whereof CB is parallel to EF, and AC perpendicular thereto; it is evident, that that force alone which is denoted by CB, acts in moving the wheel forward along EF, whilft the force denoted by AC does in the fecond wheel prefs it directly downward againft the road, and in the third lifts it directly upward; whence it follows, that if the force of a horfe be juft fufficient to move the firft wheel, it will not fuffice to ftir the fecond or third. It likewife follows, that if the wheel be fo far inlarged, as that the angle which the line of direction AB makes with the plane EF, approaches nearly to a right one, the line CB will bear a very fmall proportion to AB, whilft AC becomes nearly equal thereto; fo that almoft the whole of the horfe's force will be fpent in preffing down, and thereby increafing the load; whence it appears,

that

that notwithstanding the several advantages arising from the largeness of wheels, yet may they be so far increased, as even upon account of their magnitude to render the draft impossible. By the use of small wheels whose axles lie below the level of the horse's chest, provision has been made against the inconvenience last mentioned, and the loss of force (which by reason of the roughness and inequalities of roads cannot wholly be avoided) has been rendered as little as possible, and made to obtain chiefly in level smooth roads, where there is least occasion for the whole force; whereas upon ascents, and in passing over obstacles in rough roads, where the stress is greatest, there little of the force is lost; for

Pl. 5.
Fig. 5.

the proof of which, let the wheel be of such a size, that its axle A may be below the horse's breast at B, and let AB, as before, denote the force of the horse; if the wheel be drawn along a smooth level road as EF, CB will express that part of the force which draws the wheel along the road, and AC that part of the force which is employed in lifting up the wheel, which part is lost as to the draft, but however, is not intirely useless; because, by pulling the wheel directly upward, it eases the load, and thereby renders the draft less difficult; tho' at the same time the draft is by no means as easy as it would be, if the force of the horse was applied at G, so as to draw in the direction AG parallel to EF. If the wheel instead of moving along a smooth road, be to pass over the obstacle DH, or which is the same thing, if it be to be drawn up the ascent EHL, and if the force of the horse be applied at G, so as that the direction of the draft AG may be parallel to EF, and consequently, inclined to EHL; it is manifest upon resolving the force AG into two forces, to wit AK and KG, whereof AK is parallel, and KG perpendicular to EHL; that force alone which is expressed by AK, acts in drawing the wheel up EHL; whereas the force expressed by

KG

K G acts in preffing the wheel directly against EHL,
and thereby adds to the weight of the wheel; fo
that in this cafe, fome part of the horfe's force is
loft, and the load at the fame time increafed, both
which inconveniencies are avoided where the breaft
of the horfe is fo far elevated above the axle of the
wheel, as that the line of direction AB may be parallel to EHL; for then no part of the horfe's
force will be loft, but the whole will be employed
in drawing the wheel directly over the obftacle, or
up the afcent; fo that a lefs force will be requifite
to draw the wheel over the obftacle DH in the direction AB, than in the direction AG; and this is
fully confirmed by experiments.   For whereas the   Exp. 9.
little carriage with four wheels, each three quarters
of an inch in diameter, being fo loaded as that the
weight of the carriage and load amounted to 6685
grains, was not drawn over the little obftacle one
tenth of an inch in height, by lefs than 2850 grains
acting in an horizontal direction, it will be drawn
over, by 2450 grains, provided the direction be
made parallel to the tangents of the wheels in thofe
points which touch the obftacle; and 1950 grains
will be fufficient to draw the carriage with the larger
wheels over the fame obftacle, if the direction of
the draft be made parallel to the forementioned
tangents, whereas 2050 grains were neceffary when
the direction was parallel to the horizontal plane.
And if the direction be ftill farther removed from
the parallelifm of the tangents, which may be done
by depreffing it below the horizontal plane, the
force of 2350 grains will be but juft fufficient
to furmount the obftacle, and draw the carriage
over.

Tho' in four wheel carriages, the contrivance of
fmall wheels before has its advantages, yet is it not
intirely free from inconveniencies; for by this means
the load muft of neceffity be thrown forward, and
a greater

a greater ſtreſs laid on the foremoſt wheels ; whereby the reſiſtance that ariſes from the friction of the axle againſt the wheels will become greater in the foremoſt than in the hindmoſt wheels, in proportion to the greater weight which they ſuſtain. Beſides, as the ſpaces deſcribed by wheels in each revolution are nearly equal to the peripheries of the wheels, it is manifeſt that the foremoſt wheels muſt revolve oftener than the hindmoſt, in order to rid the ſame ground. And this frequency of turning requiſite in the foremoſt wheels joined to the greater ſtreſs upon them from the load, as alſo to the greater reſiſtance which they meet with from obſtacles in the road, is the true reaſon why they are more frequently out of order, and ſtand in need of repair much oftener than thoſe behind.

LECTURE X.

MOTION OF BODIES DOWN INCLINED PLANES.

LECT.
X.
MY deſign in this lecture is to explain the chief properties of the PENDULUM; and in order thereto, I ſhall lay down the following PROPOSITIONS concerning the motion of bodies down inclined planes and curve ſurfaces.

Pl. 5.
Fig. 6.
PROP. I. *The force wherewith a body deſcends upon an inclined plane, as* AC, *is to the abſolute force of gravity wherewith the ſame body falls freely and perpendicularly, as the height of the plane to the length thereof, that is, as* AB *to* AC.

For it has been proved, that the force requiſite to ſuſtain a body upon an inclined plane, is to the abſolute weight of the body, as the height of the

plane

plane to its length; but the force wherewith a body L ı c r.
endeavours to defcend upon an inclined plane, muft   X.
be equal to the force which is neceffary to fupport
it upon that plane; confequently, the propofition is
true.

CoROL. I. Hence it follows, that the motion
of a body defcending on an inclined plane is uniformly accelerated; for fince the force which carries
a body down an inclined plane, has every where,
and in all parts of the plane, the fame proportion
to the abfolute weight of the body, and fince the
abfolute weight remains unvaried, the other force
muft do fo too; confequently, as it acts inceffantly
in equal times, it makes equal impreffions on the
defcending body, fo as to generate equal degrees of
velocity in the motion thereof; that is, in other
words, the motion of a body defcending on an inclined plane is uniformly accelerated.

CoROL. II. On account of this uniform acceleration of the motion, the times of defcending, as
alfo the velocities acquired at the end of the defcent,
are as the fquare roots of the fpaces defcribed, as in
the cafe of bodies falling freely; that is to fay, the
time wherein a body defcends upon the inclined
plane from A to D, is to the time of the defcent
from A to C, as the fquare root of AD, to the
fquare root of AC; and the velocity of the body,
when it has defcended as far as D, is to the velocity
thereof when it arrives at C in the fame proportion
of the root of AD to the root of AC.

PROP. II. *The velocity acquired in any given*
time by a body defcending on an inclined plane, is to the
velocity acquired in the fame time by a body falling
freely and perpendicularly, as the height of the plane
to its length, that is, as AB *to* AC.

For, by the firft *corollary* of the foregoing *propo-*
fition, the motion of a body down an inclined plane

4   is

LECT. is uniformly accelerated, in the same manner as the
X. motion of a body falling freely; consequently, at
the end of any given time, the velocities acquired
must be as the accelerating forces; but by the foregoing *proposition*, the accelerating force of a body
moving down an inclined plane as A C, is to the
accelerating force of a body falling freely and perpendicularly, as the height of the plane to its
length; and therefore the velocities acquired in any
given time must be in the same proportion.

PROP. III. *The spaces described in a given time
by two bodies moving from a state of rest, whereof one
descends on an inclined plane, and the other falls freely,
are in the same ratio of the height of the plane to its
length; that is, the space described by a body moving
along* A C, *is to the space described by a body falling
down the perpendicular* A B, *as* A B *to* A C.

For where the motions are equable, the spaces described in a given time, are as the velocities wherewith they are described; if therefore the velocities
be increased in a constant uniform manner, the spaces
described will likewise increase in the same manner;
but by the second *proposition*, the velocities are augmented in such a manner as in a given time to bear
the same proportion to one another, as the height
of the plane does to its length; consequently the
spaces described in a given time must be in that
proportion.

COROL. I. If from B, the line BD be drawn
perpendicular to AC, AD will be the space described by a body moving down the plane A C, in the
same time that a body falls freely down the height
of the plane from A to B.

For, from the nature of similar triangles, A C
is to A B, as A B to A D; but by the *proposition*, as
A C is to A B, so is the space described in a given
time by a body falling freely, to the space described
by

by a body defcending upon the inclined plane AC; confequently, fince AB is the fpace defcribed by the body falling freely, AD muft be the fpace defcribed in the fame time by a body defcending along AC.

COROL. II. All the chords of a circle are defcribed in the fame time by bodies running down them. For if a circle be defcribed with the diameter AB, which is the height of the inclined plane AC, the point D, which determines the fpace AD thro' which a body defcends upon the inclined plane, whilft another falls freely from A to B, will be in the periphery of the circle, becaufe the angle ADB in the femicircle is always a right one; and for the fame reafon, if the height of the plane continuing the fame, the inclination thereof be varied, fo as that it may become AG, the point E which determines the fpace AE, thro' which a body moves along the plane AG, during the time of a body's fall from A to B, will likewife be in the periphery of the circle; confequently, in the femicircle ADB all the chords as AD and AE will be defcribed in the fame time; and as in the femicircle AFB, whatever chords as BF and BH are drawn thro' the point B, other chords as AD and AE may be drawn in the other femicircle parallel thereto and equal; it follows, that whether a' body falls freely down the diameter AB, or whether it defcends along a chord as HB or FB, it will in the fame time arrive at the loweft point of the circle; or in other words, all the chords of a circle will be defcribed in equal times by bodies running along them.

Pl. 5.
Fig. 7.

PROP. IV. *The time wherein a body moves down an inclined plane as* AC, *is to the time wherein a body falls freely down* AB *the height of the plane, as the length of the plane to its height, that is, the times are as the fpaces defcribed.*

Pl. 5.
Fig. 6.

L.   For

For by the second *Corol.* of the first *Prop.* the time of a body's motion along the inclined plane from A to C, is to the time of its motion from A to D, as the square root of AC to the square root of AD; but by the second *Corol.* of the third *Prop.* the time of a body's motion along the inclined plane from A to D, is equal to the time of the fall from A to B; and therefore the time of the motion along the plane from A to C, is to the time of the perpendicular fall from A to B, as the square root of AC, to the square root of AD; that is, because from the similarity of triangles AC, AB, and AD are in continued proportion, as AC to AB, or as the length of the plane to its height.

COROL. Hence it follows, that if several inclined planes have equal altitudes, the times wherein those planes are described by bodies running down them, are to one another as the lengths of the planes.

Pl. 5.
Fig. 7.

For the time of the descent along AC, is to the time of the fall down AB, as AC to AB, and the time of the fall down AB, is to the time of the descent along AG, as AB to AG; consequently, the time of the descent from A to C, is to the time of the descent from A to G, as AC to AG, that is, the times are as the lengths of the planes.

Pl. 5.
Fig. 6.

PROP. V. *The velocity acquired at the end of the fall by a body falling down the perpendicular height of an inclined plane as* AB, *is equal to the velocity acquired at the end of the descent by a body moving down the inclined plane, from* A *to* C.

For by the first *Prop.* the accelerating force of a body falling freely from A to B, is to the accelerating force of a body moving along the plane AC, as AC to AB; and by the fourth *Prop.* as AB is

to

to AC, so is the time of the fall from A to B, to the time of the descent from A to C; so that the forces which accelerate the bodies during their motions, are to one another, reciprocally as the times that they continue to act; consequently, at the end of those times, the velocities generated must be equal. For instance, if AB be one half of AC, the force which accelerates the body in its fall from A to B, is to the force which accelerates the body in its descent from A to C, as two to one; but the time that a body takes to fall from A to B, is but one half of the time that a body takes to descend from A to C; so that the accelerating force which acts upon the body during its motion from A to C, tho' it be but one half of the accelerating force which acts upon the body during its fall from A to B, yet does it continue to act twice as long; and therefore must in the end produce the same velocity.

COROL. Hence it follows, that the velocities acquired by bodies in falling down inclined planes, are equal where the heights of the planes are equal.

For, the velocity acquired in falling from A to C, is equal to the velocity acquired in falling from A to B, as is also the velocity acquired in falling from A to G; consequently, the velocities acquired in falling from A to C, and from A to G, are equal.

Pl. 5. Fig. 7.

PROP. VI. *If a body descends along several contiguous planes as* A B, B C, *and* C D, *the velocity which it acquires in its descent from* A *to* D, *is equal to the velocity acquired by the perpendicular fall from* H *to* D, *on supposition that the body is not retarded by the shocks it suffers in the angles* B *and* C.

Pl. 5. Fig. 8.

For drawing the horizontal lines HE and DF thro' the points A and D, and producing the planes CB and DC as far as G and E; by the *Corol.* of the

last

LECT.
X.
last *Proposition*, the same velocity is acquired in the point B, by a body in descending from A to B, as in descending from G to B; consequently, the same velocity is acquired in the point C, by a body descending from A thro' B to C, as in descending from G to C; but by the same *Corollary*, the velocity acquired in descending from G to C, is equal to the velocity acquired in descending from E to C; wherefore, the velocity in the point D acquired by the descent along the three planes AB, BC, and CD, is equal to the velocity acquired by the descent from E to D, which velocity by the foregoing *Proposition*, is equal to the velocity acquired by the perpendicular fall from H to D.

Pl. 5.
Fig. 9.
COROL. Hence it follows, that if a body descends along the arch of a circle as AB or of any other curve, the velocity acquired at the end of the descent, is equal to the velocity acquired by falling down CB, the perpendicular height of the arch.

For curves may be looked upon as composed of an infinite number of right lines inclined one to another.

Pl. 5.
Fig. 10.
PROP. VII. *If two planes as AB and BD joined together at B, have equal degrees of elevation with two other planes as EF and FH joined together at F, and if AB be to EF as BD to FH; the time of a body's fall down the planes ABD, will be to the time of the fall down EFH, as the square root of AB and BD taken together, to the square root of EF and FH taken together.*

Let AB and EF be produced till BC becomes equal to BD, and FG equal to FH. Since AB is to BC, as EF to FG, AB is to AC, as EF to EG; and since those four quantities AB, AC, EF and EG are proportional, their square roots will be so too. Again, since the planes AC and EG are
equally

equally elevated, they may be looked upon as parts of one and the same plane, and therefore, by the second *Corol.* of the first *Prop.* the time of a body's fall from A to C, is to the time of the fall from E to G, as the square root of AC to the square root of EG, or as the square root of AB to the square root of EF; but the time of a body's fall from A to B,, is to the time of the fall from E to F, as the square root of AB to the square root of EF; so that the time of the fall from A to C, is to the time of the fall from E to G, in the same proportion of the time of the fall from A to B, to the time of the fall from E to F; consequently, the time of the fall from B to C, supposing the motion to begin from A, must be to the time of the fall from F to G, supposing the motion to begin from E, in the same proportion of the root of AB to the root of EF; if the bodies after their fall from A to B, and from E to F, instead of moving along BC and FG continue their motions along BD and FH, since those two planes are equally inclined to AB and EF, and since BD is equal to BC, and FH equal to FG, whatever proportion the time of the body's motion along BD bears to the time of its motion along BC, the same will the time of the motion along FH bear to the time of the motion along FG; but it has been already proved, that the time of the motion along BC, is to the time of the motion along FG, as the square root of AB to the square root of EF; wherefore, the time of the motion along BD, is to the time along FH, as the square root of AB to the square root of EF, that is, in the same proportion with the time along AB to the time along EF; and therefore, the sums of those times will be in the same proportion; that is to say, the time of the motion along AB, added to the time of the motion along BD, is to the time of the motion along EF, added to the time of the motion along FH, as the square root of AB to the square root of EF; but

it

it has been proved, that the fquare root of **AB is**
to the fquare root of **EF**, as the fquare root of **AB**
and **BD** taken together, to the fquare root of **EF**
and **FH** taken together; and therefore, the time
of a body's fall from **A** thro' **B** to **D**, is to the time
of the fall from **E** thro' **F** to **H**, as the fquare root
of **ABD** to the fquare root of **EFH**, which was
to be proved. And what has been thus proved with
regard to two planes on each fide, is in like manner
demonftrable with regard to any number of planes,
provided thofe on one fide be proportional to thofe
on the other, and that the correfponding planes
have equal degrees of elevation.

Pl. 5.
Fig. 11.
    Coʀoʟ. Hence it follows, that if bodies defcend
thro' the arches of circles, the times of defcribing
fimilar arches fimilarly pofited, are as the fquare
roots of the arches. For inftance, if bodies move
down the fimilar arches **AB** and **CD**, which are fimilarly pofited with regard to the horizontal plane
ED, the time of defcribing **AB**, is to the time of
defcribing **CD**, as the fquare root of **AB** to the
fquare root of **CD**.

    For all circles whatever, may be confidered as
fimilar polygons, confifting of an indefinite number of fides indefinitely fmall ; and therefore, fimilar arches muft confift of an equal number of fides
proportional the one to the other; and forafmuch
as the angles which thofe fides contain are equal, if
the arches be fimilarly pofited, the correfponding
fides in each arch muft have equal degrees of elevation ; and confequently, the times of defcribing
the arches will be as their fquare roots.

    In my lecture upon gravity, I fhewed you, that if
a body be thrown directly upward, it will rife to the
fame height, whence, if it fell from a ftate of reft,
it would by the end of the fall acquire the fame velocity wherewith it is thrown up ; I likewife fhewed
you, that the time of the rife is equal to that of the
fall. I now fay,

Pʀoᴘ,

Prop. VIII. *That the same things do likewise obtain with regard to bodies thrown up obliquely, whether they ascend upon inclined planes or along the arcs of curves.*

Because the same forces which accelerate the motions of bodies descending on such planes or curves, do in the very same manner retard the motions of such bodies as ascend thereon; and therefore, whatever be the time requisite for a body to descend upon an inclined plane or thro' the arc of a curve, in order to acquire any velocity, the same must the time be, wherein that velocity is destroyed in a body ascending upon the same plane or curve, and whatever be the length of the plane or curve, thro' which a body descends in order to acquire any velocity, the same must the length of the plane or curve be, thro' which it must ascend in order to have that velocity destroyed.

Corol. Hence it follows, that if by any contrivance a body be made to descend thro' the arch of a circle as from 'C to A, and with the velocity acquired by the descent to ascend along the arch AD of the same circle, the arch AD which it describes in its ascent, will be equal to the arch CA described in the descent; and the times in which those arches are described will be equal.

Pl. 5.
Fig. 12.

' And this is the case of the Pendulum; which is a heavy body as A, hanging by a small cord as BA, and moveable therewith about the point B, to which the cord is fixed. If when the cord is stretched the weight be raised as high as C, and thence let fall, it will by its own gravity descend thro' the circular arch CA; and by the *Corol.* of the sixth *Prop.* it will have the same velocity in the point A, that a body would acquire in falling perpendicularly from E to A; and by the first LAW OF NATURE, it will endeavour to go off with that velocity in the tangent AF; but being by the force of

Exp. 1.

the

Hence it appears, that if the arch of a circle wherein in a pendulum vibrates, be so divided in the points 1, 2, 3, 4, and so on, beginning from the lowest point A, as that the chords drawn from A to the several points of division, may be to one another, as those numbers, the velocities acquired by a pendulum in the lowest point A, when let fall successively from the several points of division, will be as the numbers affixed to the respective points; and it was upon this account, that in the experiments relating to the collision of bodies, the balls were constantly let fall from such heights, as that the chords of the arches which they described in their descent, might be to one another in the same proportion with the velocities wherewith the balls were supposed to meet at the lowest point.

The times wherein pendulums of unequal lengths vibrating in similar arches, perform their vibrations, are to one another, as the square roots of their lengths; for instance, the time wherein the pendulum BA vibrates thro' the arch FG, is to the time wherein the pendulum BC vibrates thro' the arch DE similar to FG, as the square root of BA to the square root of BC.

Pl. 5.
Fig. 15.

For, by the *Corol.* of the seventh *Prop.* since the arches FA and DC are similar and similarly posited, the time of the descent thro' FA, is to the time of the descent thro' DC, as the square root of FA, to the square root of DC; but by the *Corol.* of the eighth *Prop.* the time of the descent thro' FA, is one half of the time of the vibration from F to G, and the time of the descent thro' DC, is one half of the time of the vibration from D to E; consequently, the time of the vibration thro' FG, is to the time of the vibration thro' DE, as the square root of FA, to the square root of DC; that is, because the arches FA and DC are similar, as the square root of BA to the square root of BC, that is, the times of the vibrations are

as

as the square roots of the lengths of the pendulums. And forasmuch as the times wherein pendulums perform their vibrations, are to one another inversly as the number of vibrations performed in a given time; the numbers of vibrations performed by pendulums in a given time, are to one another inversly as the square roots of the lengths of the pendulums. For instance, if the length of the pendulum B A, be to the length of the pendulum BC, as one to four, the number of vibrations performed in any given time by the shorter pendulum, is to the number of vibrations performed in the same time by the longer, as the square root of four to the square root of one, that is, as two to one; which case is experimentally confirmed by two pendulums, whereof the longer being 39.125 inches, vibrates in one second of time; and the shorter being 9.781 inches, vibrates in half a second, and performs two vibrations in the same time that the longer performs one.'

Exp. 3.

Inches.

Length of a pendulum vibrating. $\begin{cases} \text{in a 2d.} \begin{cases} 39.125 \ \textit{Halley.} \\ 39.207 \ \textit{Newton.} \end{cases} \\ \text{in } \frac{1}{2} \text{ a 2d.} \begin{cases} 9.781 \ \textit{Halley.} \\ 9.801 \ \textit{Newton.} \end{cases} \end{cases}$

The time of a pendulum's vibration is no way altered by varying the weight thereof; for since the gravity of every body is proportional to its quantity of matter, as proved in my lecture upon gravity, all bodies in the same circumstances are moved by the force of gravity with the same velocity; and therefore, if the length of a pendulum continues the same, it will perform its vibrations in the same time, whatever be the magnitude of the appending weight; which may be confirmed by the following experiment. Let two unequal weights be hung by two threads so as to constitute two pendulums equal in length, and let them at the same instant of time

Exp. 4.

fall

fall from equal heights, they will keep pace together fo as to perform their vibrations in equal times.

In the foregoing part of this lecture I fhewed
you, that the vibrations of one and the fame pendulum vibrating thro' unequal but fmall circular
arches, are performed in times that are very nearly,
but not precifely, equal. Whence it follows, that
however ufeful fuch a pendulum may be in meafuring time where great exactnefs is not requifite,
yet can it by no means be admitted as an accurate
meafure of time, unlefs by fome contrivance it be
made to perform all its vibrations in equal arches,
which, confidering the unavoidable imperfections
of all machines, is extremely difficult, if not impoffible; for it has been found by experience, that
the beft regulated pendulum clocks, wherein the
greateft care has been taken to make the pendulums
vibrate in equal arches, have notwithftanding varied in a courfe of time, fo as to ftand in need of a
new regulation, which they could not poffibly do
in cafe the pendulums, whereon the regularity of
all the other movements depends, continued conftantly to vibrate in equal arches.

In order therefore to obtain an exact unerring
meafure of time, it is neceffary to make a pendulum vibrate in fuch a manner, as that all its fwings,
whether they be thro' larger or fmaller arches, may
be performed in times exactly equal; and this may
be done by making a pendulum vibrate in the
curve of a cycloid, as I fhall now demonftrate;
but I fhall firft fhew you the manner wherein that
curve is generated, and what its chief properties
are, as alfo by what contrivance a pendulum is
made to vibrate in fuch a curve.

Pl. 5.
Fig. 16. If a circle as C E F, which touches the right line
A B in the point C, be moved along that line in
the manner of a wheel from C to D, fo as to
perform an intire revolution; the point C will by
virtue of its double motion defcribe the curve line
CID,

CID, which curve line is called a *cycloid*; and the
right line C D is called the *base*, the line I K perpendicular to the base at its middle point is called the
axis of the cycloid, and the point I the *vertex*, and
the circle C E F or K L I is called the *generating
circle*.

From any point in the cycloid as H, let a right
line as H L, be drawn parallel to the base C·D, and
continued till it meets the generating circle K L I,
described about the axis I K; and let the line H M
touch the cycloid in the point H; this being done,
the chief properties of the cycloid are these three.

First, The arch I P L of the generating circle,
intercepted between the vertex of the cycloid and
the point L, wherein the right line H L meets the
generating circle, is equal in length to the right
line H L.

Secondly, The chord I L of the circular arch
I P L, is parallel to the right line M H, which
touches the cycloid in the point H.

Thirdly, The cycloidal arch I H intercepted between the vertex and the point H, is double the
chord I L.

The demonstrations of these properties may be
seen in HUYGENS, WALLIS, COTES, and others
who have wrote of the cycloid.

The contrivance whereby a pendulum is made
to vibrate in the curve of a cycloid, is thus. A cycloid as A V B, being described on the base A B, Pl. .
let the axis V D be produced towards C, till D C Fig. 17.
becomes equal to V D; thro' the points C and A,
and C and B, let two semi-cycloids C A and C B be
drawn, each equal to half of A V B, their vertices
being at A and B; if then we suppose C A and C B
to be two plates of some breadth, and an heavy
body to hang from the point C by a string equal in
length to C V, and to vibrate between the plates
C A and C B, the upper part of the string will con-

4   stantly

ftantly apply itfelf to that plate towards which the
body moves, and by fo doing caufe it to move in
the cycloid AVB as has been proved by Huygens
the author of this contrivance; and likewife by
COTES in his treatife *de motu pendulorum,* where he
has delivered the whole doctrine of pendulums in
four THEOREMS, which I fhall here lay down and
explain.

Pl. 5.
Fig. 17. 　　THEOREM I: *If a pendulum vibrating in a cycloid as* BVA, *begins its motions downward towards*
V, *from any point taken at pleafure as* L, *and if upon a radius as* VL, *equal in length to the cycloidal
arch* VL, *a circle be defcribed; the velocities of the
pendulum in the feveral points of the cycloidal arch
will be as the right fines in the circle which are raifed
from the correfponding points in the* radius; *for inftance, if in the* radius LM *be taken equal to* LM *in
the cycloid, and from the point* M *in the* radius *correfponding to the point* M *in the cycloid, be raifed the right
fine* MX, *the velocity of the pendulum in the point* M,
after it has defcended from L, *will be as the fine* MX.

　　For the proof of which, from the points L and
M in the cycloid, let the right lines LOR and
MQS be drawn perpendicular to the axis, cutting
the generating circle in O and Q, from whence to
the vertex, let the right lines OV and QV be
drawn. By the *Corol.* of the fixth *Prop.* the velocity which the pendulum acquires in defcending
along the cycloid from L to M, is equal to the velocity acquired by a body in falling perpendicularly
from R to S; but the velocity which a body acquires in falling perpendicularly, is in the fubduplicate *ratio* of the fpace defcribed, as I proved in
my lecture upon gravity; confequently, the velocity acquired by the pendulum in its defcent from
L to M, may be expreffed by the fquare root of
　　　　　　　　　　　　　　　　　　RS;

RS; but RS, being equal to the difference between RV and SV, the velocity in the point M
may be expreffed by the fquare root of the difference between RV and SV; or, becaufe, RV multiplied into the axis DV, is to SV multiplied into the
fame DV, as RV to SV, the velocity may be expreffed by the fquare root of the difference between the
product of RV×DV and SₗV×DV; but from the
nature of the circle, the product of RV × DV is
equal to the fquare of VO; and the product of
SV×DV is equal to the fquare of VQ; wherefore,
the velocity at M may be expreffed by the fquare
root of the difference between the fquare of VO and
the fquare of VQ; but, by the third property of
the cycloid, VO is equal to one half of the cycloidal
arch VL, and VQ to one half of the arch VM;
wherefore, as VO fquare, is to VQ fquare, fo is
VL fquare, to VM fquare.; confequently, the velocity of the pendulum at M may be expreffed by
the fquare root of the difference between the fquare
of VL and the fquare of VM; but the cycloidal
arches VL and VM are by fuppofition equal to
VL and VM in the *radius* of the circle; and, from.
the nature of a right-angled triangle, the difference
between the fquare of VX, which is equal to VL,
and the fquare of VM, is equal to the fquare of
MX; wherefore, the velocity of the pendulum at
the point M, is as the fquare root of MX fquare,
that is, as MX, as was afferted in the *Theorem.*
And what has been thus proved with regard to the
velocity at the point M, is in like manner demon-
ftrable with regard to the velocity at any other
point as N; namely, that it is as the right fine
NY raifed from the point N in the *radius* corre-.
fponding to the point N in the cycloid; fo that the
velocities of a pendulum defcending in a cycloid,
are in the feveral points of the cycloidal arch, as the
right fines in a circle which are raifed from the correfponding points of the *radius*, the *radius* being
equal

equal in length to the cycloidal arch intercepted between the vertex and that point from which the
pendulum begins its motion. Thus, VM and VN
in the *radius* of the circle, being taken equal to
VM and VN in the cycloid, so as that the points
M, N, V, in the *radius*, may correspond to the
points M, N, V, in the cycloid, the velocities of
the pendulum in those points are to one another, as
the sines MX, NY, and VZ, the *radius* VZ expressing the greatest velocity at the vertex V.

Pl. 5.
Fig. 17.
' THEOREM II. *If a body be supposed to move uniformly in the curve of the circle, with a velocity equal
to the velocity acquired by the pendulum in its descent
from L to V, which velocity is, as was just now
shewn, expressed by the* radius *VZ; any arch of the
circle as* XY *taken at pleasure, will be described by the
body moving along it in the forementioned manner, in
the same time that the pendulum, which begins its motion from the point L in the cycloid, describes the cycloidal arch* MN, *corresponding to and equal in length
to* MN, *that part of the* radius, *which lies between
the sines* MX *and* NY, *which terminate at the extremities of the circular arch* XY.

Let the sine FGH, be drawn indefinitely near
to the sine MX, and let XG be drawn parallel to
MF: and let MF in the cycloid be equal to MF in
the *radius* of the circle. By the foregoing *Theorem*,
the velocity of the pendulum in the point M, is as
MX; and therefore, since F is supposed to be indefinitely near to M, the little cycloidal arch MF,
equal to MF in the *radius*, is to be looked upon as
described by the pendulum with a velocity which is
as MX; and the little circular arch XH, is by
supposition described with a velocity which is as
VZ, equal to VX; and the triangles MXV and
GXH being similar, inasmuch as the angles at M
and G are right ones, and the angle MXV equal to
GXH,

GXH, becaufe GXV is the complement of each of them to a right one; XH is to XG equal to MF, as VX or VZ to MX; that is, XH and MF are to one another, as the velocities wherewith they are defcribed; confequently, they muft be defcribed in the fame time. And what has been thus demon-ftrated of MF and XH, is in like manner demon-ftrable of the feveral correfponding parts in the cy-cloidal arch MN, and circular arch XY; confe-quently, the whole cycloidal arch MN, will be de-fcribed by the pendulum in the fame time, that the circular arch XY is defcribed by a body moving along it uniformly with the velocity expreffed by VZ; and by the fame way of reafoning, the time of defcribing any other cycloidal arch as LV, is equal to the time of defcribing the correfponding circular arch LZ.

Corol. As a *Corollary* it follows, that the time wherein a pendulum defcribes any arch of a cy-cloid as MN, may be expreffed by the correfpond-ing circular arch XY.

For, as the motion along the curve of the circle is fuppofed to be uniform, the time of defcribing any arch as XY, muft be as the length of the arch; but by the *Theorem*, the times of defcribing the cir-cular arch XY, and the cycloidal arch MN, are equal; confequently, the time in which the pendu-lum defcribes the cycloidal arch MN, is as the cir-cular arch XY.

Theorem III. *The time of one intire vibration of a pendulum moving in a cycloid, is to the time wherein a body falls perpendicularly thro' a fpace equal in length to the axis of the cycloid, as the periphery of a circle to its diameter.*

All things being fuppofed as before, the time of defcribing the femicircular periphery LZP with the

Pl. 5.
Fig. 17.

M velocity

velocity expressed by VZ, is to the time of describing the femidiameter LV with the same velocity, as the femicircular periphery to the femidiameter, or as the whole periphery to the diameter; but the time of describing the femicircular periphery LZP with the velocity VZ, is equal to the time of an intire vibration; for, by the eighth *Prop.* the time wherein the pendulum describes the cycloidal arch LV, is one half of the time wherein it performs an intire vibration; and by the fecond *Theorem,* the time wherein a pendulum describes the cycloidal arch LV, is equal to the time wherein the quadrantal arch of the circle, to wit LZ, is described with the velocity expreffed by VZ; confequently, the time of an intire vibration, is equal to the time of defcribing the femicircular periphery LZP; and the time of defcribing the femidiameter LV with the velocity VZ, is equal to the time of a body's fall down the height of the axis DV; for, by the fecond *Corol.* of the third *Prop.* the fall down the axis DV, is performed in the fame time with the defcent along the chord OV; and by the eighth *Prop.* the velocity acquired at the end of the defcent along the chord OV, will in the fame time with that of the defcent defcribe a fpace equal to twice OV; but by the third property of the cycloid, twice VO is equal to the cycloidal arch LV, which by fuppofition is equal to the femidiameter VL; and confequently, the velocity acquired at the end of the defcent along the chord OV, is fuch, as will in a time equal to that of the fall down the axis DV, defcribe the femidiameter LV; but, by the *Corol.* of the fixth *Prop.* the velocity acquired at the end of the defcent along the chord OV, is equal to the velocity acquired by the pendulum in its defcent along the cycloidal arch from L to V, which by the firft *Theorem,* is as VZ; wherefore, the time of defcribing the femidiameter LV with the velocity VZ, is

equal

4

equal to the time of the fall down the axis DV; but it has been already proved, that the time of deſcribing the ſemicircular arch LZP with the velocity VZ, is to the time of deſcribing the ſemidiameter LV with the ſame velocity as the periphery of the circle to its diameter; and it has been likewiſe proved, that the time of deſcribing the ſemicircular arch with the velocity VZ, is equal to the time of an intire vibration of the pendulum; conſequently, the time of ſuch a vibration, is to the time of the fall down the axis, as the periphery of the circle to its diameter.

Corol. From what has been proved it follows, that the time of a vibration of a pendulum moving in a given cycloid is given; or in other words, that all the vibrations of ſuch a pendulum, whether they be in larger or ſmaller arches, are performed in times exactly equal.

For, as it has been proved, that the time of the vibration which begins from the point L, is to the time of the fall down the axis, as the periphery of the circle deſcribed on a *radius* equal to the cycloidal arch VL, to its diameter; it may in like manner be demonſtrated, that if the vibration begins from any other point as M, the time thereof will bear the ſame proportion to the time of the fall down the axis, that the periphery of a circle deſcribed on a *radius* equal in length to the cycloidal arch VM, does to its diameter; but the *ratio* of the periphery to the diameter in any one circle, is the ſame with that in any other; wherefore, the times of the vibrations thro' unequal arches, have all the ſame *ratio*, to the time of the fall down the axis, and of conſequence muſt be equal.

From this equality in the times of the ſwings it is, that this kind of pendulum is preferable to ſuch as vibrate in circular arches, as being a more exact and juſt meaſure of time; a minute of mean or equal time being preciſely meaſured by ſixty ſwings

of

L e c t. of a pendulum of this kind, whofe length is èqual
   X.   to three horary feet, which anfwers to 39 inches and
one eighth of our meafure, according to Doctor
HALLEY; or to 39 inches and one fifth, according to Sir ISAAC NEWTON; and now that I have
mentioned *mean* or *equal*, otherwife called *true time*,
it will not be improper in this place, to fhew you
wherein it differs from that time, which by aftronomers is called *unequal* and *apparent time.*

   As time in itfelf does not fall under the notice
of our fenfes, and as the parts thereof go on in a
continued fucceffion one after another, no two ex-
ifting together, it is impoffible to difcover the equality or inequality of any two portions of time, by
an immediate comparifon of one with the other;
and therefore, it was neceffary for thofe who firft
thought of diftinguifhing the parts of time, to have
recourfe to fomething fenfible, and of a different
nature from time, as a meafure thereof. And as
nothing feems better fitted to ferve this purpofe,
than fuch natural apppearances as fall under every
man's notice, and at the fame time have frequent
returns, it is highly probable, that in the firft ages
of the world, men obferving the frequent rifings
and fettings of the fun, took the one or the other
for their firft meafure of time, calling that portion
of time which paffed between two rifings or fettings,
which immediately fucceeded each other, by the
name of a *day*; in like manner it is rational to fuppofe, that upon obferving the frequent returns of
the full and new moons, they made the one or the
other their fecond meafure of time, calling that
fpace which paffed between two fucceffive new or
full moons by the name of a *moon* or *month*. And
it is likely, that for fome time they contented themfelves with thefe meafures, without knowing or
confidering whether they were exact or not: but
in procefs of time, as men became better acquainted
with the motions of the heavenly bodies, they difcovered

2

covered fome irregularities in the apparent motion
of the fun, and of confequence, an inequality in the
natural days which depend on that motion; inafmuch as the portion of time, which paffes between
the fun's departure from the plane of any meridian
and its next return thereunto, is not always the
fame. By confidering the caufes of this inequality,
they were led into a method of making fuch cor-
rections in the natural days, by adding to fome,
and taking from others, as reduced them all to a
mean equal length; each day being made to confift of 24 equal *hours*, each of which is divided into
fixty equal parts called *minutes*, and each of thefe
into fixty others called *feconds*, and thefe again into
thirds, and fo on in a *fexagefimal* progreffion, the
parts of each denomination being conftantly equal
among themfelves. And thefe parts of time thus
reduced to an equality conftitute the mean or equal
time, as it ftands diftinguifhed by aftronomers,
from the unequal or apparent time, which is meafured by the apparent motion of the fun.

In order to have a conftant meafure of equal
time, Huygens contrived a method of adapting
pendulums to clocks, whereby their motions are fo
exactly regulated, as that in a clock whofe movements are rightly adjufted, the feconds, minutes,
and hours, are for fome time pointed out with the
greateft exactnefs; I fay, for fome time only, becaufe it is not poffible that any clock whatever
fhould continue exactly true for a long courfe of
time; for as the pendulums of clocks according to
Huygens's firft contrivance, and by the general
practice of clock-makers at this day, are made to
vibrate in circular arches, where the times of the
vibrations are not precifely equal, unlefs the arches
thro' which the pendulum moves be fo too. If the
wheels on account of the thickening of the oil by
frofty weather, or from any other caufe grow more

M 3   fluggifh,

sluggish, so as to give the weight, which in clocks is the moving power, greater resistance than according to the first adjustment, the force of the crown-wheel upon the palates of the pendulum will likewise be diminished, and of consequence, the pendulum being thrown less forcibly will move thro' smaller arches than before, and by so doing, will measure out smaller portions of time, the time of sixty swings not amounting to a minute, upon which account the clock must gain, and go too fast. On the other hand, whenever the parts of the movement which rub one against another do, by reason of the thinning of the oil by the heat of the weather, grow more slippery, or from their constant friction become more smooth, so as to give less resistance to the moving power than according to the first adjustment, the crown-wheel acts more forcibly on the pendulum, and causes it to vibrate in larger arches, by which means the time of each swing is inlarged, and of course the clock loses and goes too slow. To remedy these inconveniences HUYGENS thought of a second method of adapting pendulums to clocks, so as to make them perform their vibrations in cycloidal arches; by which means, tho' the force of the crown-wheel upon the pendulum should vary, so as to cause it to vibrate sometimes in larger and sometimes in smaller arches, yet will not any variation arise from thence in the times of the vibrations; as is evident from the *Corollary* of the third *Theorem*; so that in clocks whose motions are governed by pendulums vibrating in cycloidal arches, the irregularities arising from the variation of the force of the crown-wheel upon the pendulum are wholly avoided; and yet a clock of this kind will not always go true; for as the pendulum cannot vibrate in the curve of a cycloid, unless the uppermost part of the string does as often as it moves from the per-

pendicular

pendicular towards either fide, form itfelf into a
cycloidal arch; and as this cannot be done unlefs
that part of the ftring be made of filk, or fome
other foft and pliable fubftance, which as fuch is
apt to imbibe the moifture of the air; whenever
the weather becomes remarkably moift, the watery
particles which float in the air, will infinuate themfelves into the pores of the ftring, and by fo doing
caufe it to contract and fhorten; upon which account, the vibrations of the pendulum will be
quickened, as will appear from the next *Theorem,*
and the clock will gain. So that neither a clock
of this, nor of any other kind, can go exactly true
for any long courfe of time, which is a thing well
known to clock-makers, who have frequently experienced the beft regulated clocks to vary in the
compafs of a few months, fome feconds from the
equation table, fo as to ftand in need of a new regulation.

THEOREM IV. *The times wherein pendulums of* Pl. 5.
different lengths as CV *and* AB *perform their vi-* Fig. 17,
brations, are to one another in the fame proportion with 18.
the fquare roots of the lengths of the pendulums.

For, by the third *Theorem,* the time wherein the Pl. 5.
pendulum CV performs its vibrations, is to the Fig. 17.
time wherein a body falls down the axis DV, as the
circumference of a circle to its diameter; and by
the fame *Theorem,* as the circumference of a circle
is to the diameter, fo is the time wherein the pendulum AB performs its vibrations, to the time Pl. 5.
wherein a body falls down the axis EB; confequent- Fig. 18.
ly, the time wherein the pendulum CV performs
its vibrations, is to the time wherein the pendulum
AB performs its vibrations, as the time of the fall
down DV, is to the time of the fall down EB;
but, as I proved in my lecture upon gravity, the
time of the fall down DV, is to the time of the

fall

fall down EB, as the square root of DV, to the
square root of EB; or, becaufe DV is one half of
CV, and EB one half of AB, as the square root of
CV, to the square root of AB; wherefore, the
time wherein the pendulum CV performs its vibrations, is to the time wherein the pendulum AB
performs its vibrations, as the square root of CV,
to the square root of AB, that is, the times are as
the square roots of the lengths of the pendulums;
fo that if one pendulum be four times as long as
another, the fhorter will vibrate in half the time,
fo as to perform two vibrations in the fame time
that the longer performs one.

In this *Theorem*, as alfo in every thing elfe that
has been hitherto faid concerning the pendulum, the
force of gravity is fuppofed to be given; whence
it follows, that if pendulums of different lengths,
Pl. 5.
Fig. 17,
18.
as CV and AB, perform their vibrations in equal
times, the force of gravity in fuch pendulums muft
vary, and that in proportion to the lengths of the
pendulums, that is to fay, the force of gravity in
the pendulum CV, muft be to the force of gravity
in the pendulum AB, as CV to AB. For, as the
times of the vibrations are fuppofed to be equal, the
times of the perpendicular falls down the axes DV,
and EB muft likewife be equal, inafmuch as they
have been proved to be proportional to the times
of the vibrations; fince therefore, forces which act
conftantly and uniformly are to one another as the
velocities which they generate in any given time,
the force of gravity which carries a body down DV,
muft be to the force of gravity which in the fame
time carries a body down EB, as the velocity acquired at the end of the fall down DV, to the velocity acquired at the end of the fall down EB; but I proved in
my lecture upon gravity, that the velocity acquired in falling down DV, is fuch as will in a fpace of
time equal to that of the fall, carry a body thro' a fpace
equal to twice DV, that is, thro' a fpace equal to
the

the length of the pendulum C V ; and therefore,
the time being given, the velocity may be expressed
by the length of the pendulum ; and for the same
reason, the velocity acquired in falling down E B,
may be expressed by the length of the pendulum
A B ; consequently, the force of gravity which
moves the pendulum C V, is to the force of gravity
which acts upon the pendulum A B, as the length of
the former, to the length of the latter. Since therefore, it has been found by experience, that a pendulum which vibrates in a second of time under the
line, must be lengthened as it is removed from the
line, and that more and more as its distance therefrom increases ; it is manifest, that the force of
gravity is less in the æquatorial parts of the earth,
than in any other, and that it increases continually
as the distance from the line increases, so as to be
greatest under the poles ; in what proportion this
increase of gravity is made, and from what cause
it proceeds, I shewed in my lecture upon gravity.

As the several parts of the cycloidal arch L V,
have different inclinations to the plane of the horizon, it is evident, from what has been said concerning the motion of bodies upon inclined planes,
that the force which accelerates the motion of a
pendulum in its descent from L to V, must continually vary ; it being greatest in the point L, and
thence continually lessening as the cycloidal arch
shortens, till at length in the point V it intirely
vanishes ; and what is particularly remarkable in
this case is, that the accelerating forces in the several points of the cycloid, are to one another in
the same proportion with the cycloidal arches intercepted between the vertex and the respective
points ; for instance, the force which accelerates the
pendulum in the point L, is to the force which accelerates the same in the point M, as the arch L V,
to the arch M V.

Pl. 5.
Fig. 17.

For,

For, as the points L and M have the same directions with their tangents, the accelerating forces
in those points must be the same with the forces
which accelerate the motions of bodies descending
along the tangents; or because the chords OV and
QV in the generating circle are, by the second property of the cycloid, parallel to the tangents at L
and M, as the forces which accelerate bodies in
their descent upon the chords OV and QV; but
forasmuch as those accelerating forces act constantly
and uniformly, they must be to one another, as the
velocities which they generate in a given time; and
therefore, since it has been proved, that the chords
OV and QV are described in the same time, the
accelerating forces are as the velocities acquired at
the end of the descent along those chords; but it
has likewise been proved, that those velocities are
as the lengths of the chords; consequently, the
force which accelerates a body descending along the
chord OV, is to the force which accelerates a body
descending along the chord QV, as OV to QV;
but forasmuch as by the third property of the cycloid, OV is one half of LV, and QV one half of
MV, as OV is to QV, so is LV to MV; and
therefore, the accelerating force along OV, is to the
accelerating force along QV, as the cycloidal arch
LV, to the arch MV; but it has been proved, that
the accelerating force along OV, is the same with
the accelerating force in the point L, and that the
accelerating force along QV, is the same with the
accelerating force in the point M; consequently,
the force at L, is to the force at M, as LV to MV;
as what has been thus demonstrated of the forces
at the points L and M, is in like manner demonstrable of the forces at any other points, so that in
a pendulum descending in the arch of a cycloid, the
accelerating force is in every point as the length of
the cycloidal arch intercepted between the point and
the

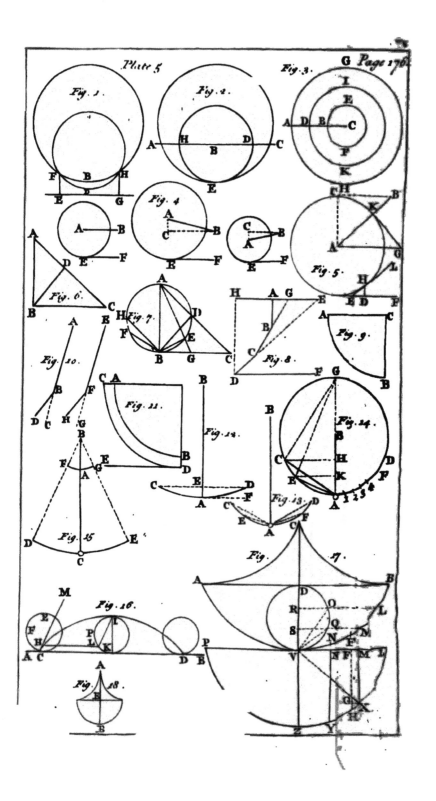

Plate 5.   G Page 176

the vertex; or in other words, the force is every where proportional to the space to be defcribed.

This then being the law of the accelerating force, and it having been proved, that the pendulum, whether it begins its motion from L or M, 'or any other point in the cycloid, will arrive in the fame time at the loweft point V; it follows, that if- feveral bodies, placed at different diftances from any point or center, begin to move towards it at the fame inftant of time, with forces that are every where proportional to the diftances from the center, they will all arrive at the center at the fame inftant of time; which I thought fit to mention in this place, in order to avoid the trouble of demonftrating the fame, when I come to treat of the motions of mufical ftrings, towards the explaining of which this property will be of ufe.

LECTURE XI.

Of the Motion of Projects.

AS the Doctrine of Projects, whereof I intend to treat in this lecture, cannot be rightly apprehended without fome knowledge of the Parabola; I fhall by way of introduction fhew the manner wherein that curve is generated, and point out fuch of its properties as I fhall have occafion to make ufe of in explaining the motion of projects, referring you for their demonftrations to thofe authors who have wrote of the conick fections.

If a cone as A BC, be touched by a plane in the right line A B, and be cut by another plane parallel to the former, the curve which arifes from the interfection of the plane with the furface of the cone is called a Parabola; being fuch as is reprefented in Fig. 2, in which the higheft point P is called the principal

LECT.
XI.

Pl. 6.
Fig. 1.

Pl. 6.
Fig. 2.

principal vertex; the right line C A P paffing thro'
the point P, and perpendicular to the tangent at that
point, is called the *axis*; a right line as D A, drawn
from any point in the curve perpendicular to the axis,
is called an *ordinate to the axis*; P A the part of the
axis intercepted between the vertex and the ordinate,
is called the *abfciffe to that ordinate*; a right line,
being a third proportional to the abfciffe and its refpective ordinate, is called the *principal parameter*,
or the *parameter to the axis*; a right line as D E H,
drawn from any point in the curve parallel to the
axis, is called a *diameter*; a right line as P E, intercepted between any point in the curve and the diameter, and parallel to B D which touches the curve
in the point D, is called an *ordinate to that diameter*;
D E the part of the diameter lying between the vertex D and the point E, is called the *abfciffe to the
ordinate* P E; and a right line, being a third proportional to the abfciffe D E and the refpective ordinate E P, is called the *parameter to the diameter*
D H, or *to the vertex* D.

The fquare of any ordinate divided by the refpective abfciffe, is equal to the refpective parameter; thus the fquare of D A divided by P A, or the
fquare of O Q divided by P Q, is equal to the principal parameter; and the fquare of E P divided by
D E, as alfo the fquare of L M divided by D L, is
equal to the parameter belonging to the vertex D.
The fquares of the ordinates to the axis, or to one
and the fame diameter, are to one another in the
fame proportion with their refpective abfciffa's.
Thus, the fquare of D A is to the fquare of O Q, as
P A to P Q; and the fquare of P E is to the fquare
of M L, as D E to D L.

In one and the fame *parabola*, the principal parameter is the leaft of all the parameters; and the
other parameters increafe, as the diftance of their
vertices from the principal vertex increafes, tho' not
in the fame proportion.

If

If from any point in a *parabola* as D, an ordinate be drawn to the axis, and if from the same point a tangent be drawn upward, it will meet the axis when produced ; and AB, the part of the axis intercepted between the ordinate DA and the tangent DB, will be bisected. by P ·the principal vertex.

These · things being premised; if a body be thrown into any direction whatever that is. not perpendicular to the plane of the horizon, it will in its motion describe a *parabola*.

For the proof of which, let AE be the direction of the projection, which in the 3d Fig. is parallel to the horizon, and in the 4th and 5th inclined thereto ; and let AE be the space which the project would describe in any given time by means of the force impressed, supposing it had no motion downward from the force of gravity ; likewise let AB be the space thro' which it would descend in the given time by virtue of its own gravity, supposing it had no other motion ; then compleating the parallelogram ABCE, it is·manifest from what was formerly said concerning the composition of motion, that at the end of the given time, the project must by virtue of its double motion, be found in the point C ; but, forasmuch as the motion impressed in the direction AE is uniform, the space described, that is AE, must be as the time in which it is described ; consequently, AE square, or BC square, is as the square of the time ; but AB or EC, which is the space described in the same time by the force of gravity, is likewise as the square of the time, as I proved in my lecture upon gravity ; consequently, AB is as the square of BC ; and therefore, from the nature of the *parabola*, the point C thro' which the project moves, must be in the curve of a *parabola*, whose diameter is A B, the vertex A, the point from whence the project begins its motion, and the

parameter

Pl. 6.
Fig. 3, 4,
5.

parameter belonging to that vertex, EC square divided by AE, or AE square divided by EC; and what has been thus demonstrated of the point C, is in like manner demonstrable of all the other points thro' which the project moves; consequently, the line which it describes is a *parabola*.

The velocity of a project in any point of the *parabola* as A, is such as a body acquires in falling down the fourth part of the parameter belonging to that point. For the velocity of the project in the point A is such, as would carry it from A to E in the same time that a body descends from E to C; and the velocity acquired in the descent from E to C is such, as in the same space of time with that of the fall, would carry a body thro' a space equal to double EC; consequently, that velocity is to the velocity of the project in the point A, as twice EC to AE, or as EC to $\frac{1}{2}$ AE; but as EC is to $\frac{1}{2}$ AE, so is the velocity acquired in falling from E to C, to the velocity acquired in falling down the fourth part of the parameter belonging to the vertex A; for, by the nature of the *parabola*, the parameter belonging to the vertex A, is equal to $\frac{AEq}{EC}$; wherefore, the velocity acquired in falling from E to C, is to the velocity acquired in falling down the fourth part of the parameter, as the square root of EC to the square root of $\frac{\frac{1}{4}AEq}{EC}$, which square roots are to one another, as EC to $\frac{1}{2}$ AE, as may appear by multiplying each into the square root of EC; so that the velocity acquired in falling thro' a fourth part of the parameter belonging to the vertex A, and the velocity of the project in the point A, have one and the same proportion to the velocity acquired in falling from E to C; consequently, from the nature of proportionals, those two velocities must be equal.

Hence

Hence it follows, that if projects move through the same or different *parabolas*, the squares of their velocities in the several points of the *parabolas*, are to one another, as the parameters belonging to the respective points; for, since the velocities in the several points are equal to the velocities acquired in falling down the fourth part of the parameters belonging to those points, and since the squares of the velocities acquired in falling down the fourth part of the parameters, are to one another as the spaces described, as I proved in my lecture upon gravity, it is evident that the squares of the velocities wherewith projects move thro' the several points of the *parabolas* which they describe, are to one another in the same proportion with the quarter parts of the parameters belonging to those points; but the quarter parts of the parameters being to one another as the whole parameters, the squares of the velocities in the several points of the *parabolas* must bear the same proportion to one another, that the parameters do which belong to those points.

Since this is the case, and since by the nature of the *parabola* the principal parameter is less than any other, and that the other parameters grow larger as the points to which they belong are more distant from the principal vertex; if a project be cast obliquely upward, as in *Fig.* 4. from A towards E, its velocity must continually decrease as it rises and approaches the uppermost point P, wherein the velocity being least must thence increase continually as the project descends and recedes from the point P; and as in one and the same *parabola*, where the distances of any two points as A and K, from the principal vertex P, are equal, the parameters belonging to those points are likewise equal; it is manifest, that a project must have equal velocities in those points; and of consequence, setting aside any difference which may arise from the resistance of the air,

Pl. 6.
Fig. 4.

air, the project will, *cæteris paribus*, strike a mark
as forcibly in the point K as it does at its first setting out in the point A.

The velocity wherewith a project is thrown being
given, the velocity thereof in any point of the curve
may be thus determined. In the *parabola* of *Fig.* 3.
let the axis BA be continued upward to D, so as that
DB may equal the height from which a body must
fall, in order to acquire the same velocity wherewith the project sets out from G; then from any
point in the curve taken at pleasure as K, let the
horizontal line KL be drawn, and the velocity of
the project in the point K, will be to the velocity
wherewith it began its motion from G, as the square
root of DL, to the square root of DB.   For, in
my lecture upon gravity, I proved, that if a body
be thrown directly upward from B towards D, with
the same velocity that it acquires in falling from
D to B, it will in any point of its ascent as L, have
the same velocity that it would acquire in falling
from D to that point ; but the velocity acquired in
the descent from D to L, is to the velocity acquired
in the descent from D to B (which velocity is by
supposition equal to the velocity wherewith the body is thrown up) as the square root of DL, to the
square root of DB ; and by the eighth *Prop.* of my
last lecture, the velocity of the project at K, is the same
with the velocity at L ; consequently, the velocity
thereof at K, is to the velocity wherewith it set out from
G, as the square root of DL, to the square root of DB.
Whence it follows, that if DB be equal to 1600
feet, and DL to 400, the velocity of the project at K,
is but one half of the velocity which it had at its setting out from G ; and if DL be equal to 900 feet,
then is the velocity at K, three fourths of the velocity at G ; so that a project being thrown obliquely upward with such a velocity as would carry it to
the height of 1600 feet if thrown directly upward,

will

will lofe a fourth part of its velocity by the time it has rifen to the perpendicular height of 700 feet, and one half of its velocity when it has rifen 500 feet more.

The velocity wherewith a project is thrown from any given place being given, as alfo the pofition of a mark, the directions wherein the project muft be thrown, in order to hit the mark, may be determined in the following manner.

Let A be the place from whence the project is Pl. 6. thrown, C the mark fituated in the line AC whofe Fig. 6. length is given, as alfo the angle CAB, which it makes with the horizontal line AB; at A erect the perpendicular AP, equal to the parameter belonging to the point A, which parameter is given, inafmuch as the velocity wherewith the project is caft from the point A is given; for it is equal to four times the height from which a body muft fall in order to acquire that velocity. Let AP be bifected by the line KH, cutting it perpendicularly in G; at A erect AK perpendicular to AC, and let it be continued till it meets KH. From the point of concourfe K, with the *radius* KA, let the circle AHP be defcribed. This being done, let a right line as BCEI, be erected perpendicular to the horizontal line AB, fo as to pafs thro' the mark C, and if poffible to cut the circle in two points as E and I; AE and AI are the two directions, in either of which, the project being caft with the given velocity, will hit the mark.

For, drawing the lines PE and PI, the angles CAE and APE are equal, from the nature of the circle; and from the nature of parallel lines, the angles CEA and EAP are equal; confequently, the triangle AEC is fimilar to the triangle PAE; and therefore, PA is to AE, as AE to EC; wherefore, multiplying the extremes and means, and dividing by EC, PA is equal to $\frac{AEq}{EC}$. In like man-

ner

ner, the triangles PAI and CAI being similar, PA is equal to $\frac{\text{AIq}}{\text{IC}}$. Wherefore, since PA is equal to the parameter at the point A, it follows, from the nature of the *parabola*, that those *parabolas* which the project describes when thrown in the directions AE and AI, must pass thro' the point C; consequently, the mark will be hit by a project thrown in either of those directions.

Whenever the mark is placed at such a distance from A on the line ACM, suppose at M, as that the perpendicular NMH which passes thro' the mark, becomes a tangent to the circle at H, the mark is then at the utmost limit on the line AM, to which a project thrown with the given velocity can reach, and there is but one direction, to wit AH, wherewith the mark can be hit; for it is evident, that any other direction must terminate in some point of the circumference above or below the point H; whence if a perpendicular be let fall to the horizontal line AN, it must of necessity fall on this side of HN with respect to A, and of consequence, cut the line AM in a point less distant from A than is the point M.

The line AH, which denotes the direction of the project, when thrown to the greatest distance possible on the line AM, bisects the angle PAM, which measures the visible distance between the zenith or vertical point P and the mark M. For, by the nature of the circle, the angle MAH is equal to the angle HPA; and forasmuch as in the triangles HPG and HAG, the sides PG and AG are equal by construction, and GH common to both, and the angles at G right ones, the angle HPG is equal to HAG, consequently, HAG or HAP is equal to MAH, that is, the angle PAC is bisected by the line AH.

2

If

"If the line ACM be fituated below the horizontal line AN, which is the cafe when the mark is feated on a defcent, let all things be conftructed as before, and the fame things will obtain; to wit, A I and AE, will be the directions neceffary to hit the mark at C; and the line AH will bifect the angle PAM, which meafures the apparent diftance between the zenith, and the mark; and the point M will be the utmoft limit on the line A M, of a project thrown with the given velocity; the demonftrations of which are exactly the fame as in the foregoing cafe.

Pl. 6.
Fig. 8.

If the mark be placed on a level, the line ACM will coincide with the horizontal line ABN, and the parameter AP, will pafs thro' the center of the circle and become a diameter, the points K and G coinciding.

In this cafe, the horizontal diftance of the mark, to wit AC or AB, is as the fine of the doubled angle of elevation; or in other words, the horizontal range, or the diftance to which a project is thrown on the plane of the horizon with a given velocity, is as the fine of the doubled angle of elevation.

For AC, the horizontal range of a project thrown with a given velocity in the direction AE, is equal to DE, the fine of the angle AKE; but, from the nature of the circle, the angle AKE is double the angle APE, which is equal to CAE, the angle of elevation; confequently, A C, the horizontal diftance of the mark, or the diftance to which a project is thrown on the plane of the horizon with a given velocity, is as the fine of the doubled angle of elevation.

Hence it follows, that in order to throw a project with a given velocity, to the greateft diftance poffible on the plane of the horizon, the direction of the projection muft be elevated in an angle of 45 degrees; for, fince the fine of twice 45 or 90 de-

grees

grees is equal to the *radius*, and of consequence, the greatest of all the fines; and fince the horizontal ranges at the feveral angles of elevation are to one another, as the fines of the doubled angles of elevation, it is manifeft, that the greateft range, or as it is ufually called by gunners, the greateft random, muft be when the project is caft in a direction whofe elevation is 45 degrees ; moreover, the greateft random is ever equal to one half the parameter at the point from which the projection is made ; for the line AM, which expreffes the greateft random, is equal to the *radius* KH, or half the diameter AP, which by the conftruction, is equal to the parameter belonging to the point A ; fo that where the velocity with which a project is thrown is given, the utmoft diftance which that project can reach on the horizontal plane, is likewife given ; for it is equal to twice the height, from which a heavy body muft fall in order to acquire the velocity wherewith the project is thrown ; the parameter belonging to the point A, having been already proved equal to four times that height.

A fecond confequence of the horizontal ranges being as the fines of the doubled angles of elevation is, that if two projects be thrown with equal velocities, in directions whofe elevations are equally diftant from 45 degrees above and below, for inftance, if the elevation of one be 60 degrees, and that of the other 30, whereof the former exceeds 45 degrees, and the latter falls fhort thereof by 15 degrees, the horizontal ranges will be equal, or, in other words, the two projects will fall on the plane of the horizon, at the fame diftance from the place of projection ; for as the fum of any two arches of a quadrant, whereof one exceeds 45 degrees, as much as the other is exceeded thereby, is equal to a quadrant, it is manifeft, that two fuch arches are complements to each other ; wherefore, fince by the nature of the circle, the fine of a doubled arch is equal

4   to

to the fine of its doubled complement, the fines of the doubled angles of two elevations equally diftant from 45 degrees above and below, muft be equal ; and fo of confequence muft the horizontal ranges which are proportional to thofe fines. And thus it would conftantly be, were it not for two caufes which do in fome meafure difturb this law of projects, fo as to make the horizontal ranges of the higher elevations to fall fhort of thofe of the lower.

The firft of thefe difturbing caufes is the *air*, which as it refifts, and thereby retards the motions of projects, muft, *cæteris paribus*, caufe a greater retardation in thofe motions which are of longeft continuance ; confequently, fince the higher the elevation of the direction is, the longer is the time of the project's motion, as fhall be fhewn hereafter ; if the directions wherein two projects are caft with equal velocities, be equally diftant from 45 degrees, the one above and the other below, the project which is thrown in the higher direction, will be more retarded than that which is thrown in the lower ; and of courfe, will fall on the plane of the horizon at a lefs diftance from the place of projection.

The fecond difturbing caufe, obtains with regard to fuch projects only as are thrown by the force of *gun-powder*. As the force of the powder acts upon the ball during its continuance in the barrel, fo does it likewife to fome diftance beyond the muzzle ; and by fo doing makes the ball to move forward in a right line, which line is commonly called the *line of impulfe of fire* ; at the end of which, the ball quitting the blaft of the powder, begins to move in the curve of a *parabola*.

Now, tho' the air gave no refiftance to projects, yet muft the horizontal ranges of a ball fhot out of the fame piece with equal charges, in two directions equally diftant above and below 45 degrees, be different on account of the different directions of the

line

line of impulse of fire; for, let us suppose a gun at A, to discharge two equal balls with equal quantities of powder, one in the direction AB, and the other in the direction AC, AB being as far above 45 degrees, as AC is below it; and let AB and AC denote the lines of impulse of fire, so that at B and C the balls will begin to move in *parabolick* curves; from which points let fall the perpendiculars BD and CE; it is manifest, that AD, which is the fine of the complement of the higher elevation, will denote that part of the horizontal range which is owing to the line of fire, when the project is thrown according to the higher elevation; and AE, the fine of the complement of the lower elevation, will be that part of the horizontal distance, which is owing to the line of impulse when the project is thrown according to the lower elevation; consequently, since the fine of the complement of a lesser angle is ever greater than that of a larger angle, the horizontal range of the lower elevation must exceed that of the higher, so that where projects are thrown with the same velocity by the force of powder, in directions equally distant above and below 45 degrees, those must range fartheft which are thrown according to the lower elevations, as well on account of the line of fire, as of the resistance of the air.

Pl. 6.
Fig. 8.

The altitude to which a project rifes, is as the versed fine of the doubled angle of elevation; for the proof of which, let AE be the direction of the projection, and let AC or AB be bisected in T, and from the point of bisection erect the perpendicular TR; since the point T is equally distant from A, where the project begins its motion, and from B, where the motion of the project ceases, TR will be the axis of the *parabola* which the project describes; and, from the nature of the *parabola*, will be bisected in V by the principal vertex;

where-

ject will rise, inasmuch as the versed sines of the doubled angles of elevation increase continually with the elevation, till at length the elevation becoming perpendicular, the versed sine of the doubled elevation becomes equal to the diameter, which being the greatest of all the versed sines, the altitude of the perpendicular projection must likewise be greatest; and it is equal to one fourth of the parameter; for, I shewed you in my lecture upon gravity, that if a body be thrown up with any velocity, it will rise to the same height, from whence if it fell from a state of rest, it would by the end of the fall acquire the same velocity wherewith it is thrown up; and in this lecture I proved, that the velocity wherewith a project moves in any point of the *parabola*, is equal to the velocity acquired by a heavy body in falling down the fourth part of the parameter belonging to that point; consequently, a project thrown up with a given velocity from the point A, will rise to a height equal to the fourth part of the parameter belonging to that point. Hence it appears, that the greatest height of the perpendicular projection, is equal to half the greatest random, inasmuch as the greatest random has been proved equal to half the parameter belonging to the point A.

The

The time of the flight of a project thrown with a given velocity, is as the sine of the angle of elevation: for instance, the time of the flight of a project thrown with a given velocity in the direction AE, is as the sine of CAE, the angle of elevation; for since the project moves thro' the curve of a *parabola* from A to C, by virtue of its uniform motion in the direction AE, and of its accelerated motion in the direction EC, it is evident, that the time of its flight thro' the *parabola*, must be equal to the time of its uniform motion from A to E; but as the velocity is given, the time of the motion from A to E; must be as the space described, that is, as AE; or by the nature of proportionals, as one half of AE; but AE being the chord of the arch AE, which measures AKE, the doubled angle of elevation, one half of AE is equal to the sine of half the arch AE, that is, to the sine of the arch which measures CAE, the angle of elevation; consequently, the time of the flight is as the sine of the angle of elevation. Hence it follows, that the greater the elevation is, the longer the time of the flight will be; as also that the time of the perpendicular flight is greatest of all, the sine of the perpendicular elevation being equal to *radius*.

If the velocity wherewith a project is thrown be required, it may be determined from experiments in the following manner; by the help of a *pendulum* or any other exact *chronometer*, let the time of the perpendicular flight be taken; then, forasmuch as the times of the ascent and descent are equal, the time of the descent must be equal to one half of the time of the flight, consequently, that time will be known; and, forasmuch as a heavy body descends from a state of rest at the rate of 16 feet in the first second of time, and that the spaces thro' which bodies descend are as the squares of the times; if we say, as one second is to sixteen feet, so is the square

of

of the number of seconds which exprels the time of
the defcent of the project, to a fourth proportional,
we fhall have the number of feet thro' which the
project fell, which being doubled, will give us the
number of feet which the project would defcribe in
the fame time with that of the fall, fuppofing it
moved with an uniform velocity, equal to that
which it acquired by the end of the fall; which laft
found number of feet, being divided by the number of feconds which exprels the time of the project's defcent, will give a quotient, exprelling the
number of feet thro' which the project would move
in one fecond of time with a velocity equal to that
which it acquired in its defcent, which velocity is
equal to the velocity wherewith the project was
thrown up; confequently, the velocity wherewith
the project was thrown up is difcovered. To illuftrate this by an inftance, let us fuppofe half the time
of the perpendicular flight to be 8 feconds; then,
as one is to 16, fo is 64, the fquare of 8 feconds, to
1024; which being doubled, and then divided by
8, gives 256 in the quotient; which fhews that the
project was thrown upward with fuch a velocity as
would carry it, fuppofing it moved uniformly, at the
rate of 256 feet in one fecond of time.

Perhaps it may be objected, that the method here
laid down for difcovering the velocities of projects,
is founded on experiments in which projects are fuppofed to move freely without any let or impediment,
whereas the air refifts and retards all projects in their
motions, fo as not to fuffer them to rife to the fame
height, or to return with the fame velocity, that
they would in cafe they moved *in vacuo*; in anfwer
to which it muft be confefled, that in the experiments here made ufe of, the air does refift and im-
pede the motions of projects, fo as to fhorten their
afcent, and to leffen the velocity of their return;
but then this does very little affect the truth of

the

LECT.
XI.
the conclusions which are gathered from these experiments concerning the velocities, wherewith projects begin their motions; for, as in the method laid down, the only thing necessary to be known from experiment, is the true time of the flight of a project, supposing it to move *in vacuo*; if that time can be had from these experiments, the velocity wherewith the project sets out may be rightly determined, notwithstanding the resistance of the air; but the time of the flight of a project thrown directly upward, is very nearly the same *in vacuo*, as in the air; for, as much as the time of a project's ascent is shortened by the resistance of the air, so much very nearly is the time of its descent lengthned by the same resistance, consequently, the whole time of the flight in air must be very nearly equal to the time of the flight *in vacuo*; and therefore, the time of the flight *in vacuo* is got, by taking the time of the flight in air.

Degrees.	Sines.	Versed sines.
30	50	13
45	70	29
60	86	51
90	100	100
120	—	148

Experiments

*Experiments concerning projects made with a small
mortar, the length of whose chase was 5¼ inches;
the diameter of the ball and chase 3¼ inches; weight
of the hollow ball 23000 grains; length of the
chamber 2 inches, and its diameter ¾ inch.*

Quantity of powder in grains.	Degrees of elevation.	Horizontal ranges in feet.	Times of the flights in half seconds.
60	30	135	5
60	45	150	6
60	60	120	8
60	90		11
90	30	200	6¼
90	45	220	8
90	60	200	9¼
90	90		11
120	30	420	9
120	45	450	11
120	60	300	12¼
120	90		16
140	45	660	15
140	90		18
180	30	1000	13
180	45	1100	17
180	60	900	21
180	90		26
240	45	1750	20
240	60	1390	25
240	90		32

LECTURE

LECTURE XII.

OF HYDROSTATICKS.

IN this lecture I shall give you an account of the gravitation and pressure of WATER, and such other FLUIDS, as are commonly called LIQUIDS.

A *fluid* in general is a body, whose parts yield to any force impressed, and in yielding are easily moved one among another.

The minute particles of fluids do not seem to differ from those of solid bodies; inasmuch as fluids and solids are frequently converted into one another. Thus water and watery fluids are by cold changed into ice; which by heat is again reduced to its fluid state. Metals of all kinds being melted become fluid, and upon cooling grow solid again. The most solid and ponderous woods, as also the hardest stones, may by the force of fire in a great measure be converted into water, as is well known to the chymists. And there are not wanting instances in nature of the grossest bodies being turned into the subtile fluids of air and light, and these again into gross bodies. Which changes can scarcely be accounted for, unless we suppose the minute particles of fluids to be of the same nature with those of solid bodies. But be this as it will, most certain it is, that fluids as well as solids consist of heavy particles, whose gravity is ever proportional to the quantity of matter which they contain. This having been found as far as experience reaches to be the universal property of matter, whatever be the form under which it appears. Most indeed of the antient naturalists, not being sensible of any weight or pressure from the air about them, or from the incumbent water when immersed therein, were of opinion, that the parts of one and the same element

did

did not gravitate one upon another; which opinion has been exploded by the moderns as erroneous; and that it is fo, will appear from the following experiment.

Let an empty phial clofe ftopped and immerfed in water, be fufpended from one end of a balance and poifed; then let the ftopple be taken out, that the water may run in, the phial upon receiving the water will preponderate, and bear down the arm of the beam from which it hangs; which evidently proves, that the parts of water retain their gravity in water, fo as to prefs and bear down upon the parts beneath them; otherwife the phial would not become heavier upon the admiffion of the water.

From the gravity of the parts it follows, that fetting afide all external impediments, the furface of a liquid contained in a veffel muft be fmooth and level; for fhould any part ftand higher than the reft, it muft defcend by the force of its gravity, and in fo doing, fpread and diffufe itfelf till it comes to be on a level with the other parts. As the gravity of the parts reduces the upper furface to a level, fo does it likewife occafion a preffure on the lower parts, greater or lefs in proportion to their depth below the furface, each part fuftaining a preffure equal to the weight of all thofe which lie above it; whence it follows, that the parts which are at equal depths below the furface, are equally preffed, and of confequence muft be at reft, contrary to the opinion of thofe, who make the nature of fluidity to confift in the conftant actual motion of the parts one among another. Should this equality of preffure at any time be deftroyed, then indeed a motion will arife in the parts of the fluid, and continue till the preffure becomes equal again, as may appear from the following experiment; whereby the truth of what has been faid concerning the preffure of the fuperior parts of fluids on thofe beneath them, will likewife be confirmed.

Take

Take a glafs tube open at both ends, and stopping one end with a finger, immerge the other in water to any depth whatever; upon the immerfion, the water will rife in the tube, but the height to which it rifes, whilft the upper orifice continues stopped, will be but fmall; but upon removing the finger, it will rife to the fame height with the water without *.

When the tube is immerfed, that portion of water which lies beneath the orifice ceafes to be equally preffed with the other portions that are at the fame depth; for that portion bears no other preffure than what arifes from the fpring of air included in the tube, (which preffure is equal, as fhall be fhewn hereafter, to the preffure arifing from the weight of the external air) whereas, the other portions do not only bear the preffure of the air, but likewife the weight of the incumbent water; forafmuch therefore, as the portion of water which lies beneath the orifice, is preffed down lefs forcibly than the adjacent portions, it muft give way and rife in the tube; but the height to which it rifes, whilft the upper orifice of the tube continues stopped, can be but fmall; becaufe, as the water rifes it compreffes the air in the tube, and thereby ftrengthens its fpring, fo as to make it prefs with greater force; and when the air is fo far compreffed by the rifing water, as that the force of its fpring, added to the weight of the elevated water, makes the fame preffure on that portion of water which lies beneath the orifice, as the joint weight of the atmofphere and external water does on the other portions, which are at the fame depth with the former, then the water ceafes to rife. Upon opening the upper orifice of the tube, by the removal of the finger, the compreffed air finding a paffage thro' that orifice, expands and dilates itfelf till it becomes of an equal denfity with

* The water is tinged of a fine blue purple colour with a few grains of Sal Armoniack and Copper.

the

the external air; by which means, the preſſure ariſ-
ing from the condenſation of the air is taken off, and of conſequence, the water which lies beneath the orifice is leſs preſſed than the adjacent portions, and for that reaſon muſt riſe, and continue ſo to do, till the elevated water in the tube gravitates as forcibly on the water beneath the orifice, as the external water does on the neighbouring portions; but this it cannot poſſibly do, till it comes to be of an equal height with the external water.

Should a lighter liquid be poured on the external water, the water within the tube will riſe yet higher than before; and the height to which it riſes above the ſurface of the external water, will be ſo much leſs than the height of the lighter liquor above the ſame ſurface, by how much the ſpecifick gravity of the water exceeds that of the lighter liquor; for inſtance, if the ſpecifick gravity of the water be to the ſpecifick gravity of the lighter liquor, as two to one, the height of the water in the tube above the level of the external water, will be to the height of the lighter liquid, as one to two; becauſe in that caſe, one part of water makes an equal preſſure with two parts of the lighter liquid. To illuſtrate this by an experiment.

Let oil of terpentine, whoſe ſpecifick gravity is to the ſpecifick gravity of water, as 83 to 100, be poured on the external water to the height of eight inches and an half, and the water will riſe in the tube to the height of 7 inches and $\frac{1}{10}$ above the level of the external water, that is, the heights of the water and oil will be in the reciprocal proportion of their ſpecifick gravities; for 7 and $\frac{1}{10}$ is to 8 and $\frac{1}{2}$, or, which is the ſame thing, 73 is to 85, very nearly, as 83 to 100.

The ſame thing is in like manner confirmed by the following experiment.

Let one end of a ſmall tube open at both ends, be immerſed in mercury contained in a larger tube,

and

and let water be poured upon the mercury in the larger tube to the height of 34 inches; the mercury will rise in the smaller tube to the height of two inches and an half above the level of the mercury in the larger tube; so that the height of the mercury in the smaller tube above the level of the mercury in the larger, will be to the height of the water above the same level in the reciprocal proportion of their specifick gravities; for 2½ is to 34, as 1 to 13½, which numbers express the proportion of the specifick gravity of water to that of mercury.

The pressure which the lower parts of a liquid sustain from the weight of those which lie above them, exerts itself every way in all manner of directions, and that equally; or in other words, whatever be the force wherewith a drop of any liquid is pressed downward by the weight of the incumbent liquid, with the very same force is that drop pressed upward, as also laterally and obliquely, and in a word, in all kind of directions whatever; otherwise the drop, which from the nature of fluidity, readily yields and gives way to any impression, must by reason of the pressure from above move out of its place; but this it cannot possibly do, because the drops all around it being equally pressed from above, do on all sides resist the motion of that drop, with the same force that it endeavours to move; consequently, the drop must continue at rest, and be pressed on all sides with the same force that it is from above; and what has been thus proved of one drop, is in like manner demonstrable of all the rest; and therefore, the pressure on the lower parts of a liquid exerts itself equally every way, as will appear from the following experiment.

Exp. 5.   Let four tubes open at both ends be immersed in water to the same depth, their upper orifices being first stopped, and let the lower orifices be so situated, as that the water in entring may move directly upward in one, and directly downward in another,

obliquely

obliquely in a third, and horizontally in the fourth; L ᴇ ᴄ ᴛ. upon opening the upper orifices, the water will rife XII. in all of them to the fame height with the external water, as being preffed in the feveral directions with a force equal to the weight of the incumbent water.

From the preffure of liquids upwards it is, that folid bodies fpecifically lighter than liquids, are made to afcend when immerfed therein. For when a folid body is immerfed in a liquid, it preffes that part of the liquid whereon it refts, with a force equal to the weight of a column compofed of the body it-felf, and that portion of liquid which lies upon it; and the water preffes upward againft the body, with a force equal to the weight of a like column of the liquid alone; which force, inafmuch as the liquid is heavier than the folid, muft overcome the force wherewith the body preffes downward, and of con-fequence, the body muft rife with the difference of thofe forces; as fhall be fhewn more fully in my next lecture. If by any contrivance the preffure of the liquid from beneath can be taken off, a body tho' fpecifically lighter will not rife in a liquid, but remain immerfed, as in the following experi-ment.

A brafs plate being joined to one end of a cylin- Exp. 6. drical piece of wood, and another plate of the fame fize and fhape being fixed in water; let the cylin-der be totally immerfed, and let its plate be laid upon the other in fuch a manner, as that no water may get between; the cylinder tho' fpecifically lighter will remain beneath the water, being preffed down by its own weight and that of the incumbent water, whilft the contrary preffure of the water from beneath, is kept off by means of the plate whereon the cylinder refts.

As bodies fpecifically lighter than liquids, are forced up, on account of the preffure from below being greater than the force wherewith the bodies

O   prefs

press downward; so on the other hand, bodies specifically heavier must sink, because the force wherewith they press downward exceeds the pressure from beneath which opposes their descent; and the force wherewith they descend is equal to the difference of those forces; as shall likewise be shewn in my next lecture. If by any contrivance those two forces can be reduced to an equality, then the bodies will not descend, but remain suspended in the liquid; as in the following experiment.

Exp. 7.　　Let a brass plate, whose specifick gravity is to that of water, as 9 to 1, be adapted to the neck of a glass vessel in such a manner, as that being immersed in water no part of the water may get upon its upper surface; let it then be immersed to the depth of nine times its own thickness, (that is, to the depth of 2 inches and $\frac{7}{10}$, the thickness of the plate being $\frac{1}{10}$ of an inch) and it will remain suspended; but upon pouring ever so little water upon its upper surface, it will immediately descend and fall to the bottom.

The plate being immersed to the depth of nine times its own thickness is pressed upward by a force equal to the weight of a column of water, whose height is nine times as great as the thickness of the plate; which weight, inasmuch as the specifick gravity of the water is to that of the plate, as 1 to 9, is equal to the weight of the plate, that is, to the force wherewith the plate presses downward; for as none of the water lies on its upper surface, it can press downward with no other force than what arises from its own gravity; consequently, in this case, the force which resists the descent, is equal to the force which promotes it, and of course, the plate must remain in its place. When a little water is poured on the plate, the weight of that added to the weight of the plate, overcomes the resisting force

Exp. 8. of the water, and causes the plate to descend. Should the plate be immersed to twice the former depth,

it

it will not defcend tho' loaded with water to the height of nine times its own thicknefs; for as in this cafe, the depth to which it is immerfed is double the former, fo likewife is the force wherewith the water preffes upward ; confequently, that force is fufficient to fupport twice the weight of the plate, and therefore will fuftain the plate when loaded with water, to the height of nine times its own thicknefs, fuch a quantity of water being juft equal in weight to the plate.

If by pouring on more water, the force wherewith the plate preffes downward be increafed, or by raifing the plate nearer to the furface of the water, the force wherewith the water preffes upward be diminifhed, the plate will fall to the bottom ; and on the other hand, if by immerfing the plate to a greater depth, the preffure of the water upward be increafed, the plate will be thruft upward againft the glafs, and would actually afcend were it not hindered by the glafs.

From what has been faid it follows, that if S be put to denote the number expreffing the fpecifick gravity of the plate, that of water being unity ; D be the depth to which the plate is immerfed expreffed in the thicknefs of the plate, and H the height of the water upon the plate expreffed likewife in the thicknefs of the plate ; D muft be equal to the fum of S and H in all cafes where the plate remains fufpended ; and if there be no water upon the plate, then D and S muft be equal ; wherefore, if in the former cafe, D be greater than S and H taken together, or in the latter, than S alone, the plate will afcend if not hindered by the glafs ; and on the other hand, if D be lefs, the plate will defcend and fall to the bottom.

The preffure which the bottom of a veffel fuftains from a liquid contained in it, whatever be the fhape of the veffel, is equal to the weight of a pillar of the liquid, whofe bafe is equal to the *area* of

the

Lect. the bottom, and whose height is the same with the
XII. perpendicular height of the liquor.

That this is the case in vessels that are equally
wide from top to bottom is plain and obvious; inasmuch as the bottom of every such vessel does actually sustain such a pillar of liquor. But that the
case should be the same in irregular vessels, is not
so easy to conceive; for instance, that in a vessel
which from a large bottom grows narrower as it
rises, so as perhaps at length to be contracted into
a tube, the bottom should bear the same pressure
when the vessel is filled, as it would were the vessel
equally wide throughout from bottom to top, seems
strange and surprizing, and yet it is what necessarily follows from the nature of fluidity; for that part
of the bottom which lies directly beneath the tube,
sustains the weight of a pillar of liquor which reaches
to the top of the tube, the vessel being supposed to
be full, and being pressed with the weight of that
pillar, reacts with an equal pressure on that portion
of the liquor which touches it; and that pressure, inasmuch as it exerts itself equally in the liquor every way, is propagated laterally thro' the
several portions of liquor which are contiguous to
the bottom of the vessel; and forasmuch as this lateral pressure does in like manner exert itself equally every way, the bottom of the vessel must be e-
qually pressed in every point; consequently, since
that portion of it which lies beneath the tube, bears
a pressure equal to the weight of a pillar of liquor,
whose height reaches to the top of the vessel, every
other equal portion must bear a pressure equal to
the same weight; and of course, the whole bottom
must be pressed as forcibly, as if the vessel continued of the same wideness to the top, and was filled with the liquor.

Exp. 9.　　To confirm this by an experiment. Let there
Pl. 6.　be two glasses open at both ends, and of such shapes
Fig. 10,　as are exhibited in the two figures, whose lower parts
11.

4.

MN

MN are cylindrical and equal, and of a capacity
just sufficient to admit the brass plate made use of
in the last experiment; which must be fitted to each
of them successively, in order to constitute two vessels of equal bottoms, but of different capacities;
and being so fitted, let it be immersed in water,
as in the last experiment, to such a depth, as that
it will be necessary to load it with water in order
to make it sink; that is, let the depth be more
than nine times the thickness of the plate, which
depth must be the same in both cases; let then
water be poured on the plate, and let it be observed what height of water is requisite to force down
the plate when the wider vessel is made use of, and
it will be found, that the same height will suffice in
the narrower vessel; consequently, the small pillar
of water in the narrower vessel, must press the plate
with a force equal to the weight of a pillar of water
of the same height, and of a base equal to the *area*
of the plate; for such a pillar does actually press
the plate in the larger vessel, as is evident from the
bare inspection of the figures, and the pressures made
on the plate in both vessels are equal, inasmuch as
they overcome equal resistances.

From what has been said it appears, that where
the base of a vessel is given, the pressures upon it
are as the perpendicular heights of the liquid, whatever be the shape of the vessel. And universally, the
pressure on any base is measured by the product of
the *area* of that base into the perpendicular height
of the liquor above it, without any regard to the
quantity of liquor contained in the vessel; so that
if we suppose a hogshead set on one end, and filled
with a liquor, and a small pipe to issue perpendicularly upward from the other end to any height
whatever, and to be filled with the same liquor,
the bottom will be as strongly pressed, and be in
as much danger of bursting out, as if the hogshead

O 3   was

LECT.
XIII.
by a force equal to the weight of such a portion of the liquid as is equal to it in bulk; consequently, if the specifick gravity of the solid be greater than that of the liquid, that is, if the solid weighs more than an equal bulk of the liquid, the body will descend with a force equal to the excess of its gravity above the gravity of the liquid; on the other hand, if the gravity of the liquid exceeds that of the solid, the body being as it were pressed upward by a force greater than that whereby it endeavours to go down, will ascend with the difference of those forces, that is, with a force equal to the excess of the specifick gravity of the liquid above that of the solid. When the specifick gravities are equal, the body will neither rise nor fall, but remain suspended at any depth; being pressed as strongly upward by the resisting force of the liquid, as it is downward by its own weight.   Hence it follows, that if by any contrivance the specifick gravity of a solid can be varied, so as to be one while greater, another less, and then equal to the specifick gravity of a liquid wherein it is immersed, the body will sink, or rise, or remain suspended according to the variation of its specifick gravity.   And this is the case in that ludicrous experiment of the little glass images in water, which are made to descend, or rise, or remain suspended at pleasure; the reason of which I shall explain to you, after you have seen the experiment.

Exp. 1.   The images being set to float on the water, the top of the vessel must be covered with a bladder closely bound about the neck of the vessel, to the end that the air, which lies upon the surface of the water, may not force its way out when it is condensed by the hand pressing on the bladder. The images themselves tho' lighter, are yet nearly of the same specifick gravity with the water, and being hollow, are full of air, which by means of small

holes

holes in their heels communicates with the air without. When the air which lies beneath the bladder is pressed by the hand, it presses on the surface of the water; and forasmuch as the pressure is propagated thro' all the water, those portions which are contiguous to the heels of the images, are thereby forced into the holes, by which means the air within is condensed, and at the same time, the weight of the images is increased by the additional weight of the influent water. And when so much water is forced in, as to render the specifick gravity of the images greater than that of the water, the images descend and fall to the bottom; where they remain as long as the pressure above continues, but when that is taken off by the removal of the hand, the condensed air in the images dilates and expands itself, and in so doing drives out the water; upon which account the images become specifically lighter than the water, and of course ascend. As the pressure on the bladder is greater or less, so must the quantity of water which is forced into the images; and therefore, whenever it happens that during the ascent or descent of an image, such a pressure is made as suffices to force in just as much water as is requisite to reduce the image to the same specifick gravity with the water, the image stops and remains suspended, upon increasing the pressure it descends, and ascends if the same be lessened. Some of the images begin to descend sooner, as also to rise later, than others, for one or both of these reasons; first, because some are specifically heavier than others; and, secondly, because the cavities in the legs are greater in some images in proportion to their magnitudes, than they are in others; upon both which accounts, a less pressure is requisite to make some descend, and to keep them down, than what is necessary to produce the same effects in others. For, first, let us suppose the specifick gravities of two

images

images to be different, but the cavities in their legs,
when taken of a given height, to be proportional
to their respective magnitudes; since the air is equally dense in both images, it is manifest, that it gives
the same opposition in both to the influent water;
consequently, the water when forced in by the pressure from above, must rise to equal heights in the
cavities of both; since therefore the cavities whose
heights are equal, are supposed to be proportional to
the magnitudes of the images, it is manifest, that
the quantities of water contained in those cavities
must be so too; consequently, each image receives
an addition of weight from the influent water proportional to its magnitude; or in other words, the
specifick gravities of the two images are equally
augmented; forasmuch therefore as one of the
images is supposed to be specifically heavier than the
other, it is evident, that when the specifick gravity
of the former has received such an addition, from
the influent water, as makes it a little exceed the
specifick gravity of the water, the specifick gravity
of the latter must fall short thereof; consequently,
the former must sink, an leave the other above.

Secondly, Let us suppose the specifick gravities of the two images to be equal; but let one
image be less in proportion to the cavity in its legs,
than the other is in proportion to its cavity, the
height of the cavities being given; since the water
does from the same pressure rise to an equal height
in both, it is plain, from what I just now said,
that the former must receive a greater quantity of
water in proportion to its magnitude, and consequently, a greater addition to its specifick gravity
than the latter, and of course must descend
sooner.

From what has been said it follows, that if the
proportion which the cavity in the legs bears to
the magnitude of the image be given, the greater
the

the specifick gravity of the image is, the more apt it will be to descend; consequently, in this case the aptitude or promptness of an image to descend is proportional to, and may be expressed by, the specifick gravity. In like manner, if the specifick gravity be given, the greater the proportion is which the cavity in the leg bears to the magnitude of the image, the more apt the image is to descend; and therefore in this case, the aptitude is proportional to, and may be expressed by, the cavity applied to the magnitude of the image. But if neither the specifick gravity of the image, nor the proportion of the cavity to the magnitude of the image, be given, the aptitude of an image to descend, is as the specifick gravity into the cavity applied to the magnitude of the image; that is, putting A to denote the aptitude, S the specifick gravity of the image, C the cavity in the leg, (the height whereof is always supposed to be given) and M the magnitude of the image; $A = \dfrac{SC}{M}$; or, substituting the absolute weight of the image applied to its magnitude, in the room of the specifick gravity, $A = \dfrac{WC}{M^2}$; that is, the aptitude an image has to descend, is as the weight of the image into the cavity of the leg directly, and the square of the image's magnitude inversly.

A solid specifically heavier than a liquid, being immersed therein, loses as much of its weight as is the weight of a portion of the liquid equal to it in bulk; for it has been already shewn, that a solid is carried down in a liquid by the excess only of its gravity, above the gravity of a portion of the liquid equal to it in bulk; consequently, the other part of its gravity is lost, as to any effect it has on the body itself; as will appear from the following experiment.

Let

LECT.   Let a small cylinder of brass, suspended at one
  XIII.   end of a balance and counterpoised, be immersed
Exp. 2.   in water; upon the immersion it will become lighter, suppose by 200 grains, which is the weight of
as much water as is equal in bulk to the cylinder;
for a cylindrical vessel, just large enough to contain
the cylinder, being hung at one end of a balance
and poised, and then filled with water, preponde-
rates with the weight of 200 grains.

Since a solid when immersed in a liquid, loses
as much of its weight, as is equal to the weight of
a portion of the liquid of the same dimensions with
the solid, it follows, that all bodies whatever, whose
magnitudes are equal, however different their specifick gravities may be, do suffer an equal loss of
weight in the same liquid. Thus a cylinder of
Exp. 3.   block-tin, equal in dimensions to the brass cylinder,
but specifically lighter, being immersed in water,
loses 200 grains, as did that of brass.

Tho' a solid loses part of its weight when immersed in a liquid, yet it must not be imagined that
the weight so lost by the solid, is actually destroyed,
but that it is imparted to the liquid, the liquid constantly gaining in weight what the solid loses. For
Exp. 4.   if the vessel with the water wherein the cylinders
were immersed, be put into a scale and poised; upon the immersion of either cylinder, it will preponderate with the weight of 200 grains, which is what
the cylinder loses.

Solids equal in weight, but of different specifick
gravities, being immersed in the same liquid, suffer
losses of weight reciprocally proportional to their
specifick gravities; for as the loss of weight which
any body suffers in a liquid, is equal to the weight
of as much of the liquid as is equal in bulk to the
solid, the loss sustained is ever proportional to the
magnitude of the body; whatever proportion therefore the magnitudes of bodies have to one another,
the same will the losses of weight have which they
                                                suffer;

suffer; but the magnitudes of bodies equal in weight, but of different specifick gravities, are to one another in the reciprocal proportion of their specifick gravities; consequently, so are the losses of weight which they suffer. Which is confirmed by the following experiment.

Let two cones, one of lead, the other of tin, **Exp. 5.** whose specifick gravities are to one another, as 112 to 74, and the weight of each 400 grains, be immersed in water, after the manner of the cylinders; upon the immersion, the lead will lose $35\frac{1}{4}$ grains, and the tin 54; but $35\frac{1}{4}$ is to 54, as 74 to 112, that is, reciprocally as the specifick gravities of the metals.

From the losses of weight being reicprocally pro- **Exp. 6.** portional to the specifick gravities, it follows, that if two bodies of different specifick gravities, which balance each other in air, be immersed in water or any other liquor, the *æquilibrium* will be destroyed, and that which has the greatest specifick gravity will descend; as will appear, by hanging the cones, one at each end of a balance, and then immersing them in water, for the lead will preponderate.

The specifick gravity of a solid specifically heavier than a liquid, is to the specifick gravity of the liquid, as the absolute weight of the solid, to the loss of weight which it suffers in the liquid; for the specifick gravities of bodies being as the absolute weights applied to the magnitudes, where the magnitudes are equal, the specifick gravities are directly as the absolute weights; if therefore we compare the solid with a quantity of the liquid equal to it in magnitude, their specifick gravities must be as their weights; but the absolute weight of such a quantity of the liquid, is equal to the loss of weight sustained by the solid; consequently, the specifick gravity of the solid, is to that of the liquid, as the whole weight of the solid, to the loss which it sustains in the liquid.

Hence

Hence we have a method of difcovering the fpecifick gravities of fuch folid bodies as are heavier
than water; I mean, of difcovering the proportions
of their fpecifick gravities to the fpecifick gravity
of water. For if we fuppofe the fpecifick gravity of
water to be unity, and put L to denote the lofs of
weight which any body, whofe fpecifick gravity we
look for, fuftains in water, and W its whole weight;
then $L : W :: 1 : \dfrac{W}{L}$; confequently, $\dfrac{W}{L}$ expreffes the fpecifick gravity of the folid, that of water
being unity; and therefore, in order to know the
fpecifick gravity of any folid heavier than water,
nothing more is requifite, but to difcover the quantities denoted by W and L, and to divide the firft
by the laft; the firft is had, by taking the weight
of the body in air, and the laft, by taking the weight
in water, and fubducting it from the weight in air;
for the remainder is the lofs of weight, which dividing the weight in air, gives a quotient expreffing
Exp. 7. the fpecifick gravity of the body. To apply this
to a particular cafe, let it be propofed to difcover
the fpecifick gravity of a piece of tin, which being
weighed in air, is found to be 300 grains, and in
water, 259$\frac{1}{4}$, which being fubducted from the former, leaves 40$\frac{1}{4}$ for the lofs of weight; fo that in
this cafe, W denotes 300, and L 40$\frac{1}{4}$; and therefore, dividing 300 by 40$\frac{1}{4}$, we fhall have 7$\frac{4}{10}$ for
the fpecifick gravity of tin, that of water being
unity. Whence it appears, that tin, bulk for bulk,
is more weighty than water, in the proportion of
74 to 10.

If the body, whofe fpecifick gravity is required,
be lighter than water; then, forafmuch as its gravity is not fufficient to caufe a total immerfion, the
lofs of weight which it fuffers in water cannot be
found out by weighing it alone in that liquid; let
it therefore be joined to fome other body fo weighty,
that the compound may fink; but firft let the lofs
of

of weight which the heavier body alone fuftains in
water be found out, as before; and then let the lofs
of weight which the compound fuftains be likewife
difcovered, whence deducting the lofs of weight
fuftained by the heavier, the remainder will exhibit
the lofs fuftained by the lighter; confequently, dividing the weight of the lighter by that remainder,
the quotient will exprefs the fpecifick gravity required; that is, putting W for the weight of the
body whofe fpecifick gravity is fought, L for the
lofs of weight fuftained by the compound, and
l for the lofs fuftained by the heavier body $\frac{W}{L-l}$
expreffes the fpecifick gravity of the body. To
apply this to a particular cafe; let the weight of a
piece of wood fpecifically lighter than water be
220 grains, and let it be joined to a piece of tin of
160 grains, whofe lofs in water is found to be 17
grains; then the compound being weighed in water, will be found to lofe 334 grains; fo that in
this cafe, W is equal to 220 grains, L to 334,
and l to 17; and L lefs l, is equal to 317 grains.
And therefore, dividing 220 by 317, we fhall
have $\frac{694}{1000}$ for the fpecifick gravity of the wood,
that of water being unity. So that that kind of
wood is bulk for bulk lighter than water, in the
proportion of 694 to 1000.

Exp. 8.

If the body whofe fpecifick gravity is fought be
diffolvable in water, then inftead of water, let fome
other liquor be made ufe of, which will not diffolve
the body; and let the proportion of the fpecifick
gravity of the body to the fpecifick gravity of that
liquor, be difcovered by the foregoing method; as
alfo the proportion of the fpecifick gravity of that
liquor to the fpecifick gravity of water, by the method which fhall be fhewn prefently. Then in whatever proportion the fpecifick gravity of the liquor exceeds or falls fhort of the fpecifick gravity of water,

in

LECT. XIII. in the same proportion let the specifick gravity of the body with regard to that of the liquor be increased or diminished, and it will give the specifick gravity of the body with respect to that of water; that is, if we put A for unity or the specifick gravity of water, B for the specifick gravity of the other liquor, and C for the specifick gravity of the body with regard to that liquor; then by saying as A is to B, so C to a fourth proportional, we shall have $\frac{BC}{A}$ for the specifick gravity of the body with respect to that of water; or rejecting the divisor A being equal to unity, and putting S for the specifick gravity of the body with respect to water, we shall

Exp. 9. have $S = BC$. To apply this, let the specifick gravity of Roman-vitriol be required; let the weight of a piece in air be 67 grains, and in spirit of wine 41 grains; consequently, its loss of weight in the spirit is 26 grans, which dividing 67, gives 2.576 for the specifick gravity of the vitriol with regard to the specifick gravity of the spirit, which in this case is supposed to be unity; but the specifick gravity of the spirit with regard to that of water is less than unity, being only $\frac{87}{100}$, as shall be shewn presently; wherefore B is $= 0.87$, and C is $= 2.576$; consequently 2.24, which is the product arising from the multiplication of those two numbers, expresses the specifick gravity of Roman vitriol with respect to that of water, which is a unity; and therefore, in whole numbers, the specifick gravity of Roman-vitriol exceeds that of water, in the proportion of 224 to 100.

The specifick gravities of liquors are discovered by taking the losses of weight sustained by one and the same solid in the several liquors; for since the loss of weight in each liquor, is equal to the weight of as much of the liquor as is equal in bulk to the body; by taking the losses of weight sustained by
the

the fame body in the feveral liquors, we get the abfolute weights of fuch portions of thofe liquors as are equal in bulk ; and by confequence, the fpecifick gravities of the liquors, the fpecifick gravities of bodies equal in bulk, being to one another as their abfolute weights ; wherefore, putting L for the lofs of weight which a body fuftains in water, and little l for the lofs of weight fuftained by the fame body in any other liquor ; then, by faying, as L to l, fo is unity to a fourth term, we fhall have $\frac{1}{L}$ for the fpecifick gravity of the other liquor, that of water being unity ; fo that to difcover the fpecifick gravity of any liquor, we have nothing more to do, but to weigh one and the fame folid, both in the liquor whofe quantity is fought, and in water, and to divide the lofs of weight which the folid fuffers in the liquor, by the lofs which it fuftains in water ; for the quotient will exprefs the fpecifick gravity of the liquor. Thus, a glafs bubble, whofe weight in air is 1727 grains, being weighed in water, is found to lofe 641 grains, and 558 in fpirit of wine ; and therefore, dividing 558 by 641, we fhall have a quotient of 0.87 for the fpecifick gravity of the fpirit, that of water being unity.

Exp. 10.

When a body fpecifically lighter than a liquid, is fet to float upon it, the part immerfed is equal in bulk to a portion of the liquid whofe weight is equal to the weight of the whole body ; for fince the body finks in part, by moving fome of the liquor out of its place, and fince the weight of the body is the power which moves the liquor, the body muft continue to fink, till it has removed as much of the liquor as is equal to it in weight ; confequently, the part immerfed muft be equal in magnitude to fuch a portion of the liquor, as is equal in weight to the whole body ; which is abundantly confirmed by the following experiment.

P

A ball

A ball of pear-tree, a wood specifically lighter than water; being set to float on water contained in a glass vessel, let the vessel be placed in a scale, and counterpoised; then, taking out the ball, let the vessel be filled up with water to the same height at which it stood when the ball was in it, and the same weight will counterpoise it as before.

From the vessel's being filled up to the same height at which the water stood when the ball was in, it is manifest, that the quantity poured in is equal in magnitude to that part of the ball which was immersed; and, from the same weight counterpoising, it is evident, that the water poured in, is equal in weight to the whole ball.

The part immersed is to the whole body, as the specifick gravity of the body to the specifick gravity of the liquid; for the specifick gravities of two bodies, being to one another as their absolute weights applied to their magnitudes, if their weights be equal, their magnitudes are in the reciprocal *ratio* of their specifick gravities; since therefore, such a portion of the liquid as is equal in magnitude to the immersed part of the solid, is likewise equal in weight to the whole solid; the magnitude of the immersed part is to the magnitude of the whole body, as the specifick gravity of the solid to the specifick gravity of the liquid.

When the same body is set to float successively in different liquors, the parts immersed are to one another in the reciprocal proportion of the specifick gravities of the liquors. For the body descends in each liquor, till the part immersed takes up the room of as much liquor as is equal in weight to the whole body; and therefore, such portions of the several liquors as are equal in magnitude to the immersed parts of the body have all equal weights, but the magnitudes of bodies equal in weight, are to one another reciprocally, as their specifick gravities; consequently, in one and the same body,

4

floating

floating in different liquors, the parts immerfed are reciprocally as the fpecifick gravities of the liquors. On this principle is founded the HYDROMETER; which is an hollow glafs ball, with a fmall hollow ftem of about 5 or 6 inches in length, oppofite to which, on the other fide of the ball, adheres a fmaller ball filled in part with mercury, or fome other weighty body, to the intent, that when the ball is fet to float in water, or any other liquor, the ftem may be kept uppermoft, and in a pofition perpendicular to the furface of the liquor; and at the fame time, that the machine may be fo far immerfed, as that the ftem only, or fome part thereof, may remain above the liquor: the ftem being graduated from top to bottom, has numbers annexed to every degree, expreffing the magnitudes of the parts which lie below the feveral degrees.

The ufe of this little machine is to difcover the fpecifick gravities of liquors, which is done in the following manner. The *hydrometer* being firft fet to float in water, the degree to which it finks muft be obferved, and the number thereto annexed; then being fet to float in any other liquor, the degree to which it finks, with the number annexed, muft likewife be noted; for as this number is to the former, fo is the fpecifick gravity of water, to that of the other liquor, as is evident from what was juft now faid. To illuftrate this in the cafe of water and fpirit of wine. The *hydrometer* being dropt into water, finks to the degree whofe number annexed is 87; and being dropt into fpirit of wine, finks to the degree whofe number is 100; whence it appears, that the fpecifick gravity of water is to that of fpirit of wine, as 100 to 87.

Tho' *hydrometers* may be ufeful in difcovering the fpecifick gravities of liquors for loofe and inaccurate computations, yet are they not to be depended on in cafes where great exactnefs is required;

Exp. 12.

and

and that for two reasons; First, because it is extreamly difficult to graduate the stems so exactly,
as that the numbers annexed shall truly express the
magnitudes of the parts below them. Secondly,
because, partly from the motion of the hydrometer
in the liquor, and partly from the rising of the liquor about the stem from the attractive force of the
glass, it is hardly possible to determine with exactness the degree to which the hydrometer sinks.
Upon both which accounts, as also because the
method of determining the specifick gravities of
liquors by means of the glass bubble is much more
easy and exact, this method by the *hydrometer* is
intirely laid aside.

LECTURE XIV.

OF HYDROSTATICKS.

Lect.
XIV.
IN this lecture I shall give you an account of the
flux of water from RESERVOIRS thro' *orifices*
and *pipes*.

If water, flowing out at an orifice in the bottom
of a vessel, be kept constantly at the same height in
the vessel, by being supplied as fast above as it
runs out below, the velocity wherewith it flows out,
is as the square root of the height of the water above
the orifice.

For if we suppose the column of water which
stands directly above the orifice, to be divided into
an indefinite number of plates of an equal but exceedingly small thickness, it is manifest, that whatever be the force of gravity, wherewith the uppermost plate presses upon the second, the second presses upon the third with a double force, and the third
upon the fourth with a triple force, and so on; so
that the plate which is next the orifice is pressed
downward

downward by the joint gravities of the several plates which lie above it, and likewise by the force of its own gravity, inasmuch as there is no other plate beneath it whereon to rest; consequently, from its own gravity, and that of the several plates above it, it does all at once receive as many equal impressions from gravity, as it would successively in falling down the height of the water; and of course, must pass thro' the orifice, with the same velocity that it would acquire in falling down that height; but I proved in my lecture upon gravity, that the velocity which a body acquires in falling thro' any space, is as the square root of the space; consequently, the velocity wherewith the water flows out, is as the square root of the height of the water above the orifice.

To confirm this by an experiment; let there be two vessels in all things alike, excepting that one is four times as tall as the other, the height of one being 20 inches, and of the other 5; let each of them have a circular orifice in the bottom, a fifth part of an inch in diameter; and being both filled with water, let them be set a running, and let the water be supplied as fast above as it runs out below; the taller vessel will discharge about twenty one ounces in the space of a quarter of a minute, and in the same time the shorter will discharge about 11 ounces. Now, forasmuch as the orifices thro' which the water flows are equal, and likewise the times of the flux, the quantities discharged are as the velocities; consequently, the velocity wherewith the water flows out of the taller vessel, is to the velocity wherewith it flows out of the shorter, as 21 to 11, that is, nearly as 2 to 1, which are the square roots of the heights of the water above the orifices.

As the pressure sustained by the lower parts of water from the weight of those above, exerts it-

Exp. 1.

P 3

self

self with the same force laterally that it does downward, it matters not whether the orifice through
which the water flows, be at the bottom or side of
a veffel; for the water will flow out of both with
the same velocity, provided they are at equal
depths below the upper surface of the water ; and
therefore, the velocity of water flowing out of an
orifice in the side of a veffel, is as the square root
of the height of the water above the orifice ; as
Exp. 2. will appear, by repeating the laft experiment with
veffels whofe orifices are in their sides ; for the
quantities difcharged will be the fame as before.

　　From what has been faid it follows, that if an
orifice in the side of a veffel be fituated as far above
an horizontal plane, as it is below the upper furface
of the water, the water will fpout from that orifice,
to the diftance of twice the height of the orifice
Pl. 6. above the plane. For inftance, if AOBC be a vef-
Fig. 12. fel full of water, O an orifice in the side, whofe
height OD above the horizontal plane DH, is equal
to OA, the diftance of the orifice from the top of the
water ; DH the horizontal diftance to which the
water fpouts, will be double of OD, the height of
the orifice above the plane. For the fpouting wa-.
ter has two motions, one uniform from the preffure
of the water in the veffel, in the direction OF perpendicular to the orifice, the other accelerated from
the force of gravity in the direction OD perpendicular to DH; which two motions do by no means
hinder one another, but by their combination caufe
the water to fpout in the curve of a *parabola*. Now,
the velocity wherewith the water moves in the direction OF, being equal to the velocity acquired by
a body in falling from A to O, or from O to D;
in the fame time that it falls from O to D, and by
fo doing, reaches the horizontal plane, it will be
carried in the direction OF, thro' a fpace equal to
twice OD, (inafmuch as all bodies whatever that
move

move uniformly, with a velocity equal to that which is acquired by a body in falling thro' any height, do in the same time with that of the fall, defcribe a fpace double of that of the fall) ; confequently, the horizontal diftance to which the water fpouts, will be equal to twice the height of the orifice above the plane. Thus, from an orifice in the fide of a Exp. 3. veffel, the depth whereof below the furface of the water is 20 inches, the water will fpout to the diftance of 38 inches on an horizontal plane, whofe diftance below the orifice is likewife 20 inches; and where the depth of the orifice below the top of the water is 5 inches, the water will fpout to the diftance of $9\frac{1}{4}$ inches on an horizontal plane fituated at the diftance of 5 inches below the orifice; fo that in both cafes the diftances to which the water fpouts, are nearly double the diftances of the planes below the orifices; and they would be exactly double, were it not that the water is retarded a little by the oppofition it meets with from the air.

The diftances to which water fpouts on an horizontal plane, from orifices in the fides of different veffels, the orifices being at equal heights above the plane, are to one another as the fquare roots of the heights of the water above the orifices.

For fince the orifices are at equal heights above the plane, the times of the defcent of the water from the feveral orifices to the plane muft be equal; confequently, the horizontal diftances to which the water fpouts, muft be as the velocities wherewith it fpouts; but thofe velocities are as the fquare roots of the heights of the water above the orifice; confequently, fo muft the horizontal diftances. Thus, if two veffels be fo placed, as that the orifices in Exp. 4. their fides fhall be 20 inches above an horizontal plane; the height of the water in one veffel being 20 inches above the orifice, and in the other 5; the water will fpout from the former, to the diftance

P 4

of

of 38 inches, and from the latter, to the diftance of 19 inches; but 38 is to 19, as 2 to 1½, therein as the fquare roots of the heights of the water above the orifices, for the heights are as 4 and 1.

The diftance to which water fpouts from an orifice in the fide of a veffel, whatever be the height of the orifice above the plane, as alfo of the water above the orifice, may be thus determined; let
Pl. 6.
Fig. 13.
BR reprefent an horizontal plane, F an orifice in the fide of a veffel at any height above the plane, and AB the height of the upper furface of the water above the plane. On AB as a diameter, defcribe the femicircle ADB, and at F fet off FE perpendicular to AB, and meeting the circle in E. The diftance to which the water fpouts on the plane BR from the orifice F, is proportional to the line FE.

For, from the nature of motion, the fpace defcribed, is as a rectangle under the time and velocity; but in this cafe, the time of the motion is as the fquare root of FB, and the velocity wherewith the water fpouts, is as the fquare root of AF; confequently, the fpace thro' which the water runs in the horizontal direction, is as the fquare root of the rectangle AFB; but, by the nature of the circle, the fquare root of the rectangle AFB is equal to FE; confequently, the horizontal diftance to which the water fpouts on the plane BR from the orifice F, is as FE.

Hence it follows, that the diftance to which the water fpouts, is as the fine of the arch AE, whofe verfed fine AF is equal to the height of the water above the orifice. And, forafmuch as any two fines, which are equally diftant from the center, are equal, it follows that the water muft fpout to the fame diftance from two orifices as F and L, whofe diftances from the center are equal; as alfo, that it muft fpout to the greateft diftance from an orifice

in

In the center, the fine CD being in that cafe equal to *radius*, and confequently the greateft.

To confirm what has been faid; let a veffel whofe height is 16 inches, and which is perforated in the middle, and likewife at the diftance of $5\frac{1}{3}$ inches above and below the middle, be filled with water, and fet upon an horizontal plane; the water will fpout from the middle orifice to the diftance of above 15 inches, and from each of the other two, to the diftance of about 10 inches.

All things being fuppofed as before, the diftances to which the water fpouts, fetting afide what little difturbance may arife from the refiftance of the air, are equal to twice the fines of the arches, whofe verfed fines are equal to the heights of the water above the orifices. For, the diftance to which the water fpouts from the central orifice C, is to the diftance to which it fpouts from any other orifice as F, as the fine CD is to the fine FE; but forafmuch as the orifice C is as far diftant above the plane as it is below the furface of the water, the diftance to which the water fpouts from that orifice is equal to twice CB, or twice CD; confequently, the diftance to which it fpouts from F muft likewife be equal to twice FE, and fo of any other orifice.

Water which fpouts perpendicularly upward fets out with fuch a velocity, as is fufficient to carry it to the fame height with the water in the veffel from which it fpouts. For the velocity wherewith it fets out, is equal to the velocity acquired in falling down the height of the water; and, in my lecture upon gravity, I fhewed, that a body thrown directly upward rifes to fuch a height, whence if it be let fall, it will by the end of the fall acquire the fame velocity wherewith it was thrown up; confequently, the water fpouts with a velocity fufficient to carry it to an equal height with the water in the refervoir; however, it cannot poffibly arrive at that height,

by

by reason of the resistance it meets with from the air; which, as it cannot be taken off, must lessen the heights of all jets whatever, so as to make them fall short of the heights in the reservoirs; besides, the water in the uppermost part of, the jet, when it has lost all its motion, rests for some time on the part next below it, and by its weight obstructs and retards the motion of the whole column, and thereby lessens its height; and so great is the resistance arising from this cause, as that the jet is frequently destroyed by it, the rising water being by fits and starts pressed down to the very orifice from which it spouts.

Exp. 7.　By giving the jet a little inclination, the uppermost parts, when they have lost their motion upward, are made to fall off from the rest, whereby the resistance which arises from their weight is taken off. And this is the true reason why, *cæteris paribus*, such jets as are a little inclined, rise higher than those whose ascents are perpendicular.

The velocity wherewith water flows out of a cylindrical pipe inserted horizontally into the side of a vessel, is as the square root of the height of the water in the vessel above the place of the pipe's insertion directly, and the square root of the length of the pipe inversly; for since the pipe is cylindrical, the velocity wherewith the water flows out at one end, must be equal to the velocity wherewith it flows in at the other; but the velocity wherewith it flows in, is in the proportion laid down; for the pressure of the incumbent water in the vessel, cannot make the water which lies next the orifice flow into the pipe, unless at the same time it drives forward all the water contained in the pipe; for which reason, the water in the pipe may be looked upon as an obstacle which resists and impedes the moving cause. Now, where a cause

acts

acts under the difadvantage of a clog or impediment, the potency of fuch a caufe is increafed, either
by diminifhing the impediment, or augmenting the
abfolute ftrength and vigour of the caufe itfelf;
where the ftrength and vigour of the caufe is given,
the potency thereof increafes in proportion as the
impediment leffens, and leffens as that increafes;
and where the impediment is given, the potency of
the caufe increafes, and leffens in proportion to the
increafe and diminution of the abfolute ftrength and
vigour of the caufe; confequently, the potency is
in a *ratio* compounded of the ftrength or magnitude
of the caufe, and of the weaknefs or fmallnefs of
the impediment; that is, it is as the magnitude of
the caufe directly, and as the magnitude of the impediment inverfly; or as the magnitude of the caufe
applied to the magnitude of the impediment. Now,
in the cafe before us, where the preffure of the water in the refervoir is the moving caufe, and the water in the pipe is the impediment, the magnitude
of the former is meafured by a rectangle under the
height of the water, and the orifice of the pipe;
and the magnitude of the latter by a rectangle under the orifice of the pipe, and the length thereof;
or rejecting the orifice as being ever the fame in
both, the magnitude of the moving caufe, is as the
height of the water, and that of the impediment,
as the length of the pipe; and therefore, putting
H for the height of the water in the refervoir above
the place of the pipe's infertion, and L for the length
of the pipe; $\frac{H}{L}$ will denote the preffure of the water in the refervoir, as leffened by the refiftance of
the water in the pipe; and putting O for the orifice of the pipe, $\frac{HO}{L}$ will exprefs the force which
drives the water into the pipe; and forafmuch as
the motion generated in any time by a force acting
constantly

conftantly and uniformly, is as a rectangle under the force and time; putting T for the time that the water continues to flow into the pipe, $\frac{HOT}{L}$ will be as the motion generated in the water flowing into the pipe; but the motion generated in the influent water, is as the quantity which flows in, multiplied into the velocity wherewith it flows; and therefore, putting Q and V for the quantity and velocity, $\frac{HOT}{L}$ is as QV; or, becaufe the quantity which flows in, is in a *ratio* compounded of the orifice, time, and velocity; by fubftituting O, T, V, which denote the orifice, time, and velocity, in the place of Q, we fhall have $\frac{HOT}{L} = OTV^2$; and ftriking out OT from both fides, we fhall have $\frac{H}{L} = V^2$; confequently, V is as $\sqrt{\frac{H}{L}}$; that is, the velocity wherewith the water flows out of the refervoir into the pipe, and confequently, the velocity wherewith it flows out of the pipe, is as the fquare root of the height of the water in the refervoir, applied to the fquare root of the length of the pipe.

Hence it follows, that if the length of the pipe be varied whilft the height of the water in the refervoir continues the fame, the quantities difcharged in any given time, will be to one another inverfly as the fquare roots of the lengths of the pipe; for fince the diameter of the pipe, and the time of the flux are given, the quantities difcharged muft be as the velocities wherewith they run out, that is, in the inverfe *ratio* of the fquare roots of the lengths of the pipe.

To confirm this by an experiment; let a pipe of 16 feet in length, and half an inch in diameter, be inferted horizontally into the fide of a veffel; and

let

let the water in the veffel be kept conftantly at the
height of 3 feet above the place of the pipe's infertion; the pipe when fet a running will difcharge
above $161\frac{1}{4}$ ounces in half a minute; let it then be
made fhorter by 12 feet, and fet a running again,
and it will in the fame fpace of time difcharge 321
ounces, that is, near twice as much as before; fo
that the quantities difcharged, will be to one another reciprocally as the fquare roots of the lengths
of the pipe, which in this cafe are as 4 and 1.

TABLE I.

L	Q	T
1	$436\frac{1}{4}$	
4	321	$\frac{3}{4}$
9	$211\frac{3}{4}$	$1\frac{3}{4}$
16	$161\frac{1}{4}$	3
25	132	5
36	87	9
49	72	14
64	65	$20\frac{1}{2}$
81	$61\frac{1}{2}$	29
100	59	42

TABLE II.

D	Q
1	$87\frac{1}{4}$
4	$88\frac{1}{2}$
9	88
16	81
25	74
36	$67\frac{1}{2}$
49	65
64	$58\frac{1}{2}$
81	56
100	54

June the 21ft, 1722, I made feveral experiments
concerning the motion and difcharge of water thro'
pipes, in the following manner.

There was a refervoir of 3 feet in height, which
was kept conftantly full during the flux of the water; at the bottom was inferted horizontally a pipe
of half an inch in diameter, whofe length when
greateft was 100 feet, but being compofed of feveral pieces, was capable of being made of ten different lengths; which lengths were the fquares of the
natural numbers. Into this pipe were inferted horizontally (as occafion was) ten other pipes, each of
them 6 inches long, and ¼ inch in diameter; the
places of their infertion into the main pipe were diftant from the refervoir the fquares of the natural
numbers in feet; the axes of the fmall pipes made
an angle of 80 degrees, with that of the main pipe;
the reafon why they were inferted in fuch an angle
was, that it had been obferved that the water flowed out of orifices made in the main pipe nearly in
that angle.

In TAB. I. L denotes the length of the main
pipe (the fmall pipes not being inferted), Q the
quantity in *troy* ounces difcharged in half a minute of time, T the time in feconds which the water took to flow from the refervoir to the extremity
of the pipe, the fame having been firft exhaufted.

In TAB. II. D denotes the diftance from the refervoir, at which the fmall pipe was inferted into
the main pipe; Q the quantity in *troy* ounces difcharged by the fmall pipe in half a minute of time,
the main pipe being ftopped. ——

TABLE

TABLE III.

1	2	3	4	5	6	7	8	9	10	P	Sum.
84¾										63¼	148
	80									59	139
		74								54½	128½
			65½							49¼	115¾
				56¼						48	104¼
					45¼					45	90¼
						41½				42	83½
							31¼			43	74¼
								26		44½	70½
									6½	59½	66
67½	57	40	27	19½	9½	5	5¼	4	1¾	8	244½
69½	63½	56½	28	17½	7¼	5½	5	5	7		264½
69¼	63½	54½	29½	12½	12¼	10¼	8¼	8			269½
68	62	50	24½	27	17¼	9¾	9				267½
69¼	62½	50¾	24¾	28½	18¾	16½					271
69¼	63½	51½	25	31	27¼						267½
69¼	64	55	32	43							263¼
72¼	69	65	43								249¼
76½	75½	75½									227½
82¼	84										166¼
87¼											87¼
	76	50½	36½	27½	13¼	6	5¼	6½	¾	3½	225¾
		55	42	30½	13¼	6½	5½	7	2	4½	166
			49	32	14¾	15¼	6¼	9	2¼	5¾	135¼
				40	30¼	17¼	8¾	9¼	3	6¼	114¾
					39	21	10½	10¼	3½	8½	92¾
						37½	16¼	11	6¼	12½	84
							25	16½	7	21	69½
								24	8¾	30½	63¼
									9¼	53	62¼
										59	59

In TAB. III. the numbers at the top denote the ten ſmall pipes, P the main pipe, and the numbers below denote the quantities in *troy* ounces diſcharged in half a minute of time, by the pipes denoted by the numbers directly above them. The blanks denote, that the pipes denoted by the numbers directly above them at the top, were ſtopped at the time that the others diſcharged.

LECTURE

lofs of weight; when I formerly made the e:
ment, the lofs of weight amounted to ten grain
the magnitude of the exhaufted air I found
34 cubick inches; for upon immerfing the l
in water, and opening the valve which covere
mouth, the quantity of water which flowed in
poffeffed the place of the exhaufted air, amou
to 8628 grains, which being divided by 253 ½
number of grains in a cubick inch of water,
34 in the quotient; fo that from this experim
is manifeft, that 34 cubick inches of that air, v
more immediately furrounds us, are equal in w
to ten grains; and that the fpecifick gravity o
fame air is to the fpecifick gravity of water, a
to 8628, or, as one to 862½; the fpecifick g
ties of bodies equal in bulk, being to one·an
as the abfolute weights of the·bodies.

As the air rifes·above the furface of the ear
grows rarer, and confequently lighter; a ł
bulk of air,. being lighter at the diftance of a
than at the earth's furface, and lighter again
                                               dif

'diſtance of two miles, and ſo on continually. And
yet notwithſtanding this diminution of gravity in
the ſuperior parts of air, ſo great is the height of the
atmoſphere, as to render the weight of the whole
very conſiderable ; as will appear from the following experiment.

Let a piece of common glaſs be placed as a cover *Exp. 2.*
on the top of the receiver ; and upon exhauſting the
air, the glaſs will at firſt be preſſed cloſe to the receiver, and at length broken by the weight of the
air, which reſts upon-it.

While the air continues undiminiſhed in the receiver, it does by virtue of its elaſticity preſs as
ſtrongly againſt the lower ſurface of the glaſs, as
does the incumbent air by means of its weight upon
the upper ſurface ; as ſhall be ſhewn hereafter ;
conſequently, as long as the air remains undiminiſhed in the receiver, the weight of the incumbent
air can have no ſenſible effect on the glaſs ; but upon leſſening the quantity, and therewith the ſpring
of the included air, the glaſs being no longer ſupported from below, is preſſed down, and broken
by the weight of the air above ; and for the ſame
reaſon, a ſquare glaſs phial when exhauſted cracks
and flies to pieces.

From the weight and preſſure of the air on the
ſurface of liquors it is, that they are made to riſe
in exhauſted tubes open at one end, as will appear
from the following experiments.

Let a glaſs veſſel with mercury be placed under a *Exp. 3.*
receiver, and let a tube open at one end be ſuſpended above the veſſel in ſuch a manner, as that the
open end may at pleaſure be let down into the mercury ; if then, the air being drawn out of the receiver, the tube be let down, the mercury will not
riſe therein as long as the receiver continues empty ;
but upon readmitting the air, it will immediately
aſcend. The reaſon of which is, that upon ex-

Q   hauſting

hausting the receiver, the tube is likewise emptied
of air; and therefore, when it is immersed in the
mercury, and the air readmitted into the receiver,
all parts of the mercury are pressed upon by the
air, except that portion which lies beneath the orifice of the tube; consequently, it must rise in the
tube, and continue so to do, until the weight of
the elevated mercury presses as forcibly on that portion which lies beneath the tube, as the weight of
the air does on every other equal portion without
the tube. But to proceed to a second experiment
of the same kind.

Pl. 6.
Fig. 14.
Exp. 4.
    Let two glass tubes as A and B, each above 30
inches long, of which A is open at one end only,
but B at both, be so contrived, as by means of
screws to be let into the little glass vessel CD, in
the manner represented in the figure, A being filled with mercury, and then screwed into the vessel,
let mercury be poured into B, till both that and
the vessel are full; let then the vessel be inverted;

Pl. 6.
Fig. 15.
and let the extremity of B be immersed in a vessel
of mercury, the mercury will descend thro' B, and
continue so to do, till A is emptied, as also so much
of the vessel CD as is above the level of the upper
orifice of B. This being done, let A be so far un-
screwed, as to permit the air to pass between the
threads of the screw into the empty part of the vessel; upon the admission of the air, the mercury
will rise in the tube A. For, from the circumstances of the experiment it is evident, that the part
of A which stands above the level of the mercury
remaining in the vessel, is perfectly void of air;
consequently, while the mercury all around the
tube is pressed by the newly admitted air, that portion which lies beneath the tube suffers no pressure
from above; and of course must rise, and continue
to rise, until the weight of the elevated mercury
becomes a balance to the pressure of the air without.

By

Plate 6. *Page 252.*

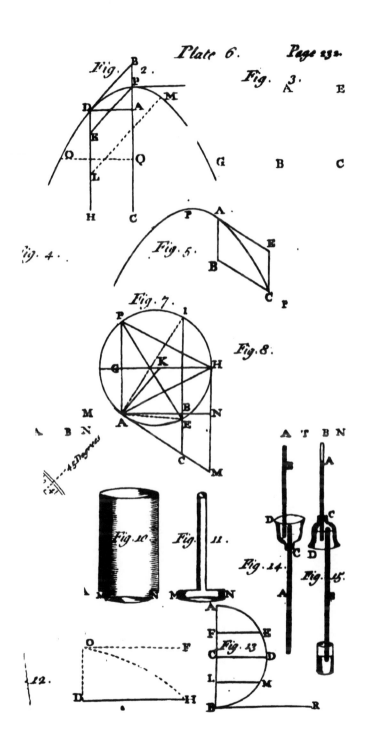

Fig. 2.

Fig. 3.

Fig. 4.

Fig. 5.

Fig. 7.

Fig. 8.

45 Degrees

Fig. 10.

Fig. 11.

Fig. 12.

Fig. 13.

Fig. 14.

Fig. 15.

By the weight and preffure of the air, water is raifed in common pumps, and fire engines, as will appear by confidering their ftructures, and the manner in which they work. A B reprefents the body of a pump, which is commonly an hollow cylinder of wood or lead, C a plug fixed near the bottom of the pump, with an hole in the middle, covered by a leathern valve, fo contrived as to open and give way to the water in paffing upward, but to fhut clofe and obftruct the paffage downward; D a fecond plug of the fame kind, and perforated in like manner with the former. This plug is commonly called the *fucker* or *pifton*, and being moveable, is drawn up and thruft down at pleafure, by means of the iron rod E to which it is faftened. The fides of the fucker are every where cafed with leather, whereby it is made to fit the cavity of the pump fo exactly, that neither air nor water can pafs between. At fome diftance above the fucker is an orifice as O in the fide of the pump, thro' which the water is difcharged at the time of working, in the following manner. The fucker being drawn up, the fpace between that and the lower plug is left void of air; then forafmuch as the water, which ftands about the pump, is every where preffed by the air, except in that part which anfwers to the hole of the plug, it muft there give way, and pafs up into the cavity of the pump; and upon depreffing the fucker again, as it cannot return downward by reafon of the valve, which fhuts clofe upon the hole, and ftops the paffage, it rifes up thro' the fucker, and lodges itfelf thereon; fo that upon the next elevation of the fucker, it is carried towards the top of the pump, and thrown out at the orifice O.

If inftead of an orifice above the fucker, we fuppofe one juft above the lower plug, with a valve opening outwardly, fo as to fuffer the water to flow out, but not to return. And if we fuppofe the ſucker to be ſolid without a perforation, the figure will repreſent a forcing pump, or fire engine, in which the water riſes above the lower plug in the ſame manner, and from the ſame cauſe, that it does in a common pump; and by the preſſure made upon it by the ſucker when thruſt down, it is forced out at the orifice, and that ſo ſtrongly, as by the help of leathern pipes to be conveyed to the tops of the higheſt houſes.

The air in any particular place does not always continue of the ſame weight, but is ſometimes heavier, and ſometimes lighter; which plainly argues a variation in the quantity, inaſmuch as the gravity of any body is proportional to the quantity of matter which it contains. From what cauſe this variation ariſes, is not eaſy to determine. Dr. HALLEY is of opinion, that the diminution of the quantity of air in any place, is the effect of two contrary winds blowing from that place, whereby the air is carried both ways from it; and of conſequence, the incumbent cylinder of air is diminiſhed; as for inſtance, if in the *German* ocean it ſhould blow a gale of *weſterly* wind, and at the ſame time an *eaſterly* wind in the *Iriſh* ſea; or if in *France* it ſhould blow a *ſoutherly* wind, and in *Scotland* a *northern*; that part of the atmoſphere which is impendent over *England* would, he thinks, be thereby carried off and diminiſhed. He likewiſe conceives, that the increaſe of the quantity of air in any place, is occaſioned by the blowing of two contrary winds towards that place, whereby the air of other places is brought thither and accumulated. And upon this foot, he endeavours to account for what is commonly obſerved in this part of the world; namely, that the atmoſphere, *cæteris paribus*, is always heavieſt upon an *eaſterly* or *north-eaſterly* wind. This happens, ſays he, becauſe, that in the great *Atlantick* ocean, on this ſide the thirty fifth degree of *north* latitude, the *weſterly* and *ſouth-weſterly* winds blow almoſt always; ſo that

that whenever the wind comes up here at *eaſt* or north-*eaſt*, it is ſure to be checked by a contrary gale, as ſoon as it reaches the ocean; for which reaſon, the air over us muſt needs be heaped up in greater abundance, as often as thoſe winds blow. To confirm this hypotheſis of contrary winds being the cauſe of the variation in the weight of the air, he obſerves, that within the *Tropicks*, where there are no contrary currents of air, this variation does not obtain; but that the atmoſphere continues much in the ſame ſtate of gravity in all kinds of weather. Now, whether this, or whatever elſe be the cauſe of it, moſt certain it is, that the weight of the air does vary; and ſo conſiderable is the variation, that the weight of the air in its heavieſt ſtate, exceeds the weight thereof when it is lighteſt, in the proportion of almoſt ten to nine.

The changes which the air undergoes as to its gravity, are obſerved by means of the *Barometer* or weather-glaſs; which, as it was the invention of Torricellius, is known among the naturaliſts by the name of the *Torricellian tube or inſtrument.* It conſiſts of a ſmall glaſs-tube, about three feet long, cloſed at one end; which being filled with mercury well purged from air, is inverted into a cylindrical box of timber, wherein ſome mercury is lodged; upon the inverſion ſome of the mercury falls out, whereby the upper part of the tube is left empty whilſt the lower part continues full. Now, foraſmuch as it has appeared from experiments, that the ſuſpenſion of the mercury in the tube is owing to the preſſure of the air on the ſtagnant mercury; the pillar of mercury which is kept up in the tube, muſt always be equal in weight to a pillar of the atmoſphere of the ſame thickneſs; conſequently, as the weight of the atmoſphere varies, the height of the mercury in the barometer muſt do ſo too; the mercury conſtantly riſing as the weight of the air increaſes, and ſinking as that

leſſens.

leſſens. That the minute variations in the height of
the mercury may be obſerved, that part of the
tube which lies between the limits of the leaſt and
greateſt height, to wit, from 28 to 31 inches, is
graduated; each inch being divided into ten or
twelve equal parts by means of a table, whereunto
the tube is fixed; whereon likewiſe are inſcribed in
their proper places ſuch conſtitutions of the air and
weather, as have been obſerved to accompany different heights of the mercury. In contriving this
inſtrument, care muſt be taken to make the box,
which contains the ſtagnant mercury, ſo large, as
that upon the riſing or falling of the mercury in the
tube, the height of that in the box may ſuffer little
or no variation; for ſhould the ſtagnant mercury
ſink upon the riſing of the mercury in the tube, or
riſe as that ſinks, which muſt be the caſe where the
box is ſmall; the riſe or fall of the mercury in the
tube will appear to be leſs than it really is; as for
inſtance, if when the mercury riſes half an inch in
the tube, it does at the ſame time fall a quarter in
the box, the riſe in the tube, which appears to be
only half an inch, is in truth three quarters; becauſe
the height of the mercury is always to be computed
from the ſurface of that in the box. So, on the
other hand, if the mercury by falling half an inch
in the tube riſes a quarter in the box, the true de-
ſcent in the tube is three quarters of an inch; inaſmuch as the height of the mercury in the tube above
the ſurface of the ſtagnant mercury in the box, is
leſs after the fall by three quarters of an inch. By
making the circular *area* of the box thirty or forty
times greater than that of the tube, (which is generally the caſe, the tubes of moſt barometers being but one fifth of an inch wide, and the boxes
an inch and a quarter) the ſtagnant mercury
in the box may be kept conſtantly at the ſame
height very nearly, the greateſt variation of the
height not amounting to more than the tenth or

twelfth

twelfth part of an inch, which is inconsiderable.

If the tube instead of being continued directly upward, be bent at the height of 28 inches, in the Pl. 7. manner here represented, it is then called an *inflected* Fig. 2. or *diagonal barometer*; in which the inclined part AB may constitute an obtuse angle of any magnitude with the perpendicular part BC; but the nearer the angle approaches to a right one, the longer must the inclined part be; for it must be continued until the perpendicular altitude thereof AH, above the horizontal line HB, becomes equal to three inches, which is the difference between the greatest and least height of the mercury in the barometer; otherwise, the mercury will not have room to rise to its utmost height, at such times as the constitution of the air requires it. This barometer shews the minute variations in the weight of the air much more accurately than the former; because the rise or fall of the mercury in the inclined part AB is very sensible, when an alteration in the perpendicular height is scarcely to be perceived. But then the box which contains the stagnant mercury, ought to be much larger in proportion in this than in the former; because in this, a much larger quantity of mercury rises into, and falls out of the tube, upon the changes of the weather.

If the lower part of the tube in the first barometer, instead of being inserted into a box, be turned up in the form of a crook, it is then called a *curved barometer*, in which the crooked part generally terminates in a large bubble open at top. The bubble Pl. 7. contains the stagnant mercury, which, as it is pres- Fig. 4. sed upon more or less by the incumbent air, is forced up to a greater or smaller height in the strait part of the tube. In this barometer the bubble ought to be so large in proportion to the tube, as that upon the greatest variation of the height of the mercury in the tube, the height thereof in the bubble may not vary above one tenth of an inch; the

necessity

neceffity there is for this, is evident from what is said concerning the magnitude of the box in the firft kind of barometer.

Befides the barometers hitherto mentioned, there is the *wheel*, as alfo the *pendant* or *conical*, *barometer*, and others of various kinds; which, however they may differ as to their ftructures, do all agree in fhewing the changes in the weight of the air, by the rifing and falling of the mercury in their tubes; wherein it fometimes, tho' very rarely, defcends as low as twenty eight inches; and at others rifes to thirty one; the mean height thereof being twenty nine inches and an half. So that a pillar of the at-mofphere, in the mean ftate of its gravity, is equal in weight to a pillar of mercury of the fame thick-nefs, and whofe altitude is twenty nine inches and an half. Whence it follows, that an inch fquare of the earth's furface, or of any other body contiguous thereto, fuftains a preffure from the incumbent at-mofphere, when in the mean ftate of its gravity, equal to feventeen pounds, eight ounces, and 374 grains; that being the weight of a fquare pillar of mercury one inch thick, and twenty nine and an half high.

From this great preffure of the air it is, that two brazen hemifpheres, whofe diameter is three inches and an half, being laid one upon another, and then exhaufted, cling fo faft together, as to require above 150 pounds to feparate and draw them afunder. And it muft be obferved, that as the globe in this experiment cannot be perfectly exhaufted, that fmall portion of air which remains within, by ex-panding itfelf, contributes to the feparation of the hemifpheres; for which reafon, they are drawn a-funder by a lefs weight than that wherewith the air preffes them together; for the diameter of the fphere being three inches and an half, the *area* of its greateft circle is nine fquare inches and three fifths nearly; confequently, the weight of that pillar of
air

air which preffes the hemifpheres together, is not left than 162 pounds, even in its lighteft ftate, when the mercury in the barometer ftands at the height of 28 inches only. If the globe, after it has been exhaufted, be hung within a receiver, upon drawing the air out of the receiver, the lower hemifphere will fall off from the other; which plainly fhews, that their cohefion is owing to nothing elfe but the weight and preffure of the air upon them.

Since the atmofphere even in its lighteft ftate is fo ponderous, as that a fquare pillar of it one inch thick weighs fixteen pounds, nine ounces, and 461 grains; it follows, that a middle-fized man, the furface of whofe body is generally allowed to contain about fifteen fquare feet, fuftains a preffure from the atmofphere, when in its lighteft-ftate, equal to the weight of 31144 pounds; which preffure on larger bodies, and in heavier ftates of the air, is ftill greater; and therefore it may well be afked, how it comes to pafs, that we are not fenfible of this preffure, great as it is. In anfwer to which it muft be obferved, that fuch preffures only are perceived by us, as do in fome meafure move our fibres, and put them out of their natural fituation. Now, the preffure of the air being equal on all parts of the body, cannot poffibly move the fibres of any one part, or force them from their fituation; but on the contrary, muft by reafon of its uniformity keep all the fibres in their proper places, and as fo doing, cannot be perceived. And that this is the cafe is evident from hence, that if the preffure of the air be taken off from one part of the body, the preffure on the neighbouring parts immediately becomes fenfible. Thus, if a man covers the top of an open receiver with his hand, upon exhaufting the receiver, and thereby taking off the preffure of the air from the palm of the hand, he will perceive

a weight

a weight on the back of his hand, and that fo great,
as to put him to pain, and almoft endanger the
breaking of his hand.

LECTURE XVI.

OF PNEUMATICKS.

LECT.
XVI.

BY the elafticity of the air, whereof I intend
to treat in this lecture, we are to underftand
that force wherewith the particles of air expand
themfelves, and recede from each other, whenever
the preffure from without, which keeps them together, is taken off. The method which I fhall
obferve in treating of this force is, Firft, to fhew
from experiments, that the air is really indued with
fuch a force; and, Secondly, to enquire into its
nature and laws.

Exp. 1.   As to the firft, if a little warmed ale, or any
other liquor fomewhat glutinous, be put into a glafs
and included in a receiver, upon exhaufting the receiver the liquor will rife in large frothy bubbles,
and run over the glafs.

As the liquor is glutinous, it retains a great number of airy particles, which upon the removal of the
outward air, and therewith the preffure which it
makes on the liquor, dilate and expand themfelves;
and forafmuch as they cannot readily extricate
themfelves from the liquor by reafon of its clamminefs, they raife it up, and carry it over in the form
of froth. And for the fame reafon it feems to be,
that meath, cyder, and moft other domeftick wines,
after they have been bottled a while, upon drawing
the cork, fpurt out and fly. For as they are all in
fome meafure glutinous, they retain a good quantity of air; which upon corking the bottle is condensed

denfed by reafon of the condenfation of the air
which is lodged in the neck of the bottle; befides,
by the flight fermentation which fuch liquors commonly undergo in the bottle, a frefh fupply of air
is generated, equal in denfity to the former. When
therefore upon drawing the cork, the extraordinary
preffure arifing from the condenfed air in the neck
of the bottle is taken off, the air which is difperfed
thro' the liquor expands itfelf with great force,
and not finding a ready paffage between the parts
of the liquor, which by reafon of their clamminefs
do not eafily feparate, drives the liquor before it
in the manner of a fpout. But to proceed;

The expanfive force of the air is likewife evident
from the following experiment. Let a glafs bottle **Exp. 2.**
of a globular form, and containing a fmall quantity of water, have a fmall glafs tube open at both
ends, inferted into it fo far, as that the lower end
may be below the furface of the water; and let the
infertion be made by means of a fcrew and a collar
of leathers, in fuch a manner as that no air may
pafs into or out of the bottle; let then the whole
apparatus be placed under a tall receiver, and upon
exhaufting the air out of the receiver, the water will
rife up thro' the tube in the form of a jet, which
will be higher or lower in proportion as the receiver
is more or lefs exhaufted; the reafon of which is,
that the air included in the bottle, by endeavouring
to expand itfelf, preffes upon the furface of the
water, which therefore muft rife in the tube, as foon
as the preffure of the outward air which keeps it
down is leffened; and the greater the diminution
of that external preffure is, the higher the water
muft be thrown.

If a bladder wherein a fmall quantity of air is **Exp. 3.**
included, be placed under a receiver, upon drawing
the air out of the receiver, the bladder will fwell,
and the fwelling will be greater or lefs in proportion as the receiver is more or lefs emptied; which
plainly

plainly argues an expansive force in the inclosed
air; as does likewise the bursting of a full blown
bladder in an exhausted receiver; as also the
cracking of a square glass phial when close stopped.

If a small siphon, having a weight fastened from
the hand of the piston, and being closed at the
end so as that upon drawing up the piston no air
can get in, be suspended in an inverted position
with the weight downward, and then covered with
a receiver; upon drawing part of the air out of the
receiver, the weight will descend, and draw down
the piston; and upon the readmission of the air it
will rise again.

- When part of the air is drawn out of the receiver, that portion which remains within expands itself, whereby its spring is so far weakened, as not to
be able to stand against and support the weight, for
which reason the weight descends; whereas, upon
the return of the air which was carried off, the elastick force is so far increased, as to become an overbalance for the weight, and upon that account
drives it up.

From this and the foregoing experiments it fully
appears, that the air is indued with an expansive
force. Whence that force arises, and what the law
of its action is, comes now to be considered.

The naturalists were formerly of opinion, that
the elasticity of the air was owing to the shape and
figure of its parts: for they supposed each particle
of air to consist of several branches, which being
of a pliable nature, were capable of being compres-
sed and squeezed together by any outward force, and
of expanding and spreading themselves abroad upon
the removal of the compressing force; and this
has been thought by some to be a full and satisfac-
tory account. But that great philosopher Sir ISAAC
NEWTON was of opinion, that the expansive force
of the air is altogether inexplicable on the foot of
this,

this, or indeed any other hypothefis, except that of the air's being indued with a repelling power, whereby the particles recede and fly from each other; his words are thefe.

" That there is a repulfive virtue, feems alfo to
" follow from the production of air and vapour.
" The particles when fhaken off from bodies by
" heat or fermentation, fo foon as they are beyond
" the reach of the attraction of the body, reced-
" ing from it, and alfo from one another with
" great ftrength, and keeping at a diftance, fo as
" fometimes to take up above a million of times
" more fpace than they did before in the form
" of a denfe body; which vaft contraction and
" expanfion feems unintelligible, by feigning the
" particles of air to be fpringy and ramous, or
" rolled up like hoops, or by any other means
" than a repulfive power."

Now, fuppofing this to be the cafe, and that the repelling power of each particle exerts itfelf on the next adjacent particles only, as Sir ISAAC feemed to imagine, I fhall fhew you what the law of this repelling power is, or, in other words, how this power is varied, by varying the diftance of the particles; and in order thereto, fhall lay down the following PROPOSITION.

PROP. *If a fluid be compofed of particles endued with a repulfive power, fo as that each particle repels thofe, and thofe only, which are next it, and if the force wherewith two adjacent particles repel each other, be in a given reciprocal ratio of the interval of their centers; that is, putting I for the interval of the centers, and P for the index of the given power of that interval; I fay, if two adjacent particles repel each other with a force that is as* $\frac{1}{I^P}$, *the force which com-*

prefjes the fluid, is as the cubick root of that power of

the

the denſity of the fluid, whoſe index is P in̄re̅a̅s̅e̅d̅ 2, or P + 2; that is, putting F for the c̅o̅m̅p̅r̅e̅s̅s̅i̅n̅g̅ force, and D for the denſity of the fluid, F is a̅s̅ D raiſed up to the power whoſe index is $\dfrac{P+2}{3}$.

Exp. 7.

Pl. 7.
Fig. 5.

For the proof of this, let a portion of the fluid be contained in a given cubick ſpace, whoſe upper ſurface is denoted by the ſquare ABCG; the compreſſing force being applied to that ſurface.

The elaſtick force of the fluid, which withſtands the compreſſing force, and is exactly equal thereto, is the force of thoſe parts only which compoſe the upper ſurface; becauſe the repelling forces of the particles are ſuppoſed to exert themſelves on thoſe particles only which lie next them, and not to extend to particles more remote. But the force of the ſuperficial parts is as the number of particles in the ſurface, and the force wherewith any two adjacent particles repel each other conjointly. Now, the number of particles in the given ſquare ſurface, is reciprocally as the ſquare of the diſtance of the centers of two adjacent parts; that is, as $\dfrac{1}{I^2}$; and by ſuppoſition, the force wherewith two particles repel each other, is as $\dfrac{1}{I^P}$; and therefore, the elaſtick force of the fluid, and of conſequence the compreſſive force, or F, is as $\dfrac{1}{I^{P+2}}$. The denſity of the fluid contained in the given cubical ſpace, is inverſly as the cube of the diſtance between the centers of the particles; that is, D is as, $\dfrac{1}{I^3}$, and I is as $\dfrac{1}{D^{\frac{1}{3}}}$; and therefore, by ſubſtituting $\dfrac{1}{D^{\frac{1}{3}}}$ in the room of I, F is as $D^{\frac{P+2}{3}}$; that is, the compreſ-

ſive

4

five force is as the cube root of that power of the
denfity of the fluid, whofe index is P + 2.

COROL. From this propofition it follows, that
if the denfity of an elaftick fluid be as the force
which compreffes it, the particles repel one another
with forces that are inverfly as the diftances of their
centers.

For fince F is as D, $\dfrac{P+2}{3}$ is equal to unity, and
fo likewife is P; confequently, the P power of I,
whofe reciprocal expreffes the repulfive force of the
particles, is equal to I.

Hence the particles of air muft repel one another
with forces reciprocally proportional to the diftances
of their centers, becaufe the denfity of the air is
proportional to the force which compreffes it; as
will appear from the following experiment.

Let an inflexed tube as AB, open at both ends, **Exp. 8.**
be filled up with mercury to fome fmall height,
fuppofe DC; then ftopping the end B, fo as that **Pl. 7.**
the air may not get out when it is compreffed, and **Fig. 6.**
meafuring the length of BC, that part of the fhorter leg that is filled with air, which air, it is evident,
is compreffed by the weight of the atmofphere;
let mercury be poured in at A, till the height thereof in the longer leg above the height of the fame in
the fhorter, becomes equal to the height at which
it ftands in the barometer, by which means the air
in the fhorter leg will be compreffed by twice the
weight of the atmofphere; let then the length of
that part of the leg which is poffeffed by the air under this double preffure be meafured, and it will be
found to be juft one half of BC; whence it appears,
that the fpaces which a given quantity of air pof-
feffes under different preffures, are reciprocally proportional to the preffures; and confequently, inafmuch as the denfities of bodies where the quantity
of matter is given are reciprocally as their magnitudes,

tudes, the denſity of the air is directly as the
preſſing force. From this property of the air,
has deduced a method for determining the
thereof at any height.; what he has deliver
cerning this matter, is contained in the 5th
of his *Harmonia Menſurarum*, which I ſhall
vour to explain to you ; and in order theret
lay before you ſuch properties of the logari
curve, as I ſhall have occaſion to make uſe.of
ring you for their' demonſtrations to the for
oned author, and others who have wrote
Pl. 7. curve. Let then BDGI be a·logarithmick
Fig. 7. AH its aſymptot, that is, a right line ſo
with reſpect to the curve, as not to meet it
drawn to an infinite, or rather indefinite
BA, DC, and GF, ordinates, that is, righ
perpendicular to the aſymptot at the points
and F, and terminating in the curve. BC
gent to the curve at the point B.   The pro
of this curve, which I ſhall have occaſion t
tion, are theſe four.

Firſt, Any portion of the aſymptot inter
between two ordinates, is the logarithm or n
of the *ratio* which thoſe ordinates bear one
other ; thus AC meaſures the *ratio* of BA t
and CF meaſures the *ratio* of DC to GF
ſo likewiſe, AF meaſures the *ratio* of BA t
And if AC, AF, and AH be in arithmetic
portion, then DC, GF, and IH are in geon
proportion ; and if any 'portion of the aſ
be a given quantity, then is the *ratio* of t
ordinates, which intercept that portion, li
given.

Secondly, That portion of the aſymptot
which is intercepted between a tangent and
dinate, drawn to the ſame point of the curv
is a given quantity, or in other words, to
ever point of the curve the tangent and ordin

drawn, the portion of the afymptot which they intercept is always of one and the fame length. The portion fo intercepted is called the *fubtangent*, and it is the module, or that which regulates the magnitudes of all the logarithms in the fame fyftem ; for they are greater or lefs in proportion to the magnitude of the fubtangent; fo that if in two logarithmick curves, the fubtangent of one be double or triple the fubtangent of the other, the meafures of the fame *ratios* are likewife twice or thrice as great in the former as they are in the latter.

Thirdly, The indefinite *areas* comprehended between the curve and the afymptot, drawn on to an indefinite length beyond HI, are to one another as the ordinates which bound them in their wideft parts ; thus, the indefinite *areas* BAHI, DCHI, and GFHI, are to one another as the ordinates BA, DC, and GF.

Fourthly, The indefinite *area* BAHI, is equal to the parallelogram BACE, comprehended under the ordinate BA, and the fubtangent AC.

Thefe things being premifed, let AB reprefent the earth's furface, and let AH be a line perpendicular thereto ; then, forafmuch as the denfities of the air at different heights, are as the preffures of the incumbent atmofphere, and the ordinates in the curve, as the indefinite *areas* which lie beyond them ; if the indefinite *area* BAHI be made to denote the weight or preffure of all the air, and AB its denfity at the furface of the earth, then, by the third property of the curve, the indefinite *area* DCHI will denote the weight or preffure of all the air which lies above C, and the ordinate DC will denote the denfity of the air at that height ; and thus it is with regard to any other height, fo that at all heights, the denfities of the air will be denoted by the refpective ordinates ; wherefore, by the firft property of the curve, the difference be-

R

tween any two heights, is the meafure of the *ratio*
which the denfities of the air bear to one another
at thofe heights ; thus C F meafures the proportion
which the air's denfity at the height C bears to its
denfity at the height F. Let us now fuppofe the
force of gravity to ceafe, and that the air is fo
compreffed by fome external force, as to be every
where from top to bottom of the fame denfity, as
it is at the furface of the earth ; its weight or preffure, which before was denoted by the indefinite
area BAHI, may now be denoted by the parallelogram BACE, inafmuch as by the fourth property of the curve, that *area* and this paralle
are equal. Since then two fluids which
each other muft have their heights inverfly
fpecifick gravities, if we put unity to d
fpecifick gravity of the air at the furface
earth, and fay, as unity to 11890, whic
fpecifick gravity of mercury with refpe
of air, fo is 2¼ feet, which is the heigh
mercury in the barometer, to a fourth nu
fhall have 29725 feet for the height of th
geneal atmofphere ; and this height is eq
fubtangent AC. For fince the preffure
homogeneal atmofphere is as its denfity
height, and likewife as the rectangle BACE
fince the denfity is denoted by BA, the h
muft be denoted by AC, the module in this
of logarithms. Hence we have a method
termining the denfity of the air at any heigh
putting H to denote the height at which the
fity of the air is required, by the fecond p
of the curve, we have this analogy, as the in
number marked A, which is the module
fyftem, is to the fractional number m
which is the module of BRIGGS's fyftem, fo
expreffed in feet, to a fourth number, whi
BRIGGS's tables is the logarithm of the *ratio*

the

the denfity of the air at the earth's furface, to its
denfity at the height H, anfwerable to which in
the tables is the natural number expreffing that
ratio.

$$\underset{\text{29725}}{\overset{\text{A}}{}} : \underset{}{\overset{\text{B}}{0.43429448}} :: H : \frac{0.43429448 \times H}{29725}.$$

$$\underset{\text{29725}}{\frac{\overset{\text{B}}{26400 \times 0.43429448}}{A}} = \overset{\text{C}}{0.385661.} \qquad \overset{\text{D}}{2.4303.}$$

Thus, for inftance, if the denfity of the air at
the height of five miles, or 26400 feet, be required,
by multiplying that number by the fractional number marked B, and dividing the product by the
integral number marked A, we fhall have the logarithm marked C; anfwerable to which in the
tables is the natural number marked D, expreffing
the ratio of the air's denfity at the furface of the
earth, to its denfity at the height of five miles;
whence it appears, that at the furface of the earth,
the air is denfer than it is at the height of five miles,
in the proportion of almoft $2\frac{1}{4}$ to one; but then,
this is on fuppofition that the force of gravity continues the fame at all heights, whereas in truth,
that force decreafes in the recefs from the earth's
center in the duplicate ratio of the diftance, which
caufes the denfities of the air at different heights to
be fomewhat different from what they would be in
cafe the force of gravity did not vary.

In order therefore to determine the denfities more
accurately, let S be the earth's center, and AB,
equal to AB in the laft figure, the earth's furface,
and let F be the height at which the denfity of the
air is required; let SK be a third proportional to
SF and SA, and at the point K, let the ordinate

Pl. 7.
Fig. 8.

KG be drawn, denoting the denſity of the air at F, then taking the point M at an indefinitely ſmall diſtance above F, let SL be a third proportional to SM and SA; and at the point L, let the ordinate LN be drawn, denoting the denſity of the air at M; this being done, it will appear, that the curve BGN, which paſſeth thro' the points G and N, is the ſame logarithmick curve with the former, but in an inverted poſition.　For ſince SL is to SA, as SA to SM, and ſince SK is to the ſame SA, as SA to SF, then by equality of *ratio*, SL is to SK, as SF to SM, and by diviſion and permutation, KL is to FM, as SK to SM; or becauſe FM is indefinitely ſmall, as SK to SF; that is, as SA² to SF²; whence reducing that analogy into an equation, and dividing by SF², we ſhall have $KL = \dfrac{SA^q \times FM}{SF^q}$,

and rejecting SA⁹, as being a given quantity, we ſhall have KL as FM directly, and SF⁹ inverſly, but FM is as the quantity of air in the indefinitely little ſpace FM, and SF⁹ inverſly is as the gravitation of the ſame air, and KG is as its denſity; conſequently, the rectangle under KL and KG, or the *area* KGNL, is as the gravitation, the quantity, and denſity of that air conjointly, that is, as its preſſure on the air beneath it; and the ſum of all the ſimilar *areas* below KG, is as the ſum of all the preſſures above F, that is, as the denſity of the air at F, or as KG, which denotes that denſity; and KGNL which is the difference of the two ſums of all the ſimilar *areas*, one of which ſums begins from the point K, and the other from the point L, is as the difference of the air's denſities at F and M, that is, as KG—LN.　Let now KL be given, that is to ſay, let the ſmall portion intercepted between KG and LN be always of one and the ſame length, in whatever parts of the line AS the points K and L are taken; then KG will be as the *area* KGNL,

and

and consequently, as KG—LN; whence by division, KG will be as LN, so that the *ratio* of KG to LN is given, and of course the given line KL will be the measure of that given *ratio*; whence by the first property of the logarithmick curve, the curve which passeth thro' the points G and N is a logarithmick curve; and it is also the same with the former; for taking AO the height above the earth's surface indefinitely small, it is evident, that the force of gravity is the same at O that it is at A, consequently, the density of the air at O will come out the same, whether the law of gravity be taken into the consideration or left out; let then the ordinate OP be drawn in the former curve, and at the same distance from A in the latter curve, let the ordinate PQ be drawn. Now, since one and the same density of the air at the earth's surface is denoted in both curves by the equal ordinates BA, it is evident, that the ordinates OP and PQ, which in the two curves denote one and the same density at O, must likewise be equal; whence it follows, that both curves have the same curvature, as also the same inclination of their tangents at the points B, and their subtangents equal; that is, the latter curve is the same with the former, but in an inverted position. Now, forasmuch as BA in the latter curve denotes the density of the air at the surface of the earth, and KG its density at F, it is evident by the first property of the curve, that in this system, AK is the measure of the *ratio* which the density at the surface has to the density at F; the first thing therefore which must be done, in order to discover the density at F, is to find out the line AK, and this is done by diminishing AF in the same proportion that the earth's semidiameter SA is less than SF, the distance of F from the earth's center; for by the construction, SF is to SA, as SA to SK; whence by division, SF : SA :: AF : AK. AK being thus obtained, let it be called H; then, by the same

process

procefs as before, we may difcover the denfity of the air at the height F.

$$ 4005 : 4000 :: 5 : \frac{4000 \times 5}{4005} = 4.99375 \overset{E}{=} 26367 \text{ feet.} $$

$$ \underset{29725}{\overset{B}{\frac{26367 \times 0.43429448}{A}}} = 0.385232. \qquad \overset{G}{2.4279.} $$

For inftance, if the denfity of the air at the height of five miles be required as before; then by faying, as 4005 miles, that is SF, is to 4000 miles, that is SA, fo is five miles, that is AF, to a fourth, we fhall have the number marked E, expreffing miles, and parts of a mile, equal to 26367 feet, which being multiplied by the fractional number marked B, and the product divided by the integral number marked A, we fhall have the fractional number of Briggs's tables marked F, anfwerable to which is the natural number marked G, expreffing the *ratio* of the air's denfity at the furface of the earth, to its denfity at the height of five miles. After the fame manner may the *ratio* of the air's denfity at the furface, to its denfity at any height be computed. The refult of fuch computations I have fet down in the annexed table; the firft column of which contains the heights of the air in *Englifh* miles, whereof 4000 make a femidiameter of the earth. The numbers in the fecond column exprefs the *ratio* of the air's denfity at the furface, to its denfity at the refpective heights, and they likewife denote the rarity or expanfion of the air at thofe heights. The third column contains the denfities and compreffions at the feveral heights. The numbers at the bottom of the fecond column included in crotchets denote, that fo many figures are to be annexed to the five preceding, and thofe included in the crotchets at the bottom of the third column denote, that so many decimal cyphers are to be prefixed to the five following figures.

Heights of the air in *Eng-lish* miles.	Rarity and expansion.	Compression and density.
0	— — 1 —	1
1/4	— — — 1.0454	0.95676
1/2	— — — 1.0928	0.91509
3/4	— — — 1.1424	0.87535
1	— — — 1.1943	0.85405
1 1/4	— — — 1.2429	0.80456
1 1/2	— — — 1.3052	0.76616
1 3/4	— — — 1.3644	0.73290
2	— — — 1.4263	0.70118
2 1/4	— — — 1.4871	0.67244
2 1/2	— — — 1.5586	0.64160
2 3/4	— — — 1.6292	0.61379
3	— — — 1.7031	0.58716
3 1/4	— — — 1.7883	0.55919
3 1/2	— — — 1.8596	0.53775
3 3/4	— — — 1.9460	0.51387
4	— — — 2.0336	0.49173
4 1/4	— — — 2.1257	0.47043
4 1/2	— — — 2.2221	0.45002
4 3/4	— — — 2.3226	0.43012
5	— — — 2.4279	0.41187
10	— — — 5.9182	0.16897
20	— — 34.288	0.029164
30	— — 198.34—	0.0050418
40	— — 1136. —	0.00088028
50	— — 6449.2 —	0.00015505
100	33584[3]— —	0. [7]26798
400	11271[24] —	0. [28]88723
4000	19316[150] —	0.[154]51770
40000	33097[276] —	0.[280]30214
400000	32859[301] —	0.[305]30433
1000000	12002[303] —	0.[307]45450
Infinite.	37311[304] —	0.[308]26802

Corol.

COROL. Since SF is by conftruction equal to
$\frac{SA}{SK}$, and fince from the nature of mufical proportion, the quotients arifing from the divifion of one
and the fame quantity by quantities in arithmetick
progreffion, conftitute a feries of mufical proportionals, it follows, that if feveral diftances from
the earth's center as SF, be taken, in mufical progreffion, their reciprocals as SK, muft be in arithmetick progreffion; and by the firft property of the
logarithmick curve, the denfities of the air as KG,
muft be in geometrick progreffion.

Since the denfity of the air is proportional to
the compreffing force, and fince the compreffing
force is equal to the elaftick force, it is manifeft,
that if the denfity of the air be increafed, the elafticity will likewife increafe in the fame proportion;
and on this principle are founded artificial fountains,
which play by means of condenfed air; they are of

Pl. 7.
Fig. 9.

two kinds, fingle and double. The fingle fountain is made of brafs, and is every where fhut, excepting that thro' the middle of the bafon BB,
there paffes down a pipe PP, whofe lower end
reaches nearly to the bottom of the fountain, and
to the upper end is fitted a ftopcock, by help of
which the pipe may be fhut or opened at pleafure.

Exp. 8.

Some part of the fountain as ADC, being filled
with water poured in thro' the pipe, a condenfing
or forcing fyphon is fcrewed to the top of the pipe
above the cock, by means whereof a great quantity
of air is thrown into the pipe; which as it cannot
return back, by reafon of a valve which fhuts clofe
upon the hole of the fyphon, forces its way thro'
the water into the upper part of the fountain, and
there remains in a ftate of condenfation, greater than
that of the outward air. When therefore the con-
denfer is taken off, and the cock opened, the included air preffing ftrongly on the water which lies
beneath

beneath it, throws it up thro' the pipe, and thereby makes a jet.

The force wherewith the water is thrown up, is proportional to, and may be expreſſed by the exceſs of the denſity of the included air above that of the external air. For if the included air be equally denſe with that without, its elaſtick force muſt be equal to the compreſſive force of the atmoſphere; conſequently thoſe two forces will balance one another, and the water will continue at reſt, being preſſed as ſtrongly downward by the weight of the external air, as it is upward by the expanſive force of the included air; but if the included air be more denſe than the external, its elaſtick force will exceed the compreſſive force of the atmoſphere, in the ſame proportion that its denſity exceeds the denſity of the outward air; conſequently, that part of the expanſive force of the included air which raiſes the water, is proportional to, and may be expreſſed by, the exceſs of the denſity of the included air above that of the external air. So that putting F for the force which raiſes the water, D for the denſity of the included air, and 1 for the denſity of the air without, F is as $D - 1$.

The height in feet to which the water riſes, ſetting aſide all impediments, is equal to the product ariſing from the multiplication of 33 into the exceſs of the denſity of the included air above that of the outward air; that is, putting H for the height of the jet, and x for 33, $H = xD - x$.

For as water which is driven out of a reſervoir by the preſſure of the incumbent water, if it ſpouts directly upward, riſes to the ſame height with the water in the reſervoir; ſo if it be driven by any other force, it muſt riſe to an equal height with a pillar of water whoſe preſſure is equal to that of the driving force; foraſmuch therefore as the atmoſphere makes an equal preſſure with a height of water of

33 feet,

33 feet, the water will be thrown to the height of 33 feet by the compressive force of the atmosphere ; wherefore if we put 1 for the pressure of the atmosphere, and say, as one is to 39 or x, so is D—1, which expresses that part of the pressure of the included air which drives out the water, to a fourth proportional, we shall have xD—x, or x × D—1, for the height to which the water is thrown ; whence it appears, that if D—1 be equal to unity, which is the case when the air within is as dense again as that without, the water will rise to the height of 33 feet ; and if D—1 be equal to 2, which is the case when the included air is thrice as dense as the external, the height of the jet will be 66 feet, and so on.

    The double fountain consists of two single fountains, whose bottoms are fastened to an hollow brass cylinder, one at each end, in the manner represent-
Pl. 7.
Fig. 10. ed in the figure, wherein AA and BB denote the two fountains with their basons ; CC the hollow cylinder, which plays upon the pins DD as upon an axle ; each has a pipe as P, whose lower end reaches nearly to the bottom of the fountain. From the bason of the fountain AA, there issues another pipe as T, which passing thro' AA, and likewise the hollow cylinder CC, without communicating with either, opens at E into the fountain BB. And in like manner, such another pipe issuing from the bason of BB, and passing thro' that fountain and the cylinder, opens into the fountain AA. The hollow cylinder being placed in an upright posture by means of the carriage which supports it, and the pipes of the lower fountain being stopped, water is conveyed into it thro' the pipe T, which issues from the bason of the upper fountain ; by the running in of the water, the air contained in the lower fountain is crowded into a smaller space, and thereby condensed ; if then both the pipes of the upper

fountain

fountain be ftopped, and the lower fountain be brought into the place of the upper, by turning the cylinder on its pins, the water which it contains will fall to its bottom, and the lower end of the pipe P will be immerfed therein, in the manner reprefent- ed in the upper fountain; fo that upon opening that pipe, the water will be driven thro' it by the expanfive force of the condenfed air; and as it falls into the bafon, it will be conveyed thence by the pipe T into the lower fountain; and when the up- per is exhaufted and ceafes to play, then ftopping its pipes, and changing the places of the fountains as before, the other may be fet a going in the fame manner.

LECTURE XVII.

OF SOUNDS.

IN this lecture I fhall firft explain to you the Lict. NATURE OF SOUNDS, and then treat of the XVII. VIBRATIONS OF MUSICAL STRINGS.

That SOUNDS have a neceffary dependance on the air, will appear from the following experi- ment.

Let a bell be placed under a receiver in fuch a Exp. 1. manner as that it may be rung at pleafure; and up- on drawing the air out of the receiver, the found of the bell will grow lefs and lefs audible in propor- tion to the degrees of exhauftion, fo as at laft al- moft to die away, and fcarcely to be heard at all; and upon re-admitting the air, the found will re- vive again, and increafe in proportion to the quan- tity of air that is taken in.

As this experiment proves the air to be neceffary to the production of founds, fo the tremblings which great guns, bells, drums, and many other founding bodies communicate, by means of the

interme-

intermediate air, to fuch bodies as are near them, plainly fhew, that founds depend on tremulous motions of the air; which therefore I fhall endeavour to explain to you, together with the caufe and manner of their production. When the parts of a bell, a mufical ftring, or any other elaftick body are fet in motion by a ftroke, they vibrate, that is, they go forward and return backward alternately thro' very fhort fpaces; in going forward they propel, and thereby comprefs and condenfe the air which lies next them; and in returning backward, they fuffer the compreffed air to recede and expand itfelf, fo that the parts of the air which are contiguous to the trembling body, go and return in the fame manner with the parts of the body; and as they are endued with a repulfive power, they muft by means thereof excite the fame vibrations in thofe parts which lie next beyond them; and thefe again muft in like manner agitate the parts beyond them, and fo on continually; fo that by one fingle vibration of an elaftick body, a motion is excited in the air, and propagated directly forward, by which fome parts go forward, whilft others return back, and that alternately, as far as the motion reaches.

That this motion may more readily be conceived, Pl. 7. let ST reprefent an elaftick ftring, ftretched and Fig. 11. made faft at both ends; and by a force applied to the middle point H, let it be drawn into the pofition SET; upon the removal of the force which inflects it, it will by virtue of its elafticity return to its former pofition SHT; and forafmuch as the reftitutive force acts conftantly upon it during the time of its motion from E to H, its motion thro' that fpace muft be continually accelerated, and the velocity thereof muft be greateft at H. When the ftring has recovered the pofition SHT, it will not remain therein; but by virtue of the velocity acquired in moving from E to H, it will be carried forward till it has moved thro' a fpace as HK, equal

to

to EH, and then its motion forward will ceafe ; for as it moves towards K, the elaftick force acts continually upon it in drawing it back ; and by fo doing, retards the motion from H to K, in the very fame manner that it accelerated the motion from E to H ; confequently, by the time that the ftring has moved from H to K, it will have loft all that velocity which it acquired in moving from E to H ; as foon as it ceafes to go forward, it will be brought back again from K to H by the force of elafticity, with an accelerated motion, in the very fame manner as it was at firft from E to H ; and when it has arrived at H, it will by virtue of the acquired velocity go on to E, with a retarded motion, in the fame manner as it did from H to K. The motion of the ftring from E to K and back again, is called a *vibration* ; and it is evident from what has been faid, that fetting afide all external impediments, a ftring which has made one vibration, muft continue to vibrate for ever thro' the fame fpace ; but, whereas it meets with continual refiftance from the air, the fpace thro' which it vibrates muft on that account grow lefs and lefs continually, and at length vanifh ; and yet, notwithftanding this variation in the fpace, the times of the vibrations are all equal, as I fhall demonftrate before the clofe of this lecture ; but I take notice of it in this place, becaufe one of the chief properties of the pulfes of the air, whereof I fhall have occafion to make mention prefently, has a neceffary dependance thereon.

When the ftring is drawn into the pofition SET, if we fuppofe A, B, C, and fo forth, to be particles of air placed in a right line one beyond another, and that the diftance of the firft particle from the ftring. at E, is equal to the interval of any two adjacent particles, as it muft needs be, on fuppofition that the particles of the air fly from other bodies with the fame force that they repel one another ; upon

letting

Lect. letting the string go, as it cannot move forward
XVII. without approaching to the particle A, it must at
the very next instant after it begins its motion, com-
pel that particle ; which for the same reason, must
in the next instant after it begins to move propel
the particle B, and that must in the same manner
propel C, and C propel D, and so on ; so that the
string, and the several particles of air taken in
their order, will begin to move forward successively
one after another, at very small intervals of time.
And whereas the string is accelerated in its motion
from E to H, and retarded in its motion from H
to K, the particle A must likewise be accelerated
in one half of its progress, and retarded in the
other; for since A is equally distant from the
string, and from B, before the vibration com-
mences, and since it begins to move forward a little
later than the string ; it is evident, that upon the
first motion of the string, the distance between that
and A, must become less than the distance between
A and B ; and forasmuch as the increments of velocity which are continually generated in the string
by the action of its elasticity, are not communicated to the particle A, in the instant of time wherein
they are generated, but a little later ; it is manifest, that the string during its motion from E to
H, must continually be nearer to A than A is to
B ; and consequently, must act more forcibly in
driving A forward, than B does in driving it backward, and by so doing accelerate its motion. After
the string has arrived at H the middle point of its
progress, and ceased to be accelerated, in the very
next moment A likewise reaches the middle point
of its progress, and ceases to be accelerated, being
driven as strongly backward by B, as it is forward
by the string. But however, by virtue of the acquired motion it continues to go forward, but with
a retarded motion ; and is at length stopped by the
repulsive

repulfive power of **B**; in the fame manner that the ftring in moving from H to K is retarded, and at laft ftopped by the action of its elaftick force. Af-ter the ftring has reached K, the utmoft limit of its progrefs, in the very next moment does A likewife reach the utmoft limit of its progrefs, and then turning back, purfues the ftring, which had like-wife turned back the moment before. And as the ftring is accelerated during its return from K to H, and retarded from H to E; fo the particle A, during the firft half of its return, being nearer to B than it is to the ftring, muft be accelerated; and during the latter half, being nearer to the ftring, is thereby retarded, and at length ftopped upon its arrival at the place from whence it fet out, which happens immediately after the ftring has returned to E; and there it continues at reft, unlefs by a fecond vibration of the ftring it be again driven forward in the fame manner as before. As this particle is made to go and return thro' a very fhort fpace, by the impulfe of the ftring, fo likewife are the feveral fucceeding particles, by the impulfes of the forego-ing; and as the ftring and the feveral particles taken in their order, begin their motions forward, fuccef-fively one after another at very fmall intervals of time, fo likewife do they begin to return in their order at the fame intervals of time; whence it fol-lows, that fome of them muft go forward, at the fame time that others return back. As the particles which go forward begin their motions fucceffively one after another, they muft neceffarily come nearer together; that is, they muft be condenfed. And it muft be obferved, that the condenfation goes for-ward continually; for in the very next inftant after any particle as D, has made its neareft approach to E, E muft make its neareft approach to F; and in the next inftant F muft make its neareft approach to G, and fo on continually; fo that the conden-

fation

sation must pass forward successively in
manner thro' the several particles of air.

But that I may explain this vibratory
the air more particularly, it must be observed, that
as the string during the first half of its progress from
E to H is continually accelerated, its distance from
the particle A, must constantly grow less; and forasmuch as during the latter half of its progress from
H to K, it is continually retarded, and that in the
same uniform manner that it was accelerated from
E to H, its distance from A must constantly be inlarged, and that in the same regular manner that
it was diminished during the progress of the string
from E to H; so that by the time it has arrived at
K, the utmost limit of its progress, it is just as far
distant from the particle A, as it was when it first
set out. Upon the return of the string, inasmuch
as it is continually accelerated from K to H, its
distance from the particle A must still be inlarged
and forasmuch at it is retarded in its motion from
H to E, in the very same manner as it was accelerated from K to H, its distance from A must constantly grow less in the same regular manner than it
was inlarged during its motion from K to H; so
that upon its return to E, it is again just as far distant from the particle A, as it was at its first setting out. From what has been said, it is evident,
that the string during the time of its progress is
always nearer to the particle A, than it was before
its motion began, and that its least distance from
the particle is at H, the middle point of its progress; it is likewise manifest, that during the time
of its return, it is always more distant from the particle than it was before its motion began; and that
its greatest distance from the particle is at H, the
middle point of its return. And what has been
thus shewn of the string with respect to the particle
A, is in like manner true of that particle with respect

spect to the particle B, and of B with respect to C, and so on of every particle, with respect to that which lies immediately beyond it, as far as the motion reaches; so that each particle with regard to that which lies immediately beyond it, is in a state of condensation during its progress, and of rarefaction during its return, its greatest condensation being at the midst of its progress, and its greatest rarefaction at the midst of its return. What proportion these rarefactions and condensations bear to the density of the air in its natural state, in every point of that small space thro' which a particle of air vibrates, shall be shewn in my next lecture, as also the law of this vibratory motion.

As the parts which go forward, do in their progressive motion strike such obstacles as they meet in their way, they are for that reason called *pulses*; and the sensations which are excited in the mind by the strokes of these pulses on the drum of the ear are called *sounds*; so that sounds as considered in their physical causes, are nothing else but the pulses of the air. In order therefore to explain the nature of sounds, I shall lay before you the chief properties of these pulses.

The first of which is, that they are propagated from the trembling body all around in a sphærical manner. For tho' the parts of the body, by whose vibrations the pulses are generated, do go and return according to certain directions, yet forasmuch as every impression which is made on a fluid is propagated every way throughout the fluid, whatever be the direction wherein it is made, the pulses must spread and dilate, so as to form themselves into concentrick sphærical surfaces, or rather thin shells, whose common center is the place of the sounding body. And hence appears the reason why one and the same sound may be heard by several persons, tho' differently situated with respect to the sounding body.

A second

A second property of the pulses is, that they grow less and less dense as they recede from the sounding body, and that in the same proportion with the squares of their distances from the body. For whatever be the force wherewith the sounding body acts on the first sphærical shell of air, with the very same force does that shell act upon the second, and that again upon the third, and so on continually; so that the force which condenses the air in the several shells is given; consequently, the condensations which it produces in those shells must be inversly as the resistances it meets with; but the resistances are as the shells; and therefore, since those increase continually in the same proportion with the squares of their distances from the center, their densities must decrease in the same manner.

By reason of this diminution in the densities of the pulses, those which are farther removed from the sounding body, make slighter impressions on the drum of the ear, than those which are less distant; and hence it is, that sounds grow less and less audible, the farther they go from the sounding body; and at certain distances become so weak as not to be heard at all.

A third property of the pulses is, that all of them, whether denser or rarer, move equally swift, so as to be carried thro' equal spaces in equal times, as I shall demonstrate in my next lecture.

From this property it follows, that all sounds, whether they be loud or low, grave or acute, move equally swift, the softest whisper making equal speed with the noise of a cannon, or the loudest thunderclap; and it has been found by experiment, and I shall likewise demonstrate in my next lecture, that sounds move at the rate of 1142 feet in a second of time or thereabouts; for the velocity is not precisely the same in all seasons of the year, but is somewhat greater in *summer* than in *winter*, on account of the heat which renders the air more elastick
tick

tick in proportion to its denfity, than it is in the
cold *winter* feafon.

2. A fourth property of the pulfes is, that all thofe which are excited by the vibrations of one and the fame body, are at equal diftances from one another. For fince each pulfe is excited by one fingle vibration of the founding body, and fince all the pulfes move with equal and uniform velocities, it is manifeft, that they muft fucceed one another at diftances proportional to the times of the vibrations; but the times of the vibrations of one and the fame body are all equal; confequently, the intervals of the pulfes are fo too. And it muft be obferved, that the interval between two pulfes, which is by fome called the *length*, and by others the *breadth* of a pulfe, is that fpace thro' which the motion of the air is carried, during the time wherein any one particle performs its vibratory motion in going forward and returning back.

On the intervals of the pulfes depend the tones of founds; and here I muft obferve to you, that all the variety there is in founds, refpects either their *ftrength* or their *tone*; with regard to their ftrength, they are diftinguifhed into *loud* and *low*; and with refpect to their tone, into *grave* and *acute*, otherwife called *flat* and *fharp*. The ftrength of any found depends on the magnitude of the ftroke, which is made by a pulfe on the drum of the ear; the greater the ftroke is, the louder is the found which it excites, and the weaker the ftroke, the lower the found; and whereas all the pulfes move with equal velocities, the magnitude of the ftroke, and confequently the ftrength of the found, muft be as the quantity of matter in the pulfe; that is, as a rectangle under the denfity and breadth of the pulfe; and fuppofing the breadth of the pulfe to be given, it muft be as the denfity.

The tone of a found depends on the duration of a ftroke; the longer a ftroke is which a pulfe makes

on

on the drum of the ear, the more grave is the
found which it produces; and the shorter the strok
the more acute is the found; but since all the pul
move equally swift, the duration of a stroke m
be proportional to the interval between two suc
five pulses; and of consequence, a sound is more
less grave or acute in proportion to the length
that interval. Hence it follows, that all the sou
from the loudest to the lowest, which are excited
the vibrations of one and the same body, are of o
tone. It likewise follows, that all those soundi
bodies, whose parts perform their vibrations in eq
times, have the same tone; as also, that those
dies which vibrate slowest, have the gravest
deepest tone; and on the contrary, those which
brate quickest have the sharpest or shrillest tone.

As there may be an infinite variety in the tin
wherein sounding bodies perform their vibratic
so may there likewise be in the tones of the sou
which depend thereon; and yet amidst this gr
variety, musicians acknowledge but seven princi
notes in an *octave*; for tho' the eighth be requi
to compleat the seven intervals in an octave,
are there in truth but seven notes; for that whic
called the *eighth*, becomes the base or ground n
in the next octave ascending; and as it stands
the limits of the two octaves, it is called the *eig*
with respect to the base note below it, and
ground or base note with respect to the 15th wh
is above it; which 15th is likewise the base in
next ascending octave; for by a repetition of no
wherein the proportions of the times of the note
the first octave are preserved, the octaves may
continued on both ways, ascending and descendi
and that in *infinitum*; and yet, notwithstanding t
infinite progression in the octaves, the number
harmonick sounds is limited. Mr. SAUVEUR is
opinion, that all the harmonick sounds, that is, f
sounds as can be heard distinctly and with pleas

end in whose tones a difference can be clearly perceived by the ear, lie within the compass of ten octaves; as also, that all sounds whatever, from the lowest harmonick found, to the highest that the human ear can well bear, are contained within the limits of two octaves more. And if this be the case, it follows, that that body which gives the shrillest sound that the ear can bear, makes 4096 vibrations in the same time that one vibration is performed by that body which gives the gravest harmonick sound; for since in every octave, the time of the eighth is $\frac{1}{2}$ of the time of the base note, if $\frac{1}{2}$ be raised up to the 12th power, it will exhibit the time of the shrillest sound, that of the gravest being unity; but the 12th power of $\frac{1}{2}$ is the 4069 part of an unite; consequently, the time of the shrillest sound that the ear can well bear, and likewise of the vibration which produces it, is to the time of the gravest harmonick sound, and of the vibration whereby it is produced, as 1 to 4096; but the times of the vibrations of two bodies are inversly, as the numbers of vibrations which they perform in a given time; consequently, the body which gives the shrillest sound performs 4096 vibrations in the same time that the body which gives the gravest harmonick sound performs one; and forasmuch as Mr. SAUVEUR has found by some experiments which he made on organ pipes, of which I shall give you an account in my next lecture, that a body which gives the gravest harmonick sound, vibrates twelve times and an half in a second, the shrillest sounding body must perform 51100 vibrations in the same time; which argues great swiftness in the vibrating parts; and yet, great as it is, it has nothing extraordinary or surprising in it, if compared with the velocity of some other motions; for if we suppose the parts in each vibration to run thro' a space equal to the 10th part of an inch, tho' it is highly probable, that the lengths they run are much shorter;

and

and if we suppose them to move with the same velocity during the whole time of their motion; it follows, that they are carried at the rate of 425 feet and 10 inches in a second; consequently, they do not move with much more than two third parts of the velocity wherewith a ball flies from the mouth of a cannon.

The fifth and last property of the pulses is, that they may be propagated together in great numbers from different bodies, without disturbance or confusion; as is evident from conforts, wherein the founds of the several inftruments are conveyed diftinctly to the ears of the audience; as they move along, some of them coincide and ftrike the drum of the ear at one and the fame time, and thereby excite a fmooth regular motion, that is pleafing and agreeable; whilft others which do not mix and unite, at leaft not frequently, ftrike the ear at different inftants of time, and thereby difturb each other's motion, fo as to render them harfh, grating, and offenfive. And hereon depends almoft the whole of concords and difcords in mufick; fuch founds, generally fpeaking, being deemed concords, as are excited by pulfes which have frequent coincidences; and on the other hand, fuch founds being called *difcords*, as arife from pulfes which coincide but rarely.

The frequency or infrequency of the coincidences, depends on the proportions which the intervals of the pulfes bear one to another; as I fhall fhew you in relation to the feveral notes in an octave; in doing of which, inftead of the pulfes and their intervals, I fhall confider the vibrations of the bodies which excite the pulfes, and the times of thofe vibrations; becaufe the number of pulfes is always equal to the number of vibrations in the founding bodies, and the intervals of the pulfes proportional to the times of the vibrations.

If two vibrating bodies begin their motions together, and vibrate in equal times, it is manifeft, that

that their vibrations muſt keep pace together, and conſtantly coincide. But if the vibrations be performed in unequal times, it is plain, that they cannot conſtantly keep pace together; for which reaſon ſome of them only will coincide; and which thoſe are may be determined from the times of the vibrations; for ſince the numbers of vibrations, which are performed in a given time, are inverſly as the times of the vibrations, if the numbers which expreſs the times of the vibrations of two bodies be taken reciprocally, they will exhibit the coincident vibrations of the reſpective bodies. For inſtance, if the time of the vibrations of one body, be to the time of the vibrations of another, as 8 to 9, which is the caſe of two bodies, whereof one ſounds a ſecond or tone major to the other, every ninth vibration of the former coincides with every eighth of the latter. So again, if the times of the vibrations be to one another, as 5 to 6, which is the caſe, where one body ſounds a leſſer third to the other, every ſixth vibration of the former falls in with every fifth of the latter.

In this ſcheme, I have ſet down thoſe fractional numbers which expreſs the proportions that the times of the vibrations of thoſe bodies, which ſound the ſeveral notes in an octave, bear to the time of the vibration of that body which ſounds the baſe note; by the help of which numbers the coincident vibrations may be readily diſcovered.

$\frac{1}{2}$ *Eight.*
$\frac{8}{15}$ *Greater ſeventh.*
$\frac{5}{9}$ *Leſſer ſeventh.*
$\frac{3}{5}$ *Greater ſixth.*
$\frac{5}{8}$ *Leſſer ſixth.*
$\frac{2}{3}$ *Fifth.*
$\frac{3}{4}$ *Fourth.*
$\frac{4}{5}$ *Greater third.*
$\frac{5}{6}$ *Leſſer third.*
$\frac{8}{9}$ *Second or tone major.*
1 *Baſe note.*

For in each fraction, the denominator exhibits the coinciding vibration of that body which ſounds the note, and the numerator the coinciding vibration of the body which ſounds the baſe note.

S 4

Having

LECT. Having thus explained the nature and properties
XVII. of found, I come now to give you an account of
the VIBRATIONS of MUSICAL STRINGS, and to
fhew you in what proportions the times of the vibrations are varied, by varying the length, thicknefs, or tenfion of the ftrings; and in order thereto, I fhall lay down the following PROPOSITION.

Pl. 7.   *Let an elaftick ftring as* AB, *faftened at* A, *and*
Fig. 13. *paffing over a fmall pin or pulley at* B, *be ftretched by*
an appending weight as P, *(which I fhall call the tend-*
ing force); and by a force applied at the middle point
C, *(which I fhall call the* inflecting force*), let it be*
drawn into the pofition ADB; *if the diftance between*
C *and* D *be exceedingly fmall in proportion to the length*
of the ftring, or, to fpeak in the mathematical phrafe,
if CD *be a nafcent quantity, the inflecting force will*
be meafured by a rectangle under the fpace CD, *and*
the tending force applied to the length of the ftring.
For fince the tending force acts upon the ftring in
the direction DB, it may be denoted by that line,
and being fo denoted, it may be refolved into two
forces, whereof one acts in pulling the ftring horizontally in the direction CB, and is therefore to be
expreffed by CB; whilft the other acts in drawing
the ftring perpendicularly upward from D towards
C, and is therefore to be expreffed by the line DC;
fo that that portion of the tending force which acts
in moving the ftring upward, is to the whole force,
as DC to DB; or, becaufe D and C are fuppofed
to be indefinitely near, as DC to CB; but the
force which acts in drawing the ftring upward, is
equal to the inflecting force, becaufe they balance
each other; confequently, the inflecting force is to
the tending force, as CD to CB; and turning this
analogy into an equation, by multiplying the extreams and means, and then dividing by CB, we
fhall have the inflecting force equal to a rectangle
under the tending force, and the line CD, applied
to

to half the length of the ſtring; and therefore, foraſmuch as whole quantities are in proportion as their halves, the inflecting force will be as a rectangle under the tending force and the line CD, applied to the length of the ſtring; ſo that putting F for the inflecting force, P for the tending force, S for the line CD, and L for the length of the ſtring, F is as $\frac{SP}{L}$. Hence it follows, that if P and L, that is, if the tending force and length of the ſtring be given, the inflecting force is as the line CD, as will appear from the following experiment.

Let a ſmall braſs wire three feet long, faſtened Exp. 2. at one end, and paſſing over a pin ſo as that when ſtretched it may be in an horizontal poſition, be tended by a weight of three pounds; and let half an ounce, and an ounce, be appended ſucceſſively to the middle of the wire; in the former caſe, the point of ſuſpenſion will be drawn down $\frac{1}{10}$th parts of an inch, and in the latter $\frac{1}{10}$.

Since the force which inflects a ſtring of a given length, and tended by a given force, is as the ſpace CD, thro' which the ſtring is bent; the force wherewith the ſtring reſtores itſelf, muſt likewiſe be as CD, becauſe the reſtitutive force is in all caſes equal to the inflecting force; conſequently, the point D is carried towards C, by a force that varies with the diſtance; and therefore, whatever be the diſtance at which it begins its motion, the time wherein it arrives at C will ſtill be the ſame; as I proved in my lecture on the pendulum. Whence it follows, that the vibrations of one and the ſame ſtring, whether they be through larger or ſmaller ſpaces, are all performed in equal times.

If L and S be given, F is as P; that is, if the length of the ſtring, and the ſpace thro' which it is bent be given, the inflecting force is as the tending force; or, in other words, one and the ſame ſtring, being

LECT.
XVII.
being tended by different forces will upon the inflexion be drawn down equal fpaces by inflecting forces, which are to one another in the fame proportion with the tending forces.

Exp. 3. Let the fame wire as before be tended by a weight of fix pounds, and it will require one ounce to draw it down $\frac{4}{10}$ths of an inch, and two ounces to draw it down $\frac{8}{10}$ths; whereas, when it was tended by a weight of three pounds only, it was drawn down the fame fpaces by half an ounce, and an ounce.

If P and S be given, F is as $\frac{1}{L}$; that is if the force which tends the ftring, and the fpace thro' which it is bent be given, the inflecting force is inverfly as the length of the ftring; or, in other words, if ftrings of different lengths be tended by equal forces, they will be drawn thro' equal fpaces by inflecting forces, which are to one another inverfly as the lengths of the ftrings.

Exp. 4. Let a fmall brafs wire a foot and an half long, be tended by a weight of three pounds, and it will require an ounce to bend it down $\frac{4}{10}$ths of an inch, whereas half an ounce was fufficient to give the fame bent to the wire which was of a double length, and under the fame tenfion.

The time of a vibration of an elaftick ftring is meafured by a rectangle, under the length and diameter of the ftring, applied to the fquare root of the tending force. For if, as in the cafe of gravity, we fuppofe the force wherewith the inflected ftring reftores itfelf to act uniformly, as we fafely may, becaufe the fpace thro' which it acts is exceedingly fmall; then the motion generated will be as a rectangle under the force and the time of its acting; fo that putting M for the motion, F for the reftitutive force, and T for the time of its acting, M is as

FT;

FT; but the motion is as the quantity of matter moved into the velocity wherewith it moves; and in this cafe, the quantity of matter is as a product under the length of the ftring, and the fquare of its diameter; wherefore, putting D, L, and V, to denote the diameter, length, and velocity, FT is as D^2LV; and dividing both fides by F, T is as $\frac{D^2LV}{F}$; but the reftitutive force of the ftring being equal to the force which inflects it, and that having been proved to be as $\frac{SP}{L}$, wherein S denotes the fpace thro' which the ftring is bent, P the tending force, and L the length of the ftring; if inftead of F we fubftitute $\frac{SP}{L}$, T will be as $\frac{D^2L^2V}{PS}$; but the velocity applied to the fpace is inverfly as the time, that is, $\frac{V}{S}$ is as $\frac{1}{T}$; and therefore, inftead of that, fubftituting this, and multiplying both fides by T, we fhall have T^2, as $\frac{D^2L^2}{P}$; and therefore extracting the root, T is as $\frac{DL}{P^{\frac{1}{2}}}$;

that is, the time of a vibration, is as a rectangle under the diameter and length of the ftring, applied to the fquare root of the tending force.

Hence it follows, that if D and P be given, T is as L; that is, if the diameter of the ftring and the tending force be given, the time of the vibrations varies with the length of the ftring; as is manifeft from the divifion of the *monochord*, wherein the parts of the chord which found the feveral notes in an octave, have the fame proportions to the whole chord, that the times of the refpective notes have to the time of the bafe note; as for inftance, one half of the chord founds an octave to the whole, whofe time is one half of the time of the bafe note; and

Exp. 5.

and $\frac{2}{3}$ of the chord found a fifth, the time whereof is $\frac{2}{3}$ of the base time of the note, and so of all the rest.

Exp. 6. If P and L be given, then T is as D; that is, if the tending force and length of the string be given, the time of the vibration is as the diameter of the string; as will appear, if two wires of equal lengths be tended by equal weights, the diameter of one being the 90th part of an inch, and that of the other the 45th part; for the former will found an octave to the latter.

Exp. 7. If D and L be given, then T is inversly as the square root of P; that is, if the diameter and length of the string be given, the time of the vibration is inversly as the square root of the tending force; as will appear, if eleven wires equal as to length and thickness, be tended by weights, whose square roots are to one another inversly as the times of the notes in an octave; for the wires so tended will found the respective notes.

Eighth	—— ——	$\frac{1}{2}$ —	240
Greater seventh	——	$\frac{8}{15}$ —	$210\frac{15}{16}$
Lesser seventh	——	$\frac{9}{16}$ —	$194\frac{2}{3}$
Greater sixth	— —	$\frac{3}{5}$ —	$166\frac{2}{3}$
Lesser sixth	——	$\frac{5}{8}$ —	$153\frac{3}{5}$
Fifth	—— ——	$\frac{2}{3}$ —	135
Fourth	—— ——	$\frac{3}{4}$ —	$106\frac{2}{3}$
Greater third	—— —	$\frac{4}{5}$ —	$93\frac{3}{4}$
Lesser third	— ——	$\frac{5}{6}$ —	$86\frac{2}{5}$
Tone major, or second	— ——	$\frac{8}{9}$ —	$75\frac{15}{16}$
Base note	—— ——	1 —	60

In the left hand column of this table, the numbers express the times of the several notes; and the numbers in the right hand column, express the weights in ounces, whereby the wires which found the respective notes are tended; the square roots of
which

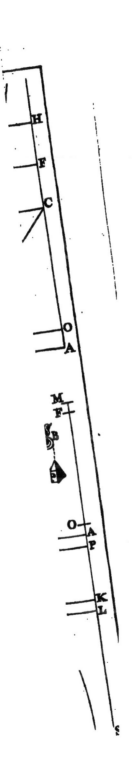

which weights, are to one another inverſly as the
imes of the reſpective notes; as for inſtance, the
weight which tends the ſtring that ſounds the octave,
is to the weight whereby the ſtring that ſounds the
baſe note is tended, as 4 to 1, whoſe ſquare roots
are as 2 to 1, that is, inverſly as the time of the
octave, to the time of the baſe note; and ſo of all
the reſt.

LECTURE XVIII.

Of the Motion of Sound.

IN my laſt lecture, wherein I treated of that motion of the air, which is productive of ſounds, I
ſhewed you, that each particle of air in going forward and returning back, is twice accelerated, and
as often retarded; but I did not then enquire into
the law of that acceleration and retardation. I likewiſe told you, that all the pulſes of the air move
equally ſwift, the demonſtration of which, I pro-
miſed to give you in this lecture.

Now Sir ISAAC NEWTON, having in a moſt
elegant manner, in the 47th *Propoſition* of the *Se-*
cond Book of his *Principles*, demonſtrated that each
particle of air, during its vibratory motion, is accelerated and retarded, in the very ſame manner as
a pendulum vibrating in a cycloid; and having likewiſe, in the 49th and 50th *Propoſitions* of the ſame
book, determined the velocity of ſound, I ſhall in
this lecture lay before you what he has ſaid, in relation both to the one and the other, in the cleareſt
light that I am able.

As to the firſt, let the line AB denote the length
of a pulſe; or that ſpace thro' which the motion of
the air is propagated, during the time that a particle performs its vibration, by going forward and
returning

Lect.
XVIII.

Pl. 8.
Fig. 1.

returning back; and let E, F and G, be three particles, or phyfical points of air fituated in the right
line at equal diftances, and at reft; and let EQ,
FR, and GT, be three equal, but exceedingly fhort
fpaces, thro' which thefe particles go and return in
their vibrations; which fpaces tho' they be here
taken of fome length, to avoid confufion in the
fcheme, are in reality fo exceedingly fmall, as to
bear no proportion to AB, the length of a pulfe.
Let x, y, and z denote any intermediate points,
in which the particles are found during their motion
forward or backward. Let EF, and FG be fmall
phyfical lines, or little portions of air, fituated in
ftrait lines between thofe phyfical points; which

Pl. 8.
Fig. 2.
Fig. 1.
lines are fucceffively moved into the places xy, yz,
and QR, RT. Let the right line PS, be drawn
equal to EQ, and on that line as a diameter, let
the circle SIPi be defcribed; and let the circumference of that circle denote the time of the vibration
of a particle, and the parts of the circumference,
the proportional parts of the time; fo as that after
any time as PH, or PHSh, if right lines as HL
and hl be drawn from the points H and h perpendicular to SP, and Ex be taken equal to PL, or

Pl. 8.
Fig. 1.
Pl, the particle E may be found at x. By this
means the particle or phyfical point E, in moving
forward thro' x to Q, and thence back again thro'
x to E, will be accelerated and retarded, in the
fame manner with a pendulum vibrating in a cycloid; inafmuch as in my lecture on the pendulum,
I fhewed you, that the fpaces defcribed by fuch a
pendulum, and the times of defcribing thofe fpaces,
are (as we have now fuppofed them to be in the
cafe of the air's motion) as the verfed fines and
arches of a circle, whofe diameter is equal in length
to the whole cycloid.

Now, in order to prove that the feveral little portions of air are agitated in the forementioned manner

by

by their elasticity, which in this case is the true
moving cause, let us suppose them to be so moved
by some cause or other, be that cause what it will;
and their elasticity will be found to be such in every point of their progress and return, as must of
necessity produce in them the same degrees of acceleration and retardation, that gravity does in a
pendulum vibrating in a cycloid.

In the circumference of the circle, let the equal
arches HI and IK, or hi and ik, be taken, bear-
ing the same proportion to the whole circumference,
that the little right lines EF and FG do to AB the
length of a pulse; and drawing the lines IM and
KN, or im and kn perpendicular to PS, inasmuch
as the points or particles E, F and G, are moved
in the same manner successively one after another,
the motion beginning with E, and each of them
performs its intire vibration, in going forward and
returning back, in the same time that the motion
is propagated thro' a space equal to AB, the length
of a pulse; if PH or PHSh denotes the time from
the beginning of E's motion, PI, or PHSi, will
denote the time from the beginning of F's motion;
and in line manner PK, or PHSk, will denote
the time from the beginning of G's motion. And
if the points E, F and G be found at x, y and z;
the lines Ex, Fy, and Gz, in the first figure, will
be respectively equal in the second, to PL, PM,
and PN, in the progress of the points; and in
their return, equal to Pl, Pm, Pn, those being the
versed sines of the arches which denote the times.
Whence it follows, that xz, which is equal to the
difference between Ex, and the sum of EG and Gz,
is in the progress of the points, equal to EG —
LN, and to EG + ln in the return; but xz is as
the expansion of the little portion of air EG, when
it is in the place xz; consequently, that expansion
is to the mean ordinary expansion, or that expan-

Pl. 8.
Fig. 1.
Fig. 2.

Fig. 1.

4   sion

fion which it has when at reft before it is
its vibratory motion, as EG—LN, to EG
that portion of air in going forward is fi
the place xz; and it is as EG + ln; or,
LN and ln are equal, as EG + LN, to EG
the portion of air in returning back, is fo

Pl. 8.
Fig. 2.

the fame place. Let now ID be drawn fr
point I, perpendicular to HL, and the nuff
angle HID, will be fimilar to the triangle
becaufe the angles at D and M are right on
the angles at I are equal, as being each of th
complement of one and the fame angle DI(
right one; confequently, DI, or its equal
to HI, as IM to the *radius* OI, equal to O
double LM equal to LN, is to double HI a
HK, as IM to OP; and by the conftructic
is to EG, as the circumference of the circle
or putting R for the *radius* of a circle, wh
cumference is equal to AB, as OP to R;
reducing thefe two analogies into equations,
have $\frac{LN}{HK} = \frac{IM}{OP}$, and $\frac{HK}{EG} = \frac{OP}{R}$; whe
multiplying thefe equations together, we fha
$\frac{LN}{EG} = \frac{IM}{R}$; and refolving this into an a
we fhall have LN : EG :: IM : R;
courfe, by fubftituting IM and R, in the pl
LN and EG, the expanfion of the fmall
of air EG, or of the phyfical point F, when
place xz or y, is to its mean ordinary exp
as R—IM to R, in its going forward,
R + im to R, in its returning; and forafm
its elafticity is inverfly as its expanfion, i
ticity, when at the point y, is to its of
elafticity, as $\frac{1}{R-IM}$ to $\frac{1}{R}$ in its progrefs,
its regrefs in the fame point, as $\frac{1}{R+im}$

and by the same way of arguing, the elastick forces of the physical points E and G, when in going forward they are found at x and z, will be to their ordinary elasticity, as $\frac{1}{R-HL}$, and $\frac{1}{R-KN}$ to $\frac{1}{R}$; and by subducting the latter of these quantities from the former, the difference of those forces will be as $$\frac{HL-KN}{R^2-R+HL-R+KN+HL \times KN} \text{ to } \frac{1}{R};$$ or, rejecting all the terms of the divisor except the first, as being indefinitely small with respect to that, as $\frac{HL-KN}{R^2}$ to $\frac{1}{R}$; or, multiplying both sides by R^2, as HL—KN to R; but forasmuch as R is a given quantity, HL—KN is as unity; consequently, the difference of the forces is as HL—KN. But from the similarity of triangles, HL—KN is to HK, as OM to OI or OP; consequently, since HK and OP are given, HL—KN is as OM; or because SP and EQ are equal, if EQ be bisected in C, as cy. And by the same way of reasoning, the difference of the elastick forces of the same points, when in their return they are found at x and z, is as the same cy; but that difference, or the excess of the elastick force of the point x above the elastick force of the point z, is the force by which the little line or portion of air xz, which lies between those points is accelerated in its progress; and on the other hand, the excess of the elastick force of the point z above that of the point x, is the force by which the same little line or portion of air is accelerated in its return; so that the force by which portion is accelerated, is every where as its distance from C, the middle point of its vibration; consequently, during its vibratory motion, it must be accelerated and retarded in the same manner with a pendulum vibrating in a cycloid; inasmuch as I proved in my lecture on the pendulum, that the

T

force

force which agitates the pendulum in the fo
tioned manner, is every where as its difhanc
the middle or lowest point of the vibration,
what has been thus proved of the little portio
is in like manner demonstrable of every othe
portion of air, thro' which the motion is
gated.

As to the velocity of sound, or what amou
the same thing, of the pulses of the air, if a pen
be made equal in length to the height of an
geneal atmosphere, whose weight is equal t
of our atmosphere, and its density the sam
that of the air at the surface of the earth;
height is, as I shewed you in a former lecture,
to 29725 feet, and which I shall now den
the letter H; in the same time that such a
lum performs an intire vibration by going f
and returning back, a pulse of the air will
thro' a space equal to the circumference of

Pl. 8.
Fig. 2.
described with the *radius* H. For if the littl
tion of air EG, vibrating thro' a small sp
PS, be acted upon at P and S, the extremi
the space thro' which it vibrates by an elasticl
equal to its gravity, it will perform its vib
in the same time that it would in a cycloid
length is equal to PS; because equal forces n
necessity move equal bodies thro' equal space
qual times. Since then, the times of vibratio
in the subduplicate *ratio* of the lengths of th
dulums, and the length of any pendulum is e
half of the cycloid, wherein it vibrates; the t
which the small portion of air would vibrate
force of its gravity in a cycloid equal in len
PS, must be to the time of the vibration of
dulum whose length is H, in the subduplicate
PO to H. But the elastick force which ac
the little portion of air in the extream poi
S, was proved to be to its whole or ordin

5

force, as HL—KN to R; that is, in the cafe before us, where the point K coincides' with P, as
HK to R; for upon the coincidence of K and P,
KN vanifhes, and HL, which in this cafe is their
difference, becomes the fine of HK, and equal to
it, inafmuch as HK is a nafcent arch. And the
whole elaftick force of that little portion of air, or,
which is the fame thing, the weight which compreffes it, is to its own weight, as the height of the
homogeneal atmofphere or H, to the fmall line
EG; whence putting e to denote the elaftick force,
which agitates the fmall portion of air in the extream
points of its vibration P and S, and w for its weight,
W for the whole elaftick force, or the weight
of the compreffing atmofphere, and reducing the
two laft analogies into equations, we fhall have
$\frac{HK}{R} = \frac{e}{W}$, and $\frac{W}{w} = \frac{H}{EG}$; whence multiplying
the two middle terms together, and likewife the
extreams, we fhall have $\frac{e}{w} = \frac{HK \times H}{R \times EG}$; and by
fubftituting PO and R for HK and EG, to which
they are proportional, $\frac{e}{w}$ is equal to $\frac{PO \times H}{R^2}$; that
is, by refolving this equation into an analogy, the
elaftick force which agitates the little portion of air
in the extream points of the fpace thro' which it
vibrates, is to its weight, as PO × H to R²; fince
then, from the nature of motion, the times wherein equal bodies are moved thro' equal fpaces, are
reciprocally in the fubduplicate *ratio* of the moving
forces, it follows, that the time wherein the little
portion of air performs its vibration by virtue of
the elaftick force denoted by e, is to the time
wherein it can vibrate thro' an equal fpace by the
force of its gravity, in the fubduplicate *ratio* of R²
to PO × H, and of courfe, to the time of the vibration of a pendulum whofe length is H, in a *ratio*

T 2 com-

compounded of the laſt mentioned ratio, and o
ſubduplicate ratio of PO to H ; that is, as R²
to H¹× PO; that is, by dividing by PO,
ing the ſquare roots in the ſimple ratio of R t
But in the time that the little portion of air
forms one vibration by going forward and re
ing back, the pulſe is carried thro' a ſpace eq
AB ; conſequently, the time in which a pulſe
from A to B, is to the time in which a pend
whoſe length is H, ſwings forward and back
as R to H, or as BC, the circumference of
cle whoſe *radius* is R, to the circumference
circle whoſe *radius* is H ; but the time of the p
motion from A to B, is to the time in whi
moves thro' a ſpace equal to the circumferen
a circle whoſe *radius* is H, in the ſame propor
wherefore, in the ſame time that a pendulum
length is H, ſwings forward and backward, a
will move thro' a ſpace equal to the circumfe
of a circle whoſe *radius* is H, which was the
to be proved.

As a *Corollàry* it follows, that the pulſes
with ſuch a velocity as a heavy body acqui
falling down half the eight denoted by H ;
the ſame time with the fall they will, with a
city equal to that acquired by the fall, deſcr
ſpace double that of the fall, that is, a ſpace
to H ; and of conſequence, in the time tha
pendulum vibrates forward and backward,
will run thro' a ſpace equal to the circumferen
a circle whoſe *radius* is H.   For, in my lectu
the pendulum, I ſhewed you, that the time o
fall thro' half the length of the pendulum, is t
time of one vibration, as the diameter of a
to its circumference ; and of courſe, to the ti
a double vibration, as the *radius* to the circu
rence.   Since then it has been proved, tha
pulſes move with ſuch a velocity as carries
thro' a ſpace equal to the circumference of a

whofe *radius* is H, in the fame time that a pendulum whofe length is H, performs a double fwing; and fince it appears that the velocity acquired by a heavy body in falling down half the height H, will carry the pulfes thro' the fame fpace in the fame time, it is manifeft, that they move with that velocity.

As a fecond *Corollary* it follows, that the velocity of the pulfes is in a *ratio* compounded of the fubduplicate *ratio* of the air's elafticity directly, and of the fubduplicate *ratio* of its denfity inverfly; for fince the velocity wherewith they move, is fuch as a body acquires in falling down half the height H, and fince the velocities acquired by falling bodies, are in the fubduplicate *ratios* of the heights from which they fall, it is manifeft, that the velocity of the pulfes is as the fquare root of H, but the height H, is directly as the air's elafticity, and inverfly as its denfity; confequently, the velocity of the pulfes is in the fubduplicate *ratio* of the air's elafticity directly, and the fubduplicate *ratio* of its denfity inverfly. Whence it appears, that the velocity of the pulfes is given, forafmuch as, *cæteris paribus*, the elafticity is as the denfity. In the *winter* time indeed, the motion of the pulfes is fomewhat flower than in *fummer*, becaufe the coldnefs of that feafon does in fome meafure weaken the elafticity, and at the fame time increafe the denfity. From what has been faid, the fpace thro' which found moves in any given time, may readily be determined; for fince it is known by experience, that a pendulum $39\frac{1}{5}$ inches long, performs a double vibration by going forward and returning back in two feconds of time, a pendulum whofe length is H, that is 29725 feet long, will perform a like double vibration in $190\frac{3}{4}$ feconds; confequently, in that time found will move thro' a fpace equal to the circumference of a circle whofe *radius* is 29725 feet; that is, it will move thro' 186768 feet, which being di-

T 3

divided by 190½, gives a quotient of 979 feet, for the space thro' which sound moves in one second of time. But it must be observed, that in this computation no regard has been had to the thickness of the solid particles of air, thro' which sound is propagated in an instant; if that therefore be allowed for, the velocity of sound will come out greater in the proportion of about ten to nine; for since the specifick gravity of air is to that of water, as 1 to 870, if we suppose the particles of air to be equally dense with those of water, and that the greater rarity of air is owing to the greater intervals between its particles, it follows, that that interval is about nine times as great as the diameter of a particle; consequently, a tenth part of the space thro' which sound is propagated is possessed by the particles of air; if therefore to 979 feet, which is the space thro' which sound would move in a second, in case the particles of air had no magnitude, we add a ninth part, or 109 feet more on account of the thickness of the particles, we shall have 1088 feet for the space thro' which sound is carried in a second of time. Besides, as there are vapours dispersed thro' the air, which being of a different tone and elasticity, do not partake of that motion of the true air by virtue whereof sound is propagated, the moving cause having on that account fewer particles of matter to agitate, must of necessity give them a greater velocity.; and from the nature of motion it is evident, that the velocity will be greater in the inverse subduplicate *ratio* of the quantity of matter to be moved; that is to say, if we suppose the atmosphere to consist of ten parts of true air, and one part of vapours, the motion of sound will be quicker in such an atmosphere, than in an atmosphere consisting intirely of true air, in the subduplicate *ratio* of 11 to 10, or in the simple *ratio* of about 21 to 20. If therefore the velocity last found be augmented in that proportion, we shall have

1142

1142 feet for the space thro' which sound moves in one second of time; and this agrees with the most accurate experiments that have been made, for discovering the velocity of sound.

The space thro' which sound moves in a second of time being thus discovered, the length of the pulses excited by the vibrations of a sounding body may likewise be found, provided the number of vibrations performed by the sounding body in a given time, can by any method be determined; for since each vibration excites a new pulse, all that is requisite to be done, is to divide 1142 by the number of vibrations which the sounding body performs in a second, and the quotient will express the length of a pulse in feet. Now, the number of vibrations which a sounding body performs in a given time, has been determined by Mr. SAUVEUR, in the following manner; " Musicians having frequently " observed, that if two organ pipes which are near- " ly unisons, be made to sound together, there are " certain instants of time, and those, as well as " they can be judged of by the ear, at equal inter- " vals, wherein their joint sound is stronger, than " in the intermediate times." This, Mr. SAUVEUR with great appearance of reason, thinks is owing to the coincidence of their vibrations at those instants ; for when by the coincidence of their vibrations, they strike the ear at one and the same instant, they must needs make a stronger impression upon it, than when they strike it separately one after another. Taking this for granted, he by the help of a pendulum, took the time between two successive coincidences in the vibrations of two pipes of considerable lengths, and nearly of the same tone; he made choice of long pipes, because the coincidences of their vibrations are rarer, and consequently, the intervals between the coincidences are more easily measured, in long pipes than in short ones. Having thus found the time which passed between two

T 4

suc-

fucceffive coincidences, he readily found the n
ber of vibrations performed by each pipe in
fame time, they being inverfly as the numbers
preffing the proportion of the tones of the pipes
for inftance, if the time between two fucceffive
incidences was found to be the fixth part of
cond, and the numbers which expreffed the pro
tion of the tones of the pipes were 45 and 46,
longer pipe performed 45 vibrations, and the
ter 46, in the fixth part of a fecond. From
experiments he found, that a pipe, whofe let
was about five *Parifian* feet, had the fame
with a ftring that vibrates an hundred times in
cond; confequently, of the pulfes excited by
founding of fuch a pipe, there are about one t
dred in the fpace of 1142 *Englifh*, or 1070 *Par*
feet; and of courfe, the length of one pulfe is al
10 *Parifian* feet and $\frac{7}{10}$ths, that is about twice
length of the pipe; whence it is probable, that
lengths of the pulfes excited by the founding
open pipes, are in all cafes equal to twice the let
of the pipes.

In a former lecture, fpeaking of the incr
which motion received by being communic
from a fmaller elaftick body to a larger, I took
cafion to give a reafon for the augmentatior
found in fpeaking trumpets; I fhall clofe this
ture, by accounting for it from the nature of
pulfes of the air. From what has been faid in
lation to the properties of thofe pulfes, it is mani
that the greater their condenfation is, the ftro
is the found which they excite; now, when
voice acts upon a portion of air confined with
trumpet, it muft neceffarily make a ftronger
preffion upon it, and of courfe condenfe it m
than when it acts upon it in an unconfined ftate; i
much as in the former cafe, the force of the voi
wholly imployed in giving motion to that fmall
tion of air which lies within the trumpet, whe

in the latter cafe, not only that portion of air is put in motion by the force of the voice, but likewife all that body of air which immediately furrounds it; the air then in the trumpet being by reafon of its confinement, more ftrongly agitated and more clofely condenfed, than it would otherwife be, muft at the *exit* of the trumpet, communicate to the air without greater degrees of condenfation; and of confequence, produce a louder found, than could poffibly be excited by the fame force of the voice, were it immediately impreffed on the unconfined air.

LECTURE XIX.

Of LIGHT.

LIGHT, whereof I intend to treat in this lecture, is a moft fubtile fluid, confifting of particles exceedingly fmall, but of different magnitudes, as fhall be fhewn hereafter, which are thrown off from luminous bodies by the vibrating motions of their parts, with a velocity furprizingly great; for they do not fpend above feven or eight minutes of an hour in paffing from the fun to the earth, as was obferved firft by Mr. ROMER, Profeffor of Aftronomy to the late King of *France*; and after him by others, by means of the *eclipfes* of the *fatellites* of JUPITER; for thefe eclipfes, when the earth is between the fun and JUPITER, are obferved to happen about feven or eight minutes fooner than they ought to do by the aftronomical tables; and on the contrary, when the earth is beyond the fun with refpect to JUPITER, they happen about feven or eight minutes later than they ought to do; fo that in the latter fituation of the earth, they are obferved to happen fourteen or fixteen minutes later than in the former; forafmuch therefore as the fatellites

lites

lites cannot difappear, but muft continue vifible to
the eye of an obferver, till all that light which they
reflect before their immerfions has paffed by the
place of obfervation, it follows, that the reflected
light of the fatellites fpends fourteen or fixteen minutes in paffing from one end of the diameter of
the earth's orbit to the other; and confequently,
half that time in moving from the fun to the earth.
Hence, if the diftance of the fun from the earth be
70 millions of miles, as it muft be on fuppofition
that its horizontal parallax is twelve feconds of a
degree, and fuch the moft accurate obfervations of
the lateft aftronomers make it; then light moves
at the rate of about 150 thoufand miles in a fecond
of time, and its velocity exceeds the velocity of
found, in the proportion of above feven hundred
thoufand to one.

The motion of light is in its own nature rectilineal, as is evident from the fhadows which all opaque
bodies caft when placed in the light of the fun, or of
any other luminous body; and yet the beams or
rays of light in paffing out of one tranfparent body
or medium into another of a different denfity, are
bent and turned out of their way; or to fpeak
more properly, they are made to change the direction of their motion; and this bending or change
of direction is commonly called *refraction*; and it
has been found by experience, that the rays in paffing out of a rarer *medium* into a denfer, are bent
in fuch a manner as to be brought nearer to a line
drawn perpendicular to the refracting furface at the
point of incidence; and on the contrary, in their
paffage out of a denfer medium into a rarer, they
decline from the perpendicular.

Pl. 8.
Fig. 3.
For the illuftration of which, let AB reprefent
a ray of light moving in air from A to B, and paffing into water at B, and let HK be perpendicular
to the furface of the water at the point B; when
the ray goes into the water, it does not continue its
motion

motion ſtrait forward in the line BC, but in ſome other line as BD, which is more inclined to the perpendicular BK. And on the other hand, if the line DB be ſuppoſed to be a ray of light moving in water from D to B, and there paſſing into air, inſtead of continuing its motion in the direction BE, it goes on in ſome other direction as BA, which being leſs inclined to, is more diſtant from, the perpendicular BH; as will appear from the following experiment. Let an empty veſſel as BCDE, have a ſmall object as A, placed at its bottom; and let it be ſo ſituated as that the ſight of the object may be intercepted by the ſide of the veſſel from an eye placed at Q; let then the veſſel be filled with water, and the ray AB, which before the pouring in of the water, moved in a right line from A to K, and by ſo doing paſſed above the eye, will upon its emerſion out of the water be bent downward, ſo as to ſtrike upon the eye, and thereby render the object viſible.

This bending of the rays in their paſſage out of one medium into another, ſeems to be owing to the attractive force of the denſer medium acting upon the rays at right angles to the ſurface, as may appear by conſidering the conſequences of ſuch an attraction.

Let then AC be a ray of light moving from A to C, and there entring into a denſer medium, the ſurface which ſeparates the two mediums being denoted by the line HK. The motion of the ray in the direction AC, being reſolved according to the known method into two, one in the direction AD, and the other in the direction AB or DC, whereof the former is parallel, and the latter perpendicular to HK; it is manifeſt, that as the ray enters into the denſer medium at C, its perpendicular motion muſt be accelerated by the attraction, whilſt its parallel motion continues the ſame; let then the line CG be taken in the ſame proportion to CD, that
the

Pl. 8. Fig. 5.

the velocity of the perpendicular motion after refraction has to the velocity thereof before the refraction; and forasmuch as the parallel motion is
the same before and after refraction, let CE be taken equal to. AD or BC, and letting fall EF equal
and parallel to CG, and drawing the diagonal CF,
the ray after refraction will describe the line CF in
the same time that it moved from A to C before
the refraction; and forasmuch as GF is equal to
AD, LM, that is, the sine of the angle MCL,
must be less than AD, the sine of ACD; consequently, by the attraction of the denser medium,
the ray in passing into that medium is brought
nearer to the perpendicular.

Again, let FC denote the motion of a ray in the
denser medium from F to C, and let this motion
be resolved into two others, one in the direction
FG or EC, and the other in the direction FE or
GC, the former being parallel, and the latter perpendicular to HK; when the ray passes into the rarer *medium* at C, the parallel motion does not suffer any change from the attraction; but the perpendicular motion is retarded by the attractive
force, which in this case acts in direct opposition to
it; let then CD be to GC, as the perpendicular velocity of the ray in the rarer *medium*, to the perpendicular velocity thereof in the denser; and let
DA be drawn equal and parallel to FG, in order to
denote the parallel motion of the ray after refraction; and the diagonal CA will be the line described by the ray after refraction, in a space of time equal
to that wherein it described the line FC before refraction; and forasmuch as AD is equal to GF, it
must be greater than LM; consequently, the angle
ACD is greater than FCG; and therefore, the ray
in passing out of a denser *medium* into a rarer, is by
the attraction of the denser *medium* bent from the
perpendicular; so that in both cases, the refraction
seems to be owing to the attractive force of the
denser

denser medium, acting upon the rays at right angles
to its surface; and what farther confirms this opinion is, that the denser any medium is, and consequently the stronger its attraction, the greater,
cæteris paribus, is its refractive power; thus oil of
vitriol, whose density exceeds the density of water
in the proportion nearly of three to two, acts more
forcibly than water on the rays of light, in bending
and turning them out of their way; as will appear
from the following experiment; let the sixth figure
represent a quadrant, whose *radius* A B is parallel Pl. 8.
Fig. 6.
to the horizon; and let A be a small coloured ob-
ject, placed on the limb of the quadrant at the extremity of the horizontal *radius*; this being viewed
thro' an empty glass vessel as C, of a prismatick
form, placed at the center of the quadrant, with
its refracting angle downwards, will appear in its
real place at A.   Let then the vessel be filled with
water, and let the object be raised on the limb of
the quadrant as high as D, that is to say, to the
height of fifteen degrees and twenty minutes, and
the rays as D B, which go from it towards the prism,
will be so bent in passing thro' the water as to enter the eye in a direction parallel to the horizon,
and represent the object as if placed at A.   And
the same thing will happen when the vessel is filled
with oil of vitriol, excepting only that the object
must be raised to a greater height, suppose to E, so
as to have an elevation of twenty degrees and eight
minutes; which plainly shews, that the rays are
more bent, and suffer a greater refraction under
the same circumstances from oil of vitriol, than
they do from water.

Refractions taken by the Quadrant and prismatick veffel.			
	Denfity.	Degrees and min.	Sines.
Water	1	15.20	2644342
Oil of vitriol	1.497	20.8	3442060
Salt water	1.2	17.52	3068029
Spirit of hartfhorn	1.011	16.	2756374
Spirit of wine	0.835	17.	2923717
Oil of turpentine	0.869	22.34	3837582
Oil of linfeed	0.939	22.57	3889277

The denfer medium begins to attract the rays at fome diftance from its furface, and it acts upon them more and more forcibly in proportion as their diftance from its furface leffens; but however, in what follows I fhall fuppofe the attractive force to act with the fame vigour in all parts of the fpace thro' which it extends itfelf; becaufe, as that fpace is indefinitely fmall, no fenfible error will arife from fuch a fuppofition. If then CD be the furface of the denfer medium, and AB the fpace thro' which the attractive force extends itfelf from A to B; a ray of light in paffing from B to A will be accelerated in fuch a manner, as that the perpendicular velocity thereof at the point A will be equal to the fquare root of the fum of the fquare of the perpendicular velocity of the ray at its incidence on the point B, and of the fquare of the perpendicular velocity

Pl. 8.
Fig. 7.

locity which it would have at A, suppofing it began its motion at B, from a ftate of reft. For fince
the attractive force is fuppofed to act uniformly thro'
the fpace BA, the motion which it generates will
as to its properties correfpond with the motion arifing from gravity; if therefore the triangle EGH Pl. 8.
be taken to denote the fpace BA, GH will exprefs Fig. 8.
the velocity of a ray at A, on fuppofition that from
a ftate of reft it begins its motion at B; but if at B
it has a velocity exprefled by any right line as IK,
parallel to GH, let the triangle be continued on till
the portion IFLK becomes equal to EGH, and
FL will exprefs the velocity of the ray at the point
A; and forafmuch as the triangle EFL, is equal
to the fum of the two triangles EGH and EIK, FL
is equal to the fquare root of the fum of the fquares
of GH and IK; that is, the perpendicular velocity
of the ray at A, is equal to the fquare root of the
fum of the fquare of the perpendicular velocity of
the ray at its incidence on the point B, and of
the fquare of the perpendicular velocity which it
would have at A, on fuppofition that it began its
motion at B from a ftate of reft. And this being
fo, the courfe and velocity of a ray of light after
refraction, in paffing out of a rarer *medium* into a
denfer, may be determined in the following manner. Let Z be a rarer *medium*, and X a denfer, fe- Pl. 8.
parated by the common furface EF, on which let Fig. 9.
a ray of light as AC, fall obliquely, and let AC
meafure the velocity of the ray in the rarer *medium*;
which velocity is the fame, whatever be the inclination of the ray. From the center C with the
radius CA, let a circle be defcribed, in which let
NM be drawn thro' the center perpendicular to
EF, and from A let fall AQ perpendicular to EF,
as alfo AO perpendicular to NC. The motion of
the ray in the direction AC being refolved into two
others, one in the direction AO or QC, and the
other in the direction AQ or OC; the line OC
will

will meafure the velocity of the perpe̶n̶d̶i̶c̶u̶l̶a̶r̶
tion; and therefore, if CP be taken to d̶
perpendicular velocity generated by t̶h̶e̶
tion of the denfer medium, the line PO w̶i̶l̶l̶
fure the perpendicular velocity of the ray
denfer medium; and forafmuch as the velo̶
the parallel motion is no way altered by the
tion, if CV be taken equal to QC, and V̶
drawn parallel to CM, and equal to PO, it i̶s̶
dent, that the ray after refraction, will defcri̶b̶e̶
line CB, and that the velocity of its motion wil̶l̶
meafured by that line.

As a *Corollary*, from what has been prove̶d̶
follows, that the velocity of the refracted ray i̶n̶
denfer medium is no way varied by varyin̶g̶ t̶h̶e̶
clination of the incident ray; for the fquare o̶f̶
being equal to the fum of the fquares of B̶
CV, or of PO and AO, and the fquare of P̶
ing equal to the fum of the fquares of CO a̶n̶d̶
the fquare of CB is equal to the fum of the fqu̶
of AO, CO, and PC; but the fquares of AO
CO are equal to the fquare of CA or CN; co̶n̶-
quently, the fquare of CB is equal to the fum
the fquares of PC and CN, which quantities c̶o̶n̶-
tinue unvaried, whatever be the inclination of
incident ray; and therefore PN or CB is a g̶i̶v̶e̶n̶
quantity; that is, the meafure of the velocity,
of confequence, the velocity wherewith the r̶a̶y̶
move after refraction in the denfer medium, i̶s̶ a̶l̶-
ways the fame, however differently inclined t̶h̶e̶
rays may be to the furface of the denfer mediun̶
their incidence thereon.

The angle ACN, which the line defcribed by t̶h̶e̶
incident ray contains, with the perpendicular to t̶h̶e̶
refracting furface at the point of incidence, is ca̶l̶l̶e̶d̶
the *angle of incidence*; and the angle BCM, w̶h̶i̶c̶h̶
the line defcribed by the refracted contains, with t̶h̶e̶
perpendicular to the refracting furface at the p̶o̶i̶n̶t̶
of incidence, is called the *angle of refraction*.

As a second *Corollary*, from what has been proved it follows, that the sines of these angles are to one another in a given *ratio*; or, in other words, that whatever proportion the sine of any one angle of incidence bears to the sine of the corresponding angle of refraction, the same does the sine of any other angle of incidence bear to the sine of the respective angle of refraction. For since CB is cut by the circle in the point T, if from B and T, BS and TR be drawn perpendicular to the *radius*, BS will be equal to AO, which is the sine of the angle of incidence, and TR will be the sine of the angle of refraction; and from the nature of similar triangles, BS is to TR, as CB to CT; that is, the sine of incidence is to the sine of refraction in the same proportion with two standing quantities; consequently, that proportion is given, whatever be the declination of the incident ray. And what has been thus proved, with respect to the sines of incidence and refraction, when rays pass out of a rarer medium into a denser, is in like manner demonstrable of those lines, when the rays move out of a denser medium into a rarer, with this difference only, that whereas in the former case the angle of incidence exceeds the angle of refraction, in the latter it is exceeded by it; for as the attraction of the denser medium by accelerating the perpendicular velocity of the rays in their passage from a rarer medium turns them out of their way, so as to bring them nearer the perpendicular, so on the other hand, by retarding their perpendicular velocity in their passage into the rarer medium, it turns them out of their way so as to remove them farther from the perpendicular, as has been already shewn; and forasmuch as the rays are turned out of their way in both cases by one and the same cause acting in the same uniform manner, it is manifest, that in both cases, they must be equally bent; consequently, as

Pl8.
Fig. 9.

U   much

much as the angle of incidence exceeds the angle of refraction when a ray passes out of the rarer medium, into the denser, so much must it be exceeded by it, when the passage of the ray is made the contrary way.

Now that the sine of the angle of incidence is to the sine of the angle of refraction in a given ratio, whatever be the inclination of the incident ray, may be proved experimentally in the following manner. Let a brass quadrant graduated on both sides, and fixed at its center to a perpendicular pillar in the manner represented, have two indices as A and B, one on each side, moveable on the center C; and let the index A, whereof the stem D is a continuation, be made to point to the 15th degree, and the index B to the 15th minute of the 20th degree; let then the pillar be immersed in water, so far as that CE the horizontal edge of the quadrant may touch the surface of the water, and upon viewing the stem D which lies within the water, it will by reason of the refraction seem to have changed its situation, and appear to lie in the same plane with the index B. And the same thing will likewise obtain, if the index A be set at the 30th degree, and B at the 30th minute of the 42d degree. Now in both these cases, the angle of incidence is equal to the angle contained between FC, the perpendicular edge of the quadrant, and the index A; and the angle of refraction is the angle made by the perpendicular edge of the quadrant, and the index B; so that one of the angles of incidence is 15 degrees, and the other 30, and the corresponding angles of refraction are nineteen degrees fifteen minutes, and 41 degrees 30 minutes; and 25, which is the sine of the lesser angle of incidence, is to 33, the sine of the corresponding angle of refraction, as 50, the sine of the greater angle of incidence, to 66, the sine of the angle

Pl. 8.
Fig. 10.

2   of

of refraction, which corresponds thereto; as in the following TABLE.

	Angles of incidence.	Sines.	Angles of refraction.		Sines.
Out of water into air.	d. 15.	2588	d. 19.	m. 15	3296
	d. 30.	5000	d. 41.	m. 30	6626
Out of oil of turpentine into air.	d. 15.	2588	d. 22.		3746
	d. 30.	5000	d. 47.		7313

LECTURE XX.

Of COLOURS.

NATURALISTS were formerly of opinion, that LIGHT was in its own nature simple and uniform, without any difference or variety in its parts. And that COLOURS, which are to be the subject of this lecture, were nothing else than certain changes or modifications of light caused by *refractions*, *reflections*, and *shadows*. But Sir ISAAC NEWTON, to whom we are indebted for almost every thing that we know with certainty concerning the nature of light, has shewn from experiments, that notwithstanding the uniform appearance of light, the particles whereof it is composed are of different colours; and that the colour of each particle is lasting and permanent, so as not to be changed either by refraction or reflexion. He has likewise shewn, that those particles which differ

LECT.
XX.

as

as to colour, differ also in degrees of refrangibility;
by means whereof, the rays of different colours may
be separated from each other, and exhibited apart.
Let a beam of the fun's light pass into a darkened
chamber thro' a round hole as H, about the fix-
teenth or twentieth part of an inch wide, so as to
fall directly on the middle of a double convex lens
as L, ground to a *radius* of five or six feet, and
placed at the distance of ten or twelve feet from
the hole; by which means the image of the hole
will be projected to I, on the other side of the *lens*,
at the distance of ten or twelve feet more, and there
appear white and round. Let then a prism of solid
greenish glass as P, be placed close behind the *lens*,
and in such a posture as that the beam of light may
fall upon it perpendicular to its axis, which is an
imaginary strait line, running thro' the middle from
one end to the other parallel to its edges; this being done, the image of the hole, instead of being
round and white, and projected to I, will be long
and coloured, and cast sidewise from I; and the
colours of the image taken in their order from that
which lies nearest to I, will be *red, orange, yellow,
green, blue, purple,* and *violet*; as in the image
MN, where the several colours are denoted by
their initial letters.

From the lengthening of the round image by
the refraction of the prism, it is evident, that of
the particles of light which form the image, some
are more refrangible than others; for were they all
alike refrangible, the distances to which they are
thrown sidewise from their first situation at I, would
be all equal, and of consequence, the second image
would be round as the first.

As in the coloured *spectrum* the *red* lies nearest to,
and the *violet* farthest from I, it is manifest, that
the red particles in their passage thro' the prism,
are pushed out of their way less, and the violet
more,

more than any other; and consequently, that the red particles have the smallest degree of refrangibility, and the violet the greatest; and that the particles of intermediate colours have intermediate degrees of refrangibility, greater or less in proportion as they lie nearer to the one or the other of the two extreams.

This difference of refrangibility in the particles of light, argues a difference likewise in their magnitudes; for since one and the same cause, to wit, the attraction of the glass, acting upon them all with equal force, and under like circumstances, produces unequal changes in the directions of their motions, it must needs be that they move with unequal forces, and consequently, that their quantities of motion are unequal, which inequality of motion can arise from nothing else but the different size of the particles, in case they all move equally swift, as they are generally supposed to do; and that they are all perfectly solid, as their power of penetrating and dissolving the densest bodies, without suffering any change themselves, seems to require; consequently, the particles of light which differ as to colour, differ also in magnitude; those of violet being smallest, and the particles of other colours increasing continually one above another, as they are more and more removed from the violet, and approach nearer to the red, whose particles are largest of all; and here it will not be improper to observe, that as the red particles are of all others the largest, they must on that account act with the greatest force, and excite the strongest vibrations in the nervous fibres of the eye; which may be one reason why reds are found to be more offensive to the eyes, than any other colour whatever.

The seven colours whereof the long image is composed are permanent and lasting, and cannot possibly be changed, either by refraction or reflexion, as will appear from the following experiments.

Let

Exp. 2.

Let a small hole be made in the paper whereo
coloured image is formed, thro' which, let ea
the seven colours pass successively, and falling
a prism, be again refracted, and they will be f
to continue the same, without the least chang
alteration; thus, the *red* when refracted, will
tinue totally of the same red colour as before;
ther *orange, yellow, green, blue,* nor any other
colour, will arise from the refraction; and the
constancy and immutability will be found in
other six colours, when refracted singly and a
from the rest. And as these colours are not cha
able by refraction, so neither are they by reflex
for if bodies of different colours be placed in th
light, they will all appear red, and in the blue li
they will appear blue, in the green light, green,
so of the other colours; in the light of any on
lour, they will all appear totally of that same
lour, with this difference only, that in some
colour will be more strong and full, in others
faint and dilute, every body appearing most f
did and luminous in the light of its own co

Exp. 3. Thus for instance, if a deep red as *carmine,*
full blue as *ultramarine,* be held together in th
light, they will both appear red; but the *car*
will appear of a strongly luminous and resplen
red, and the *ultramarine* of a faint obscure and

Exp. 4. red; and on the other hand, if they be held t
ther in the blue light, they will both appear b
but the *ultramarine* will appear of a strongly l
nous and resplendent blue, and the *carmine*
faint dark blue.

Since the colours of the rays are not capabl
being changed either by refraction or reflexio
is manifest, that if the sun's light consisted of
one sort of rays, there would be but one colou
the world; and by consequence, that the varie
colours depends upon the composition of light
is likewise manifest, that the permanent colou

natural bodies arife from hence, that fome bodies reflect fome fort of rays, and others other forts more copioufly than the reft, and upon that account appear of this or that colour. Thus *minium*, and other red bodies, reflect the red rays moft copioufly, and thence appear red; *violets*, and all other bodies of the like colour, reflect the violet rays in greater abundance than the reft, and thence have their colour; and fo of other bodies, every body reflecting the rays of its own colour more copioufly than the reft, and deriving its colour from the excefs and predominancy of thofe rays in the reflected light; for tho' all bodies appear of the fame colour, when placed together in the light of any one colour, yet every body looks more fplendid and luminous in the light of its own colour than in that of any other, which puts it paft difpute, that every body reflects the rays of its own colour in greater abundance, than it does the reft, and thence has its colour.

As natural bodies appear of divers colours, accordingly as they are difpofed to reflect moft copioufly the rays originally indued with thofe colours, fo from the different proportions which the predominant rays bear to the reft of the reflected light, arife different fhades or degrees in thofe colours. Where the predominant rays are very numerous in proportion to the reft, the colour appears ftrong and full; but as the excefs of the predominant rays leffens, the colour, from the mixture of the other rays, abates of its livelinefs, and becomes more faint and dilute; and when all the rays are equally reflected, fo as that no one kind predominates, the colour becomes white; for whitenefs is a mixture of all the colours, and it is more or lefs intenfe in proportion as the reflected rays are more or fewer in number; all *grays*, *duns*, *ruffets*, *browns*, and other dark and dirty colours, down to the deepeft *black*,

being

LECT.
XX.

being but so many lesser degrees of white, resulting from perfect whiteness on no other account
that they consist of a lesser quantity of light, and
consequently appear less glaring and luminous.

The reason why bodies reflect this or that kind
of ray more copiously than the rest, and consequently appear of this or that colour, depends altogether on the size and density of the particles
whereof the bodies are composed. Particles of
coloured bodies reflecting rays of different colours according to their different magnitudes and
densities, as has been fully proved by Sir Isaac
Newton, from experiments and observations made
on the colours of thined bodies of air, water, and
glass; by the help of which he has in the second
book of his *Opticks*, given us a table containing
seven orders or *series* of colours, together with the
thicknesses of the particles of air, water, and glass,
which exhibit the several colours in each order,
which thicknesses are expressed in parts, whereof
ten hundred thousand make an inch. The first
part of that table is here laid before you; and by
inspection thereof it will be found, that in each
order of colours, the *red* is reflected by particles of
the greatest thickness, and that the thicknesses of
the particles which reflect the other colours, grow
less and less, as the colours which they reflect are
more and more removed from the *red*. It is likewise manifest from the same table, that among the
particles which reflect one and the same colour,
those which have the greatest density, have the
least thickness; thus for instance, the thickness of
a particle of glass which reflects the *scarlet of the
second order*, is but 12½; whereas the thickness of
water which reflects the same colour, is 16⅓, and
that of air still greater, to wit 19⅔; so that the
thicknesses of the particles which reflect any colour,
increase as their densities lessen; for which reason,

particles

partides of the fame thickneſs may reflect different L e c т.
colours, provided their denſities be unequal; thus XX.
the particles of air which reflect the *violet* of the *ſe-*
cond order, have very nearly the ſame thickneſs with
particles of water which reflect the *green*, as alſo
with the particles of glaſs which reflect the *orange* of
the ſame *order*.

			Air.	Water.	Glaſs.
				Thickneſſes of	
		Very Black	$\frac{1}{2}$	$\frac{1}{3}$	$\frac{10}{16}$
		Black	1	$\frac{3}{4}$	$\frac{20}{11}$
The co-		Beginning of black	2	$1\frac{1}{2}$	$1\frac{1}{7}$
lours of		Blue	$2\frac{2}{5}$	$1\frac{4}{5}$	$1\frac{11}{16}$
the *firſt*		White	$5\frac{1}{4}$	$3\frac{7}{8}$	$3\frac{2}{5}$
order.		Yellow	$7\frac{1}{9}$	$5\frac{1}{4}$	$4\frac{1}{3}$
		Orange	8	6	$5\frac{1}{7}$
		Red	9	$6\frac{3}{4}$	$5\frac{4}{5}$
		Violet	$11\frac{1}{6}$	$8\frac{3}{4}$	$7\frac{1}{7}$
		Indico	$12\frac{5}{6}$	$9\frac{1}{8}$	$8\frac{6}{11}$
Of the *ſe-*		Blue	14	$10\frac{1}{3}$	9
cond *or-*		Green	$15\frac{1}{8}$	$11\frac{1}{2}$	$9\frac{5}{8}$
der.		Yellow	$16\frac{2}{7}$	$12\frac{1}{2}$	$10\frac{1}{3}$
		Orange	$17\frac{2}{9}$	13	$11\frac{1}{6}$
		Bright-red	$18\frac{1}{3}$	$13\frac{3}{4}$	$11\frac{2}{3}$
		Scarlet	$19\frac{2}{3}$	$14\frac{3}{4}$	$12\frac{2}{3}$

From what has been ſaid concerning the colours
of natural bodies, it follows, that if any change be
made in the ſize or denſity of the particles whereof
a body is compoſed, the colour of the body will
likewiſe be changed; for which reaſon, if two co-
lourleſs diquors be mixed together, they may in the
mixing ſuffer ſuch changes in the ſize and denſity
of their parts from their mutual actions one upon
another, as to become opaque and coloured; and
ſuch

:ure of roſes, and ſpirit of urine,   *Blue.*
ion of copper, and ſpirit of ſal ar-
niack,   *Purple.*
ion of ſublimate, and ſpirit of ſal
noniack,   *White.*
ion of ſugar of lead, and the ſo-
ion of vitriol,   *Black.*

*ariſing from the mixture of ſuch liquors as are
coloured.*

low.	Tincture of ſaffron	}	*Green.*
l.	Tincture of red roſes	}	
e.	Tincture of violets	}	*Crimſon.*
wn.	Spirit of ſulphur	}	
l.	Tincture of red roſes	}	*Blue.*
wn.	Spirit of hartſhorn	}	
e.	Tincture of violets	}	*Violet.*
e.	Solution of copper	}	
e.	Tincture of violets	}	*Purple.*
e.	Solution of Hungarian vitriol	}	
e.	Tincture of cyanus	}	*Green.*
e.	Spirit of ſal armon. coloured	}	

                             7. *Blue.*

Plate 8.

Fig. 4.

7. { *Blue.* Solution of Hungarian vitriol } *Yellow.*
   { *Brown.* Lixivium

8. { *Blue.* Solution of Hungarian vitriol } *Black.*
   { *Red.* Tincture of red rofes

9. { *Blue.* Tincture of cyanus } *Red.*
   { *Green.* Solution of copper

Colours changed and reftored.

1. A folution of copper, which is *green*, by fpirit of nitre is made *colourlefs*, and is again reftored by oil of tartar.

2. A limpid infufion of galls, is made *black* by a folution of vitriol, and *tranfparent* again by oil of vitriol, and then *black* again by oil of tartar.

3. Tincture of red rofes, is made *black* by a folution of vitriol, and becomes *red* again by oil of tartar.

4. A flight tincture of rofes, by fpirit of vitriol becomes a fine *red*, then by fpirit of fal armoniack turns *green*, and then by oil of vitriol becomes *red* again.

5. Solution of verdegreafe, from a *green* by fpirit of vitriol becomes *colourlefs*, then by fpirit of fal armoniack turns a *purple*, and then by oil of vitriol becomes *tranfparent* again.

Among the various *Phænomena* of colours, there is none more remarkable than that of the *rainbow*, which is an appearance obfervable in thofe places only where it rains in the funfhine, and where the fpectator is placed in a due pofition between the fun and the rain, with his back to the former; for which reafon it is generally allowed, that the bow is made by the refraction of the fun's light in drops of falling rain; the manner wherein it is formed, has in fome meafure been explained by Antonius de Dominis, archbifhop of *Spalato*, and after him by Des Cartes; but as neither of them underftood

                the

Lect. the true origin of colours, it was imposll
XX. them not to be defective in their accounts
therefore Sir Isaac Newton, after he had
vered the true nature and rise of colours, so
felf to the confideration of this fubject, and t
the latter end of the firft book of his *Optid*
given a full and fatisfactory account of the
matter; the fubftance of what he has there d
ed concerning the rainbow is as follows.

Pl. 9.
Fig. 1.
Let a drop of rain, or any other fpherica
fparent body, be reprefented by the fphere B
and let AN be one of the fun's rays, inciden
it at N and thence refracted to F, where
either go out of the fphere by refraction t
V, or be reflected to G; and there let it eit
out by refraction to R, or be reflected to H,
let it go out by refraction towards S, cutting
cident ray in Y; let AN and RG be produ
they meet in X. Parallel to the incident ra
let the diameter BQ be drawn, and let B
quadrant, on every point of which let us fup
ray to fall parallel to BQ; as the point of inc
removes from B towards L, the angle AXR
the rays AN and RG contain, will firft in
and then decreafe; and on the other han
angle AYS, contained between the rays A
YS, will firft decreafe and then increafe. Tl
ing fo, if we fuppofe N to be that point
quadrant BL, whereon if the incidence ra
falls, it makes the greateft angle with the ra
which emerges after one reflexion, then all tl
which fall on each fide at a very little diftanc
N, and go out after one reflexion, will emei
rallel or very nearly parallel to GR, where
which fall on the quadrant at greater diftance
N, will notwithftanding their parallelifm
their incidence be fcattered, and diverge fro
another after their emergence. If therefore
the Siniated in the direction of the forme

which go out parallel, they will enter it fo copiously as to exhibit the image of the fun in the drop of rain which reflects them; but if the eye be fo placed as to receive the latter rays which go out diverging, thofe which enter the eye will be too few to excite any fenfation; and of confequence, the image of the fun will not appear in the drop to an eye fo fituated.

. If N be the point, whereon if the incident ray AN falls it makes the fmalleft angle with the ray HS, which emerges after two reflexions; then, as before, all the rays which are incident near N, and which emerge after two reflexions, will go out parallel, and for that reafon will exhibit the fun's image to an eye fituated in their direction; but thofe rays which are incident at any fenfible diftance from N, and which emerge after two reflexions, will be fcattered as they go out, and upon that account will be too few, and confequently too feeble to excite any fenfation in the eye of the fpectator.

Now, forafmuch as the rays which are of different colours have likewife different degrees of refrangibility, the greateft angle AXR which can be made by the incident rays, and thofe which go out after one reflexion, will be of different magnitudes in rays of different colours; fo likewife will the fmalleft angle AYS, that can be made by the incident rays, and thofe which go out after two reflexions; and it has been found by computation, that in the leaft refrangible or red rays, the greateft angle AXR, is 42 degrees and two minutes; and the leaft angle AYS, 50 degrees and 57 minutes; and in the moft refrangible or violet rays, the greateft angle AXR, has been found to be 40 degrees and 17 minutes; and the leaft angle AYS, 54 degrees and 7 minutes.

. Suppofe now that O is the fpectator's eye, and OP a line drawn parallel to the fun's rays; and let

Pl. 9. Fig. 2.

POE

POE be an angle of 40 degrees and 17 minutes, POF of 42 degrees 2 minutes, POG of 50 degrees 57 minutes, and POH an angle of 54 degrees 7 minutes; and these angles turned about their common side, shall with their other sides OE, OF, OG, and OH, describe verges of two rainbows AFBE and CHDG. For if E, F, G, and H, be drops of rain placed any where in the conical surfaces described by OE, OF, OG, and OH, and be illuminated by the sun's rays SE, SF, SG, and SH, the angle SEO being equal to the angle POE, or 40 degrees and 17 minutes, shall be the greatest angle in which the most refrangible rays can after one reflexion be refracted to the eye; and therefore, all the drops in the line OE shall send the most refrangible rays most copiously to the eye, and thereby strike the senses with the deepest *violet* colour in that region. And in like manner, the angle SFO being equal to the angle POF, or 42 degrees 2 minutes, shall be the greatest in which the least refrangible rays after one reflexion can emerge out of the drops; and therefore, those rays shall come most copiously to the eye from the drops in the line OF, and strike the senses with the deepest *red* colour in that region. And by the same argument, the rays which have intermediate degrees of refrangibility, shall come most copiously from drops between E and F, and exhibit the intermediate colours in the order which their degrees of refrangibility require, that is, in the progress from E to F, or from the inside of the bow to the outside in this order, *violet, indigo, blue, green, yellow, orange,* and *red.*

Again, the angle SGO being equal to the angle POG, or 50 degrees and 57 minutes, shall be the least angle in which the least refrangible rays can after two reflexions emerge out of the drops, and therefore the least refrangible rays shall come most copiously to the eye from the drops in the line OG,

and

and ftrike the ·fenfe with the deepeft *red* in that region. And the angle SHO being equal to the
angle POH, or 54 degrees and 7 minutes, fhall
be the leaft angle, in which the moft refrangible
rays, after two reflexions, can emerge out of the
drops; and therefore, thofe rays fhall come moft
copioufly to the eye from the drops in the line OH,
and ftrike the fenfes with the deepeft *violet* in that·
region. And by the fame argument, the drops in
the regions between G and H, fhall ftrike the fenfes
with the intermediate colours, in the order which
their degrees of refrangibility require, that is, in
the progrefs from G to H, or from the infide of the
bow to the outfide in this order, *red, orange, yellow,*
green, blue, indigo, and *violet.* And fince thefe four
lines OE, OF, OG, and OH, may be fituated any
where in the abovementioned conical furfaces, what
is faid of the drops and colours·in thefe lines, is to
be underftood of the drops and colours every where
in thofe furfaces. Thus then fhall there be made
two bows of colours, an interior and ftronger by
one reflexion in the drops, and an exterior and·
fainter by two (for the light becomes fainter by
every reflexion), and their colours fhall be in a contrary order to one another, the *red* of both bows
bordering upon the fpace GF, which is between the ·
bows. The breadth of the interior bow meafured
crofs the colours, fhall be one degree and 45 minutes, and the breadth of the exterior, fhall be three
degrees 10 minutes, and the diftance between them,
fhall be 8 degrees 55 minutes; the greateft femidiameter of the innermoft, or the angle POF, being 42 degrees and 2 minutes, and the leaft femidiameter of the outermoft, or the angle POG, being 50 degrees and 57 minutes. And thefe are
the meafures of the bows, as they would be were the
fun but a point; for by the breadth·of his body,
the breadth of the bows will be increafed, and their .
diftance leffened by half a degree; and fo the

breadth

breadth of the interior will be 2 degrees 15 minutes,
and that of the exterior 3 degrees 40 minutes, and
their distance 8 degrees 25 minutes; the greatest
semidiameter of the interior bow 42 degrees 17 minutes, and the least of the exterior 50 degrees 42 minutes ;: and such Sir ISAAC NEWTON says he has
found the dimensions of the bows in the Heavens,
when he measured the same. This explication of
the rainbow is confirmed by the following experi-

Pl. 9.
Fig. 3.

ment; let a glass globe filled with water, as AB, be
hung up in the sun-shine, with a black cloth placed
behind it, and let IS be one of the sun's rays incident thereon; let the eye of a spectator whose back
is to the sun, be placed at O, and let it be directed to such a point in the lower part of the globe
suppose C, as that a strait line drawn from the eye
thro' that point, and continued on till it meets the
incident ray likewise produced, may therewith make
an angle OXI, of 42 degrees 2 minutes; and the
spectator shall then see a full *red* colour in that side
of the globe opposed to the sun as at F; let then
the eye be raised up gradually to P, till the angle
PZI becomes equal to 40 degrees and 17 minutes;
and as the eye rises, it will perceive other colours,
to wit, *yellow, green,* and *blue,* successively in the
same side of the globe.

Again let the eye be placed at Q, and let it be
directed to such a point in the upper part of the
globe suppose D, as that a strait line, drawn from
the eye thro' that point and meeting the incident ray
protracted, may therewith make an angle QSI of
50 degrees and 57 minutes, and there will appear
a faint red colour in that side of the globe towards
the sun ; let then the eye be gradually depressed to
R, till the angle RTI is 54 degrees 7 minutes, as
the eye sinks, the *red* will turn successively to the
other colours, *yellow, green,* and *blue,* as in the
former case upon the raising of the eye.

LECTURE

LECTURE XXI.

Of Dioptricks.

INTENDING in my next lecture to enquire into the NATURE OF VISION, where I shall have occasion to take notice of *Defective Eyes*, I shall in this lecture, by way of preparation, lay before you some of the chief properties of such *lenses* or glasses as are most commonly in use for assisting defective eyes; and they are of two sorts, First, such as are equally convex on both sides, and secondly, such as are on both sides equally concave. The former sort is represented in the fourth figure, and the latter in the fifth.

Let ABC be an object placed before the double Pl. 9. Fig. 6. *convex lens* HK at any distance greater than the *radius* of the sphere, whereof the *lens* is a segment; the rays, which issue from the several points of the object, and fall upon the *lens*, will in their passage thro' it be so bent by the refractive power of the glass, as to be made to convene at so many other points behind the *lens*, and at the place of their concourse they will form an image or representation of the object; and this image will be inverted, because the rays which flow from A, the uppermost point of the object, are united at F, the lowermost point of the image, whilst those which flow from C, the lowerest point of the object, are brought together again at D, the highest point of the image. So likewise those rays which issue from the right side of the object, are united in the left side of the image, whilst those which proceed from the left side of the object, concur in the right side of the image; as will appear by placing a lighted candle before a double *convex lens*, at such a distance

X · tance

tance as that the image thereof may be formed
a piece of white paper placed at a due distance
hind the *lens*; for the flame will appear invert
with its point downward; and if either side of t
flame be intercepted by the interposition of a da
body, the contrary side of the image will be o
scured.

With regard to this experiment, I must obser
to you, that tho' there is one certain distance,
which the paper must be placed, in order to ex
bit the image with the greatest distinctness, yet m
the distance be a little varied without rendring
image confused; and it is remarkable, that wh
the image is projected on the paper at the near
distance that it can with any degree of distinctne
it appears bordered all around with red; which re
ness continually decreases, as the paper is more
more removed from the *lens*; and when it is
moved to such a distance as is requisite to give t
image the greatest advantage in point of distin
ness, the redness intirely vanishes, and leaves
image equally white all over; but upon a fartl
removal of the paper, the edges of the image whi
at the nearest distance were tinged with red,

Pl. 9.
Fig. 7. now appear tinged with blue. If a candle, wh
is placed at A before the *convex lens* CD, has
image projected on a paper at EF, supposing t
to be the least distance at which it can be proje
distinctly, its edges will appear red, but upon
removal of the paper to GH, they will beco
white; and when the paper is removed to I
they will appear blue; the reason of these diffe
appearances is this, the rays of light as AC and A
which flow from the candle, being compounded
particles of different colours, whereof the red
least refrangible, and the blue most so, upon pass
thro' the *lens*, the blue rays are made to conv
soonest, and the red latest; as in the figure wh
the blue are denoted by the pricked lines, and

ed by the continued; so that an image is formed
t EF, by the concurrence of some of the more
efrangible rays, and it is tinged around its edges
y the red rays, which converging more slowly than
he rest lie outermost.

· After the blue rays have concurred, they crofs
me another, and go on diverging towards GH,
where meeting with the red rays which have not
et concurred, and there mixing with them and
he rays of other colours, they produce a white
mage, whiteness resulting from a due mixture of
ll the colours; as they proceed forward toward
K, they, by reason of their greater divergence,
pread themselves on all sides beyond the other rays,
nd by so doing, tinge the outlines of the image
blue.

On the formation of pictures by means of a
convex lens, depend the appearances of the
camera obscura, which is a small square box with a
tube iffuing horizontally from one side, at the extremity whereof is fixed a double *convex lens*; with-
n the box is placed a looking-glafs in a flanting
position, so as to be at half right angles with the
bottom of the box, which is parallel to the horizon.·
On the top of the box is placed horizontally a plate
f glafs rough on one side, whereon the pictures of
objects are reprefented in the following manner.·

Let A B be an object placed before C D, the *lens*
fixed in the tube which iffues from the box; G H
he looking-glafs inclined to the bottom of the box,
n an angle of 45 degrees, L M the plate of rough
glafs covering the top of the box horizontally. The
ays which flow from A, the uppermoft point of
he object, after they have paffed the *lens*, converge
owards F, and would actually meet at that point,
but that they are intercepted by the looking-glafs

H, which reflects them, and throws them upward; and forafmuch as the inclination of the rays
towards one another is no way altered by the reflexi-

Pl. 9.
Fig. 8.
Exp. 2.

X 2
on, .

on, they muſt meet at ſome point, as K, as far diſtant above the *ſpeculum*, as the point F is behind in.
In like manner, the rays which flow from B, the
loweſt point of the object, and which after they have
paſſed the glaſs are tending towards E, being reflected upward by the *ſpeculum*, are made to convene at I, whoſe diſtance above the *ſpeculum* is
equal to the diſtance of E behind the *ſpeculum*; and
as the rays from the extream points A and B, are
made to convene at K and I, ſo thoſe which flow
from the intermediate points of the object, are
brought together at correſponding points between
K and I, whereby the image is projected horizontally, but with its right and left ſides correſponding to the contrary ſides of the object; as may appear by placing a man before the *lens*, and cauſing
him to ſtir one of his hands; for in the image the
other hand will appear to move.

The diſtance of the image behind the glaſs is always varied by varying the diſtance of the object
before the glaſs; the image approaching as the object recedes, and receding as that approaches. For
if we ſuppoſe A and C to be two radiating points,
from which the rays AH, AK, and CH, CK fall
upon the *lens* HK, it is manifeſt, that the rays from
the more diſtant point diverge leſs than thoſe from
the nearer point, the angle at A being leſs than that
at C; conſequently, when they paſs thro' the glaſs
they muſt be brought together ſooner, and muſt
convene at ſome point as B, leſs diſtant from the
lens, than is the point D, whereat the more diverging rays from the point C are made to convene.

Pl. 9.
Fig. 9.

Where the diſtance of the object, and the *radius*
of the *lens*'s convexity are given, and where the
thickneſs of the *lens* is but ſmall, as is commonly
the caſe; the diſtance of the image from the *lens* is
determined very nearly, by ſaying, as the diſtance
of the object from the *lens*, leſſened by the *radius*

of

of the *lens*'s convexity, is to the *radius*, so is the distance of the object from the *lens*, to the distance of the image from the *lens*; that is, putting D for the distance of the object, R for the *radius* of the convexity, and F for the distance of the image,

$$D-R : R :: D : F;$$ consequently, $$F = \frac{RD}{D-R}.$$

The truth of this rule is demonstrated by the writers of DIOPTRICKS; but as all the demonstrations which I have hitherto met with are tedious and intricate, I shall not at present trouble you with them, but shall proceed to confirm the rule by experiments.

Exp. 3.

Let then the flame of a candle be placed at the distance of twelve feet and an half from a double *convex lens*, the *radius* of whose convexity is four feet two inches; that is, let the distance of the flame from the glass be equal to thrice the *radius*, and the image will be projected behind the *lens* at the distance of six feet three inches, that is, at the distance of a *radius* and an half; for in this case, R being put equal to unity, RD is three, which being divided by D — R, that is, by two, gives one and an half in the quotient.

Exp. 4.

If the flame be brought nearer to the *lens*, the image will move farther from it, and when the distance of the flame becomes equal to twice the *radius* of the *lens*'s convexity, the distance of the image will be equal to that of the flame, the *lens* standing in the midway between them; for in this case D—R is equal to R, and of consequence, F is equal to D.

Exp. 5.

The flame being placed at the distance of the *radius*, the distance of the image becomes infinite. For in this case D—R is nothing, and F is equal to $$\frac{DR}{0},$$ which expression denotes an infinite quantity; so that in this case, there will not be any image of

the

the flame; but the rays of light which flow from
the candle, after they have paffed thro' the lens,
will go on parallel to one another; and by fading-
ing, form a bright circular image, equal in fize to
the *lens*, and the magnitude thereof will remain the
fame at all diftances from the glafs.

Where the diftance of the flame is lefs than the
radius of the convexity, D—R becomes a negative
quantity, and fo of confequence does the quotient
arifing from the divifion of DR by D—R; which
fhews, that the place at which the rays meet, is on
the fame fide of the *lens* with the flame; or to fpeak
more properly, that the rays after they have paffed
the *lens*, proceed diverging from one another in
fuch a manner, as if they had flowed from a point
before the *lens*, more diftant than the place of the
flame. For the eafier underftanding of which, let
Pl. 9.
Fig. 10.
the rays AB and AC flow from the point A, whofe
diftance from the *lens* BC, is lefs than the radius
of the *lens*'s convexity; after they have paffed the
glafs, they will not continue to go on in the directions BD and CE, but in the directions BF and CG,
as if they had proceeded from fome point as H,
more diftant from the *lens* than is the point A, from
which they really flow; fo that in this cafe, the rays
after they pafs the glafs, go on diverging from one
another, but however they do not diverge as much
as they did before they paffed the glafs.

When the diftance of the flame from the glafs is
fo great as that neither the breadth of the *lens*,
nor the *radius* of its convexity bears any fenfible
proportion to it, then D—R is equal to D; and of
confequence, F is equal to R; that is, the diftance
of the image is equal to the *radius* of the glafs's
convexity, and this is the leaft diftance at which an
image can be projected by fuch a *lens*; and forafmuch as the rays of the fun, which by reafon of
the immenfe diftance of his body are always united

at

at the smallest distance, are apt to burn at the place of their union; that place is usually called the *focus*, or *burning point*, and sometimes the *absolute focus*, in contradistinction to those places whereat the images of less remote objects are formed, and which are frequently called the *respective foci*.

The length or breadth of an object, is to the length or breadth of its image, as the distance of the object from the *lens*, to the distance of the image from the *lens*. For if AC be the length or breadth of an object, and DF the length or breadth of its image; AB, which is one half of AC, is to FE, which is one half of FD, as BL to EL, the triangles ABL and FEL being similar. Hence it follows, that the nearer an object approaches the *lens*, the larger is its image, the image receding, and consequently inlarging, as the object approaches; and thus it appears to be from experiments; for the flame of a candle being placed at a distance greater than the diameter of the *lens's* convexity, in which case the distance of the image is less, appears larger than the image, but being brought within the distance of the diameter, the image, which in that case is at the same distance, becomes equal to it; and upon bringing the flame still nigher, the image becomes larger in proportion to the square of its greater distance.

Pl. 9.
Fig. 6.

Exp. 6.

The same thing is likewise evident from the magick lantern; which is a lantern out of which issues an horizontal arm, capable of being lengthened or shortened at pleasure, by means of one part sliding in and out of the other; to the extremity of the moveable part is fitted a double *convex lens*; and to that part of the arm which joins the lantern is adapted a glass, plane on one side, and *convex* on the other, the plane side looking towards the lantern; in the body of the lantern there is placed a candle, whose distance from the *plano-convex* glass

Exp. 7.

X 4
is

is fomewhat lefs than the focal diftance; fo that the light which paffes thro' that glafs, is thrown very ftrongly upon little images painted in dilute colours on pieces of plane thin glafs; which being fixed in a flider that moves to and fro acrofs the arm, are placed at a fmall diftance behind the *plano-convex* glafs in an inverted pofition, and by means of the *lens* in the moveable part of the arm, are projected in an erect pofition, on a paper or white cloth placed at a proper diftance; if by drawing out the moveable part of the arm, the pictures be removed to a greater diftance from the *lens*, the lantern muft be brought nearer to the cloth, in order to a diftinct reprefentation; becaufe, as the object recedes from the *lens*, the image approaches, and at the fame time the images will be diminifhed. But on the other hand, if by thrufting in the arm the pictures be brought near the *lens*, the lantern muft be removed farther from the cloth, and in this cafe the images will appear larger.

As convex glaffes caufe the rays of light to converge and unite, fo thofe which are concave make them feparate and diverge; for which reafon, if diverging rays fall upon a concave *lens*, they will diverge more after they have paffed thro' it, than they did before; and fuch rays as converge before their incidence, will after their paffage converge

Pl. 9.
Fig. 11. lefs; for inftance, if the rays A B and A C, which diverge from A, pafs thro' the concave *lens* B C, they will not go on in the directions B D and C E, but in fome other directions as B H and C G, fo as to widen fafter than before. On the other hand, if H B and G C be two rays converging towards K, after they have paffed thro' the glafs, they will not go on towards K, but towards a more diftant point as A, fo as to converge more flowly than before. All which is fully confirmed by experiments. For

a candle being placed before a *convex lens*, fo as to

have

have its image projected on a white paper, placed
at a due diſtance behind the *lens*, if a concave glaſs
be placed between the *convex* and the image, ſo as
that the rays which are converging towards the
image may paſs thro' it, the image will thereby be
thrown to a greater diſtance behind, the rays being
made to converge more ſlowly, and of conſequence,
to meet at a greater diſtance than they did before
the concave was interpoſed ; and it muſt be obſerved, that as the image is thrown to a greater
diſtance, it muſt for that very reaſon be inlarged ;
and foraſmuch as the larger image is compoſed of
the ſame number of rays, or rather fewer, ſome of
the rays being reflected by the concave *lens*, it muſt
on that account appear leſs bright and luminous
than the ſmaller. If by the removal of the *convex-
lens*, the rays which flow from the candle be ſuf-
fered to fall diverging on the concave, and a white
paper be placed cloſe behind the glaſs, there will
appear thereon a dark circle of ſome breadth, occaſioned by the ſhadow of the hoop which contains
the glaſs ; and the circular *area* contained within the
ſhadow will be inlightened by the rays which paſs
thro' the glaſs ; and becauſe all the rays which fall
upon the glaſs do not paſs thro' it, ſome of them
being reflected, the circular *area* will appear ſome-
what darker than the other parts of the paper,
which are expoſed to the light of the candle, without the interpoſition of the glaſs ; upon removing
the paper gradually from the glaſs, the circular *area*
will gradually inlarge, and as that inlarges, the
ſhadow which environs it will grow narrower, and
at length vaniſh ; and upon the vaniſhing of the
ſhadow, if the paper be removed a little farther,
there will ariſe a bright circle all around the circular *area*, which will grow broader, but leſs bright,
as the paper is more and more removed from the
glaſs ; and at the ſame time, the circular *area* will
continue to widen, and grow darker. All which
appear-

appearances are the natural and necessary confequences of the divergency or spreading of the rays, occasioned by their passage thro' the glass; for the farther they go from the glass, the more they must diverge, and by so doing, must on all sides spread themselves into the place of the shadow, and render it equally luminous with the rest of the *area*; and when they have spread themselves a little beyond the limits of the shadow, they fall upon such parts of the paper as were before inlightened, and there, by their additional light, exhibit that bright circle which surrounds the darker *area*; and the bright circle, by the farther spreading of the rays, as the paper is more and more removed from the glass, grows broader and less luminous; as does likewise the circular *area*, from the spreading of the rays wherewith it is inlightened.

Tho' concave glasses do not collect the rays of light, and consequently, have not a real *focus*; yet inasmuch as the rays after they have passed thro' such glasses, do flow in such a manner as that they either tend to some point behind the glass, or appear to flow from some point before it, those points are usually called the *foci*; and in double concaves of equal concavities, the *foci* for converging rays are found, by saying, as the *radius* of the glass's concavity lessened by the distance of the point of convergence from the glass, is to the *radius*, so is the distance of the point of convergence to the *focus*. And the *foci* for diverging rays are found, by saying, as the sum of the *radius* and the distance of the point of divergence from the glass, is to the *radius*, so is the distance of the point of divergence to the *focus*. So that putting F for the *focus*, R for the *radius*, and D for the distance of the point of convergence, or divergence, $F = \dfrac{RD}{R \overline{x} D}$; the negative sign being to be prefixed to D when the rays converge, and the affirmative when they diverge,

The

The demonſtration of this *Theorem* I ſhall for the preſent omit, on account of its tediouſneſs and intricacy, and ſhall cloſe the lecture with this ob-ſervation; that if rays which are converging to-wards a *focus* be intercepted by a concave *lens*, whoſe diſtance from the *focus* is equal to the *radius* of its concavity, after they have paſſed thro' the glaſs, they will ceaſe to converge and become parallel, for R and D being equal, R — D is o; conſe-quently, F is infinite; that is, the point to which the rays converge, is at an infinite diſtance, and the rays of courſe muſt be parallel.

L E C T U R E XXII.

Of Vision.

MY deſign in this lecture, is to explain the manner of Vision with the naked eye; and likewiſe to ſhew you, what aſſiſtances the ſight re-ceives from glaſſes; and in order thereto, I ſhall give you a ſhort deſcription of the eye.

Lect. XXII.

If a ſmall portion be cut off of a globe, and in the room thereof a portion of a ſmaller globe, but of an equal circular baſe, be ſubſtituted, the com-pound will exhibit the true figure of the eye; for it is of a globular form, but more *convex* before than in any other part. It conſiſts of ſeveral mem-branes which lie contiguous one to another, of which the outermoſt is called the *tunica adnata* or *conjun-ctiva*; it has its riſe from that membrane which in-veſts the ſkull, and it covers the whole ball of the eye, except the foremoſt tranſparent part; that portion of it which is viſible, is called the *white of the eye*. Beſides, this membrane, which is not reckoned among the proper coats of the eye, there are three others, which conſtitute the proper coats; the firſt of which is called the *ſclerotica*, it is a tough membrane

…a to call the transparent part *cornea*; this part is presented by ABF.

The second membrane, called *tunica choroides*, is rived from the *pia mater*, and transmitted likewise from the brain along with the *optick nerve*; is is much thinner and tenderer than the former, d tinged on the hinder part with a black liquor. he fore part is called the *uvea*, and sometimes the is, from its variety of colours. In its middle is a all hole called the sight or pupil; the *iris* consists several circular concentrick muscular fibres, hich are cut across at right angles by other strait res in the manner of so many *radii*; by the contraction of the former the pupil is lessened, and is larged by the contraction of the latter.

The third coat is usually called the *retina*, and metimes the *nervous coat*, being nothing else but e *optick nerve*, which spreads itself in the form of membrane over the bottom of the eye, over against the sight. These coats lying contiguous, rm a *capsula* or bag, wherein are contained the ree humors of the eye, called the *aqueous*, the *crystalline*, and the *vitreous*.

At a little distance behind the pupil is placed the crystalline humor, which is *convex* on both sides, t somewhat flatter before than behind; it is supported by small muscular fibres, called the *ciliary ligaments*, which are inserted into the edges of the crystalline humor at one end, and at the other, to the *tunica choroides*, and being closely united, rm a kind of membrane, whereby the cavity of e eye is divided into two parts; in the foremost which is lodged the *aqueous humor*, so called, because

becaufe in confiftence and colc
fembles water, being almoft c
tranfparent. In the hindmoft is
bumor, which has its name fron
is fuppofed to bear to melted gl

It has been generally thought
the humors of the eye are of
and that the chryftalline is muc
either of the other two ; but D
informed us in his lecture on
weighing thefe humours. in ai
lance, he found the *aqueous* and
nearly of the fame fpecifick gra
fpecifick gravity of the chryf
ceed the fpecifick gravity of the
er proportion than that of elev
it follows, that the chryftalline
ufe in bringing the rays toge
forming on the *retina* the pictu
jects, as it has been commonly
optical writers ; for tho' in fhape
ble *convex lens*, and on that acco
the rays converge, yet forafmu
between two humors, which
fame denfity with itfelf, it can l
on the particles of light ; for th
perience, to be refracted very
of one *medium* into another, wl
the denfities of the *mediums* is b

Behind all the coats and hun
optick nerve, which paffes out c
fmall hole in the bottom of the c
the eye. O reprefents the *optic*
rotica or outermoft coat, whof
rent part ABF, is the *cornea*, c
the fore part whereof AP, and
uvea or *iris*, with the pupil PP
is the *retina*, AD and FE the c

the *chryſtalline humor*, VV the *vitreous humor*, and WW the *watry humor*.

Underneath the white of the eye are inſerted into the *ſclerotica* ſix muſcles, which take their riſe from different parts of the orbit, and are diſtinguiſhed by different names, taken from the different motions which they give the eye; their tendons ſpread themſelves over the *ſclerotica*, ſo as to terminate in the confines of the *cornea*; by which means, when the ſix muſcles act together, they preſs the ſides of the eye towards each other, whereby the eye is lengthened, and at the ſame time the convexity of the *cornea* is increaſed; both which effects are in ſome caſes abſolutely neceſſary in order to diſtinct viſion, as will appear preſently.

Having given this ſhort account of the conſtituent parts of the eye, I now proceed to lay before you, the *manner of viſion*. If an object as A B, be placed at a convenient diſtance before the eye, the rays which flow from the ſeveral points of the object, and falling on the *cornea* paſs thro' the pupil, will be brought together by the refractive power of the eye on ſo many correſponding points of the *retina*, and there paint the image or repreſentation of the object, in the ſame manner as the images of objects placed before a *convex lens* are exhibited on white paper, placed at a proper diſtance behind.

Pl. 9.
Fig. 13.
Thus the rays which flow from the point A, are united on the *retina* at C, and thoſe which iſſue from B, are collected at D; and in like manner, the rays which proceed from the intermediate points of the object, are again united at ſo many intermediate points on the *retina*. On this union of the rays at the bottom of the eye, depends diſtinct viſion, for ſhould they be united before they arrive at the *retina*, or ſhould the point of their union lie beyond the *retina*, it is evident, that the rays from each point muſt take up ſome ſpace on the *retina*, and

Pl. 9.
Fig. 14.
Pl. 10.
Fig. 1.

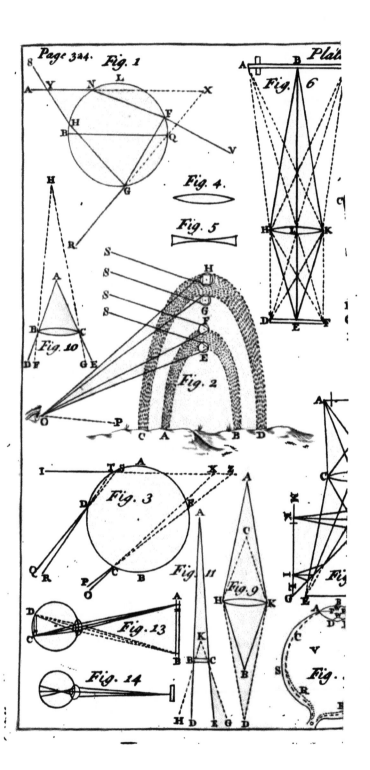

Fig. 1

Plat

Fig. 6

Fig. 4.

Fig. 5

Fig. 10

Fig. 2

Fig. 3

Fig. 11

Fig. 9

Fig. 13

Fig. 14

Fig.

of confequence, thofe which flow from conti-
ous points of the objeft will be mixed and blend-
together on the fund of the eye, fo as to exhi-
a confufed reprefentation of the objeft.

Now forafmuch as the rays which fall upon the
from radiating points, whofe diftances from the
are different, have different degrees of diver-
ce, the divergency of the rays increafing as the
tance of the radiating point leffens, and leffen-
as that increafes; and whereas thofe rays which
e greater degrees of divergence, require a ftrong-
refraftive power to bring them together at a
en diftance, than what is requifite to make thofe
et which diverge lefs, it is manifeft, that in or-
to fee objefts diftinftly at different diftances,
eye muft have a power of increafing and leffen-
its refraftive force, and thereby of adapting it-
to the different diftances of objefts; and this
oes by means of the fix mufcles which are in-
ted into the *fclerotica*; for when a radiating point
placed fo near, as that the rays which iffue from
fall upon the eye with a confiderable degree of
ergence, the mufcles aft ftrongly on the eye,
ereby the *cornea* is rendered more convex, and
confequence refrafts the rays with greater force;
fides by the lengthening of the eye from the joint
ion of the mufcles, the *retina* is removed to a
eater diftance from the *cornea*, by which contriv-
ce, the rays are made to convene at the *retina*,
twithftanding the great degree of divergence
erewith they enter the eye. As the radiating
int recedes from the eye, and the divergency of
rays of courfe grow lefs, the mufcles relax
mfelves in order to leffen the convexity of the
nea, and to fhorten the eye, a lefs convexity of the
nea, as alfo a lefs diftance between the *cornea* and
ina, being requifite to diftinft vifion in greater dif-
ces of the objeft than in fmaller.

Tho'

Tho' moſt mens eyes are ſo framed as to be abl
to ſee diſtinctly at different diſtances, yet ſome ther
are which are defective in this point, as being unabl
to ſee any thing diſtinctly but when placed ve
near; and this is the caſe of their eyes who a
called *myopes*, purblind, or ſhort-ſighted; in ſuc
the *cornea* is too convex in proportion to the lengt
of the eye; for which reaſon, all thoſe rays whic
iſſue from diſtant points, and of conſequence diverg
but little when they enter the eye, are made to con
vene before they reach the *retina*. As theſe me
advance in years, their eyes like thoſe of other ol
men, for want of a due ſupply of humors, abat
of their convexity and grow flatter; upon whic
account they begin to ſee objects diſtinctly at a di
tance, without the help of ſpectacles, and are ſo
that reaſon deemed to have the moſt laſting eyes.

By the help of concave glaſſes, purblind perſon
may ſee diſtant objects diſtinctly; for as it is th
property of ſuch glaſſes to make the rays diverg
if the rays which flow from a diſtant point, and fal
upon the eye with a ſmall degree of divergence, b
made to paſs thro' a concave *lens* of a proper con
cavity, they will thereby be made to diverge
much, as that the eye, notwithſtanding the grea
convexity of the *cornea*, ſhall not be able to brin
them together till they arrive at the *retina*.

Pl. 10.
Fig. 2.

If CD be a concave *lens*, and if B be the *focus* o
the rays which flow from the point A; that is, i
the rays which diverge from A, paſs thro' the glaſs
and by the refraction which they ſuffer in their paſ
ſage, proceed in ſuch a manner as if they had di
verged from B; and if the diſtance at which
purblind perſon ſees diſtinctly with his naked eye
be equal to the diſtance of B from the glaſs, ſuc
a perſon will by the help of the glaſs CD, be able t
ſee the point A diſtinctly; becauſe the rays whic
flow from A, after they paſs thro' the glaſs, fal
upo

upon his eye with the same degree of divergence, **L e c t. XXII.**
as if they had issued from B, the point of distinct
vision. Hence it follows, that if in the *Theorem* laid
down in my last lecture, for finding the *focus* of
double concaves exposed to diverging rays, namely

$F = \dfrac{RD}{R+D}$, wherein F denotes the *focus*, D the

distance of the point of divergence, and R the *radius* of the concavity, we suppose F to denote the
distance at which the purblind person sees distinctly
without a glass, and D the distance at which he sees
distinctly by the help of the glass, by clearing R

we shall have $R = \dfrac{FD}{D-F}$; that is to say, the

radius of the concavity of a double concave of
equal concavities, which enables a purblind person
to see an object distinctly, when placed beyond the
reach of his naked eye, must be equal to a rectangle,
under the distance at which he sees distinctly with
his naked eye, and the distance at which it is required he should see distinctly by the help of the
glass, divided by the difference of those distances.
For instance, if a person with his naked eye can
read at the distance of three inches only, and it be
required to find the *radius* of such a glass as shall
enable him to read at the usual distance of eighteen
inches; in this case, F being equal to three inches,
and D to eighteen, their product is 54; which being divided by their difference, which is 15, gives
three and ⅗ in the quotient, which shews, that
the *radius* of the glass must be three inches and
⅗ths nearly.

Where the distance at which it is required the
purblind person shall see distinctly is infinite, or in
other words, where it is so great, as that the distance
to which the power of his naked eye reaches, bears no
sensible proportion to it, there D — F becomes equal
to D, and of course, R becomes equal to F; so

that

that in order to fee fuch objects as are very remote, purblind perfons muft make ufe of concave glaffes, whofe *radii* are equal to the diftances at which they fee diftinctly with their unarmed eyes.

As purblind perfons cannot fee remote objects diftinctly, fo on the other hand, thofe who are old cannot, generally fpeaking, fee fuch as are nigh; the reafon of which is, that in old men the *cornea*, for want of a due fupply of humor to plump out the eye, has not a degree of convexity fufficient to bring the rays together on the *retina*, when they fall upon the eye with a confiderable degree of divergence; as is the cafe of all thofe rays which flow from points fituated near the eye. The proper remedy for this defect is a *convex lens*, becaufe it leffens the divergency of the rays, and brings them nearer to a parallelifm. If with refpect to the *convex lens* CD, A be the *focus* of the rays which diverge from B; that is to fay, if the rays which flow from B and pafs thro' the *lens*, do afterwards proceed in fuch a manner as if they had diverged from A, and if the diftance at which an old man can fee diftinctly with his naked eye, be equal to the diftance of A from the glafs, he will be able by the affiftance of the glafs, to fee the nearer point B diftinctly; becaufe the rays which iffue from that point in paffing thro' the glafs acquire the fame degree of divergence, with thofe which flow from A, the point of diftinct vifion, and of confequence, may as eafily be brought together on the *retina*, by the refractive power of the eye; hence, if we take the *Theorem* laid down in my laft lecture for finding the *foci* of double *convexes* of equal convexities, and fit it to the cafe before us, where the *focus* is imaginary, by making $F = \dfrac{RD}{R-D}$, if then we fuppofe F to denote the diftance at which an old eye fees diftinctly, and D the nearer diftance at which it

is

is required to make it see diftinctly with the affist-
ence of a glafs, by clearing R, we shall find it
equal to $\dfrac{FD}{F-D}$. So that the *radius* of such a double convex of equal convexities as enables an old
man to see a nigh object diftinctly, muft be equal
to a rectangle under the diftance at which he fees
diftinctly with his naked eye, and the diftance at
which he is to fee by the help of the glafs, divided
by the difference of thofe diftances. To illuftrate
this by an example; fuppofe an old man cannot
with his naked eye read at a lefs diftance than of
four feet, and it is required to affign the *radius* of
fpectacles which fhall enable him to read at the diftance of a foot and an half; in this cafe, F is four
feet, and D is one and an half, and their product is
fix, which when divided by their difference, to wit,
two and an half, gives $2\frac{4}{10}$ in the quotient; which
fhews, that the fpectacles muft be ground to a *radius* of two feet and four tenths.

If F be infinite, which is the cafe where the eye
can fee nothing but what is extremely remote, then
F—D is equal to F, and of confequence, R is equal
to D; fo that where an old man can fee no objects
diftinctly but fuch as are very far off, in order to fee
diftinctly at nearer diftances, he muft for each
diftance ufe fuch fpectacle glaffes as have their *radii*
equal to the diftance.

If D be given, then R becomes equal to $\dfrac{F}{F-1}$;
and forafmuch as the proportion of F to F—1 increafes as F leffens, R muft do fo too, which fhews,
that where the diftances at which two old eyes when
unarmed can fee diftinctly are different, in order to
make them fee diftinctly at any leffer given diftance,
the eye which can fee at the fmaller diftance muft
be furnifhed with a glafs of a greater *radius* than
the other. And herein lies the whole fecret of
younger and older fpectacles, thofe being deemed

the

the youngest, which are ground to the largest radius.

Having shewn you of what use both convex and concave glasses are in assisting defective eyes, I shall now lay before you the alterations which they produce in the appearances of objects; and First, as to *convexes*; if an object be viewed thro' a *convex lens*, at a less distance than the *focus*, it appears more remote and bigger than it does to the naked eye. That it must appear more remote, will be evident, if we consider what has been already proved in a former lecture, namely, that where rays fall upon a *convex lens*, from a point less distant than the *focus*, after they have passed the glass, they proceed in such a manner as if they had issued from a more distant point; and since this is the case of the rays which flow from each point in the object, the object must of consequence seem to be more distant than it is; and it must likewise appear greater; for if A B be an object exposed to a naked eye at O, its extream points A and B will be perceived by the eye by means of the rays AO and BO, which flow directly from those points to the eye, but if a *convex lens* as C, be interposed, the eye will no longer perceive the extremities by means of the rays AO and BO, because as they are refracted by the *lens*, they are made to concur before they can reach the eye, the eye therefore must now perceive those points by means of some other rays as AE and BD, which falling upon the glass at a greater distance from each other, are by the refractive power of the glass thrown into the directions EO and DO, and made to concur at O; so that continuing those lines directly backward as far as the object, to wit, to I and H, the eye at O will perceive the extream points of the object as situated at I and H; that is, it will perceive the object magnified. And if the eye be farther removed from the glass suppose to P, the object will appear still greater, its extremities in

Pl. 10.
Fig. 4.

that

that cafe appearing at L and K in the lines PG and PF produced. And on the other hand, if the eye continuing in its place, the object be farther removed from the *lens*, it will appear larger; for whereas at the nearer diftance the eye perceives the extream points of the object by means of the rays AE and BD, which fall upon the *lens* at E and D, and are thence refracted to O; when the object is at the greater diftance, its extremities cannot be feen by means of the rays incident on the glafs at E and D; for fince the interval between the extremities continues the fame, the rays which flow from them and fall upon the *lens* at E and D, will diverge lefs at a greater diftance of the object than at a fmaller; confequently, they will concur before they reach the eye; and therefore in this cafe the extream points of the object muft be conveyed to the eye by fome rays as aG and bF, which diverging more than the former, fall without them at G and F, whence they are refracted to the eye at O, in the lines GO and FO, which being continued backward as far as the nearer diftance of the object, to wit, to L and K, fhew that the object which at the nearer diftance appeared to extend itfelf only from I to H, does at the greater diftance feem to reach from L to K, and of confequence, appears more magnified.

If the object be removed beyond the *focus*, it will appear ftill greater; but whereas before it paffes the *focus* it appears diftinct, as alfo more and more diftant the farther it is removed from the glafs, when it gets beyond the *focus* it appears confufed, and the farther it is removed from the glafs, the more confufed it appears, and the nearer it feems to approach the eye, provided its diftance from the glafs be not fo great as to make it project its image between the eye and the glafs.

This feeming approach of the object at a time when it really recedes, and in a cafe where, according

Y 3 ing

ing to the received principles of *Dioptricks*, it ought
to appear at a diſtance, if poſſible more than infinite, has very much puzzled the writers of *Opticks*,
and was looked upon as an inſuperable difficulty,
till Doctor BERKELEY took it into conſideration
in his *Eſſay upon Viſion*, wherein, among other difficulties which he has cleared up relating to viſion,
he has given us a natural and ſatisfactory account
of this. The ſubſtance of what he has there delivered concerning this matter is, that by cuſtom and experience we are taught to judge thoſe objects near
which appear confuſed, becauſe, according to the
ordinary courſe of nature, thoſe objects, and thoſe
only, appear confuſed which are brought very near
the eye, and therefore if an object ſhall at any
time appear confuſed, tho' from another cauſe, the
mind will immediately connect nearneſs of diſtance
in the object, with that confuſion in the appearance,
as having always experienced them to go together,
and the greater the confuſion is, the nearer it will
judge the object to be, becauſe it has always obſerved the neareſt diſtances to be attended with the
greateſt confuſions: now if in the caſe before us,
Pl. 10.
Fig. 6.
we ſuppoſe A to be an object placed before the
convex lens BC, at a greater diſtance than the *focus*,
the rays after they have paſſed thro' the glaſs will
converge towards ſome point as D; if then an eye
be placed at a little diſtance behind the glaſs, ſuppoſe at E, it will perceive the object confuſed, becauſe as the rays fall upon it converging, they will
be made to meet before they arrive at the fund of
the eye, and conſequently, will be ſcattered on the
retina, and thereby render the appearance confuſed;
if the eye be moved gradually backward to F, G,
and D, or which is the ſame thing, if by carrying
the object forward, the rays be made to fall upon
the eye at leſs and leſs diſtances from the *focus*, they
will be ſcattered more and more upon the *retina*, becauſe the convergency wherewith they fall upon the
eye

eye is by so much the greater, by how much the nearer the eye is placed to the *focus* or the point D; consequently, the object as it is more and more removed from the glass, will appear more and more confused; for which reason, the mind which has been used to connect nearer distances with greater degrees of confusion, will in this case judge the object to approach, tho' in reality it recedes; and what fully confirms this is, that if by placing a concave glass at a proper distance between the eye and the convex, the convergency of the rays be taken off, and the appearance thereby rendered distinct, the object will then appear at its due distance.

If an eye be removed from a *convex lens* beyond the place where the image is projected, that is, if the eye be farther from the *lens* than is the point D, the object will appear in an inverted position, and seem to be situated between the eye and the glass; for in this case, the eye sees only the image or representation of the object, which, as I shewed in a former lecture, is projected at D in an inverted position; upon looking at the image with both eyes, it appears double, and upon shutting either eye, the image on the contrary side disappears; the reason of which is this, the eye at O perceives the image by means of the rays ODC, and therefore sees it on the same side with C, whereas the eye at P perceives it by means of the rays PDB, and on that account sees it on the same side with B; as the head is moved farther back, the distance between the two images must decrease, and at length vanish; for since the interval between the eyes continues unvaried, the rays which exhibit the image to each eye, will diverge less and less as the head is more and more removed from D, as is evident from the bare inspection of the scheme; consequently, the distance between the two images must continually decrease, and at last become so small as to be insensible.

Pl. 10.
Fig. 6.

As

As to concave glasses, since it is their property
to make the rays which flow from any point
to diverge, in such a manner as if they had issued
from a point less distant, it is evident, that an object seen thro' a *concave lens* must appear nearer than
it really is, and it must likewise appear diminished;
Pl. 10.
Fig. 7. for the extream points of the object A B, are seen
by the naked eye by means of the rays AO and
BO, which when the *concave lens* C D is interposed,
are made to diverge, so as not to meet at O, consequently, upon the interposition of the glass, the
eye will not perceive the extremities of the object by
those rays, but by some others as AK and BL,
which falling within the former, are by the refractive power of the glass made to proceed in the
lines KO and LO, so as to meet at O; wherefore
continuing OK and OL backward to the object,
the extremities of the object will be seen at E and
F, that is, the object will appear to be less than it
really is; and by the spreading of the rays in their
passage thro' the glass, some of them are made to
escape the eye, which if the glass were removed,
would fall upon the pupil; for which reason, the
object must appear less luminous; so that the property of concave glasses is to make objects appear
smaller, nearer, and more faint and obscure, than
they do to the naked eye.

LECTURE XXIII.

Of Catoptricks.

Lect.
XXIII. IN this lecture, wherewith I shall close this course,
I shall explain to you the *Doctrine* of Catoptricks, or that part of *Opticks* which treats of the
reflexion

reflexion of light; in doing of which, I shall first say something concerning the cause of that reflexion; Secondly, I shall lay down two principles, which are the chief foundation of *Catoptricks*; and lastly, I shall lay before you the most remarkable properties of plain and spherical mirrors.

As to the first, before Sir Isaac Newton published those wonderful and surprising discoveries which he made, concerning the nature and properties of light, it was an opinion generally received by the writers of *Opticks*, that the rays of light were reflected in the manner of other bodies, by striking on the solid and impervious parts of bodies; but that great Philosopher has fully proved this opinion to be erroneous; and has shewn, that the particles of light are turned back before they touch the reflecting body, by some power of the body which is equally diffused all over its surface; what he has delivered concerning this matter, is to be met with in the eighth *Proposition* of the *second Book of his Opticks*, wherein, after he has offered several reasons to prove, that light is not reflected by striking against bodies, he at last expresses himself in the following manner; " Were the rays of light reflected
" by impinging on the solid parts of bodies, their re-
" flexions from polished bodies could not be so re-
" gular as they are; for in polishing glass with sand,
" putty, or tripoly, it is not to be imagined, that
" those substances can, by grating and fretting the
" glass, bring all its least particles to an accurate po-
" lish, so that all their surfaces shall be truly plain,
" or truly spherical, and look all the same way, so
" as together to compose one even surface. The
" smaller the particles of those substances are, the
" smaller will be the scratches by which they con-
" tinually fret and wear away the glass until it be
" polished; but be they never so small, they can wear
" away the glass no otherwise than by grating and

2   " scratching

"scratching it, and breaking the protuberances,
"and therefore polish it no otherwise than by bring-
"ing its roughness to a very fine grain; so that the
"scratches and frettings of the surface become too
"small to be visible. And therefore, if light were
"reflected by impinging on the solid parts of the
"glass, it would be scattered as much by the most
"polished glass, as by the roughest. So then it re-
"mains a *Problem*, how glass polished by fretting
"substances can reflect light so regularly as it does;
"and this *Problem* is scarce otherwise to be solved,
"than by saying, that the reflexion of a ray is effect-
"ed, not by a single point of the reflecting body,
"but by some power of the body, which is evenly
"diffused all over its surface, and by which it acts
"upon the ray without immediate contact."

Now taking it for granted, that this repelling
power is the true cause of reflexion, if it be supposed to act upon the rays of light in lines perpendicular to the surface of the reflecting body; it
will thence follow, that the angle of incidence, or
the angle contained between the incident ray, and a
line drawn perpendicular to the reflecting surface at
the point of incidence, is equal to the angle of reflexion, or the angle contained between the same
perpendicular and the reflected ray. For if we suppose a ray of light to move in the direction AC,
towards the reflecting surface BCD; and if we suppose that motion to be resolved into two, one in the
direction AE, parallel to BD, and the other in the
direction AB, perpendicular to BD, it is manifest,
that of those two motions, the latter only is opposed to the repelling force; and of consequence, the
ray after reflexion, will go on in the parallel direction, with the same velocity it did before; and forasmuch as the repelling force which opposes the perpendicular motion, acts incessantly, it no sooner de-
stroys the motion of the ray towards the body, but

Pl. 10.
Fig. 8.

it

it gives it an equal degree of motion the contrary way; that is, it throws it back with the same perpendicular velocity wherewith it approached. If therefore EG be taken equal to AE, and from G be let fall GD equal and parallel to AB, EG will exprefs the parallel motion of the ray after reflexion, and DG its perpendicular motion; and the diagonal line CG, will be actually defcribed by the ray, by virtue of its compound motion; and from the nature of fimilar triangles, the angle of incidence ACE, muft be equal to ECG, the angle of reflexion; and this is the firft of thofe principles whereon the doctrine of *Catoptricks* is founded. The fecond is, that every radiant point when feen by reflexion, appears in that place where the reflected ray meets the perpendicular, drawn from the radiant point to the reflecting furface; for inftance, if from a radiant point as R, placed before the plain *speculum* AB, be let fall the line REM, perpendicular to the plane of the *speculum*; and if RC and CD be fo drawn, as that the former may denote the incident ray, and the latter the reflected; and if DC be continued on, till it meets the perpendicular REM; an eye at D will perceive the radiant point, as placed at M, the point of interfection of the reflected ray, and the perpendicular; and thus it is in all cafes of reflexion, except two, wherein this principle feems to fail; one whereof relates to plain glafs *speculums*, and the other to concave fpherical mirrors; the latter has been obferved by TAQUET, Doctor BARROW, and others; but the former has not been mentioned by any one of the *optick* writers that I know of; I fhall take notice of each in its proper place, and proceed now to confider the chief properties of mirrors, and firft, of fuch as are plain.

Pl. 10.
Fig. 9.

When an object is feen by reflexion from a plain *speculum*, its image appears as far behind the *speculum*, as the object is before; for the proof of which,

let

let R be an object placed before the plain *speculum*
AB, and let it be seen by reflexion from the point
C, by an eye situated somewhere in the line CD;
then producing CD, till it meets the perpendicular
REM, the image will, by the second principle,
appear at M; now the angles of incidence and reflexion being equal, their complements are so too,
that is to say, the angle RCE is equal to DCB or
MCE; so that in the two right-angled triangles,
the angles at C being equal, and the side EC
common to both, the triangles must be equal,
and the side ME, that is, the distance of the
image behind the *speculum* must be equal to RE,
the distance of the object before the *speculum*; and
the same thing is in like manner demonstrable,
tho' the point of reflexion be taken different from
C; for the reflected ray will constantly meet the
perpendicular in the point M; whence it follows,
that however the situation of the eye with respect
to the mirror may be changed, yet if the object
and mirror remain unmoved, the image will always appear in the same place; it likewise follows, that there cannot appear more than one
image of one and the same object; but then this
is to be understood with respect to such mirrors, as
being opaque, have but one reflecting surface; for
in looking-glasses, which by reason of their tranf-
parency, have a double reflexion in some certain
positions of the eye and object, several images
Pl. 10.
Fig. 10.
Exp. 1. may be seen. Thus if AB be a looking-glass, R
the flame of a candle, placed at a small distance
before AH, the plane of the glass produced, an
eye being placed at Q, shall see several images
standing at small distances one beyond another, in
the same position with the letters, C, D, E, F,
whereof the first and second appear bright and luminous, and the rest but faint and obscure; for the
several images taken in their order from the second,
grow more and more dark and obscure, till at length
they

they become too weak and feeble to affect the fight, and of confequence vanifh.

In order to account for this multiplicity of images, let A B C D be a looking-glafs, whofe near-eft furface, or that which lies next the eye is A B, and its farther or filvered furface is D C, R the place of the candle, and Q the place of the eye, R S a line drawn from the candle perpendicular to A O and D Y the two furfaces of the glafs pro-duced; the angle R E A being made equal to Q E B, and the line Q E being produced till it cuts the per-pendicular R S in T, the eye fhall fee the firft image at T, by means of the reflexion from the outward furface A B, the ray R E being reflected to the eye from the point E. Let a fecond ray as R G, pafs into the glafs at G, and being refracted to the point H of the farther furface, let it thence be reflected to K, and there paffing out of the glafs, let it by refraction be carried to the eye; let then Q K be pro-duced, and the eye fhall fee a fecond image fituated in that line, and that at a little diftance beyond the perpendicular R S; for if the rays fuffered no re-fraction in paffing in and out of the glafs, the fe-cond image would not be feen by means of the ray R G, but by means of the ray R H, which paffing directly from R to H, is thence reflected directly to Q, and being produced till it cuts the perpen-dicular in X, would exhibit the fecond image at X; but forafmuch as the place of the image is changed by the refraction, and brought nearer to the glafs, if we fuppofe the line Q X to be moved upward about the point Q, till it coincides with the line Q V, in which the fecond image really appears, the point X muft necessarily fall beyond the perpen-dicular, and fo of confequence, muft the place of the image. Let now a third ray as R F, pafs into the glafs at F, and be refracted to L, and from thence let it be reflected to E, and from E to M, and from

Pl. 10.
Fig. 11.

M to

M to N, where let it go out, and be refracted to the eye at Q; then producing QN to W, a third image will appear in that line somewhere beyond the perpendicular; for were there no refraction, the ray which after three reflexions exhibits the third image, would when produced cut the perpendicular in the point S; and therefore, since the line QS is raised up by the refraction, and made to coincide with the line QW, the point S, that is, the place of the third image, must fall beyond the perpendicular.

As a third image is seen by means of three reflexions, so is a fourth by five reflexions, a fifth by seven, a sixth by nine, and so on, according to the progress of the odd numbers, every succeeding image being seen by two reflexions more than the preceding; and this is the true reason why, setting aside the first and second, which being seen each by one single reflexion, appear almost equally bright, every succeeding image appears more dim and faint than the foregoing, the rays of light being rendered more weak and feeble by reflexion.

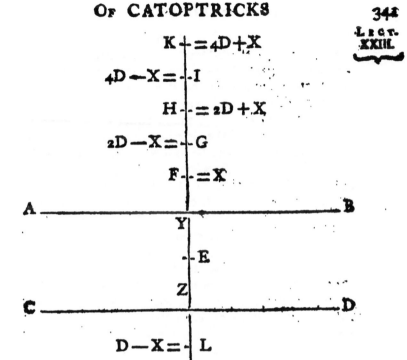

$$K \dotplus = 4D + X$$

$$4D - X = \cdot I$$

$$H \dashv = 2D + X$$

$$2D - X = \cdot G$$

$$F \dashv = X$$

A ———————————————— B

Y

$\dashv E$

Z

C ———————————————— D

$$D - X = \cdot L$$

$$M \dashv = D + X$$

$$3D - X = \cdot N$$

$$O \dashv = 3D + X$$

$$5D - X = \dashv P$$

If two plane *fpeculums* as AB and CD, be fet parallel to one another, and an object be placed any where between them as at E, the rays of light which iffue from the object and fall upon each *fpeculum*, will be reflected backward and forward from one to the other a great number of times; by which means, there will appear in each *fpeculum* a great number of images fituated one behind another, in a right line perpendicular to the *fpeculums*, and paffing thro' the object at E; in order to determine

the

LECT. XXIII. the distances of the several images from the *speculums*, let EY, the distance of the object from the *speculum* AB, be denoted by X, and let ZY, the interval of the glasses, be denoted by D; and let us first consider the reflexion which begins from the *speculum* AB; if YF be taken equal to X, then F will be the place of the first image; and forasmuch as the image at F may be looked upon as an object placed before the *speculum* CD, if ZM be taken equal to FZ, that is, to D + X, there will appear an image at M; which being considered as an object with respect to the other *speculum* AB, and YH being taken equal to MY, or 2D + X, another image will be seen at H; and for the same reason, if ZO be taken equal to HZ, or 3D + X, there will another image appear at O, and so on; again, if we consider the reflexion which begins from the *speculum* CD, by taking ZL equal to EZ, or D—X, we shall have L for the place of an image in the *speculum* CD; and by making YG equal to LY, or 2D—X, we shall have the place of another image in the *speculum* AB; and again, by taking ZN equal to GZ, or 3D—X, we shall have the place of another image in the *speculum* CD, and so on. From this manner of determining the places of several images, it is evident, that if D, which stands for the distance of the glasses, be multiplied into each of the even numbers taken in their order, and if X, which denotes the distance of the object from AB, be subducted from each product, and likewise, added to each, the differences, and the sums taken in their order, will express the distances of the several images from the *speculum* AB, the distance of the first being X; that is to say, the distance of the second image will be 2D—X, of the third 2D + X, of the fourth 4D—X, of the fifth 4D + X, and so on, according to the first Table.

TABLE

TABLE I.

Distances of the several images from the speculum
AB.

$$1 = X$$
$$2 = 2D - X$$
$$3 = 2D + X$$
$$4 = 4D - X$$
$$5 = 4D + X$$
$$6 = 6D - X$$
$$7 = 6D + X$$
$$8 = 8D - X$$

&c.

If D be multiplied into each of the odd numbers taken in their order, and if X be deducted from each product, and likewise added to each, as before, the differences and the sums taken in their order, will express the distances of the several images from the *speculum* CD, as in the second Table.

TABLE II.

The distances of the several images from the speculum
CD.

$$1 = D - X$$
$$2 = D + X$$
$$3 = 3D - X$$
$$4 = 3D + X$$
$$5 = 5D - X$$
$$6 = 5D + X$$
$$7 = 7D - X$$
$$8 = 7D + X$$

&c.

If by moving the object nearer to AB, X becomes less, then all those images whose distances are expressed by those symbols wherein X is affirmative, will come nearer to the *speculums*, whilst those whose distances are expressed by symbols

Z   wherein

wherein X is negative, move farther off; th[e]
the *speculum* AB, the first, third, fifth, sev[e]
and so-on, will approach, and the second, fo[u]
sixth, eighth, and so on, will recede; so that th[e]
veral images, beginning from the first, will
proach and recede alternately; and on the

Exp. 2. hand, those in the *speculum* CD will recede and
proach alternately, beginning from the first; h[e]
if a man puts his hand between the two *specu*
and moves his palm towards one of them, a
son looking into the other, shall see several pai[r]
hands, palm to palm, approaching each other.

Pl. 10.
Fig. 12.
Exp. 3.
If two plane *speculums* as AC and BC, b[e]
clined to one another, so as to meet in an a[n]
angle at C; and if an object be placed any w[ay]
between them, suppose at F, an eye looking
either, shall see several images standing in the
cumference of a circle, whose center is at C the
course of the *speculums*, and its *radius* equal to
the distance of the object from the concourse;
if from F be drawn FD perpendicular to the *s*
lum CA, and KD be made equal to FK, D
be the place of an image in the *speculum* CA;
if from D be drawn DE perpendicular to the
culum CB, and produced till HE is equal to
E will be the place of an image in the *speculum*
and thus by drawing perpendiculars continually
the place last found to the opposite *speculum* m[a]
places of all the images be found which are see[n]
means of those reflexions, the first whereof is
from the *speculum* CA; and in the same ma[n]
by drawing the perpendiculars FG, GL, and s[o]
may the places of all those images be found w[hich]
are seen by means of the reflexions whereof the
is made from the *speculum* CB. Now that
points D and E are in the circumference of
circle whose *radius* is CF, I thus prove in the
angles CFK and CDK, the sides FK and DK
equal by the construction; and CK is commo[n]

th, and the angles at K are right ones, where-
re the two triangles are equal, and of consequence
is equal to CF; again, the triangle CDH is
ual to the triangle CEH; the sides DH and EH
ing by construction equal, as are also the angles
H, wherefore CE is equal to CD, which is equal
CF, consequently, a circle described on the cen-
C, with the *radius* CF, will pass thro' the points
and E; and by the same way of reasoning it
ll be found to pass thro' G and L, and thro' the
tremities of all the other perpendiculars; and
:refore the several images must of necessity appear
the circumference of a circle whose center is at
concourse of the *speculums*, and whose *radius* is
ual to the distance of the object from that con-
urse. From what has been said it follows, that
the distance of the object from the concourse of
speculums be given, the images will still appear
the circumference of the same circle, notwith-
nding any alteration that may be made in the
gle whereat the *speculums* meet; if that be inlarg-
, the images will be fewer in number, and at
:ater distances from one another; and on the
her hand, if it be made less, the images will be
re in number, and stand closer together, but the
cle in whose circumference they appear, will be
same in both cases; for that is not to be lessen-
or inlarged otherwise, than by lessening or in-
ging the distance of the object from the con-
urse of the *speculums*.
Having laid before you the chief properties of
ain *speculums*, I come now to consider such *spe-
ums* as are spherical; and they are of two sorts,
:cave and *convex*; concerning which it must be
served, that as all the rays which fall upon them
m a radiating point, are reflected in such man-
r as to meet the perpendicular very nearly in one
d the same point; in order to find out the *focus*,

Z 2

or the place where the reflected rays cross one another, nothing more is necessary, but to determine the point wherein any one reflected ray meets the perpendicular; which may be done in the following Pl. 10.
Fig. 13. manner; let A be a radiating point, exposed directly before the concave glass FG, whose center is C, AB a perpendicular from the radiating point to the *speculum*, which likewise denotes the distance of the radiating point from the *speculum*, AD a ray falling on the *speculum* at D, whose distance from B is indefinitely small, DE the reflected ray meeting the perpendicular in E, CD a *radius* drawn to the point of incidence, and of consequence bisecting the angle ADE in the triangle ADE; since the angle at D is bisected by the line DC, which cuts the opposite side, AD is to DE, as AC to CE; but forasmuch as the points D and B are supposed to be indefinitely near, AD is equal to AB, and ED is equal to EB, wherefore AB is to EB, as AC is to CE; that is, the distance of the radiating point from the *speculum*, is to the distance of the point E, where the reflected ray cuts the perpendicular, commonly called the *point of interfection*, as the distance of the radiating point, lessened by the *radius*, is to the *radius*, lessened by the distance of the point of interfection; that is, putting D for the distance of the radiating point, F for the distance of the point of interfection, and R for the *radius*, D : F :: D—R : R—F; consequently, reducing this analogy into an equation, and clearing F, F will be found equal to $\frac{DR}{2D-R}$; that is, the distance of the point of interfection from the *speculum*, and consequently, the distance of an image formed by reflexion from a concave *speculum*, is equal to a rectangle under the distance of the object from the *speculum*, and the *radius*, divided by twice the distance of the object lessened by the *radius*.

2

HENCE

Hence it follows, that if an object be placed before a concave *speculum*, at an infinite distance, that is, if the distance be so great as that the *radius* of the *speculum* bears no sensible proportion to it, the image will appear on the same side of the *speculum* with the object, at the distance of one half the *radius* from the *speculum*; for in this case, D being infinite, 2D—R becomes equal to 2D, and of consequence, F is equal to R divided by 2; so that one half the *radius* is the least distance at which an image can be projected from a concave *speculum* on the same side with the object; and forasmuch as the sun's image, which, by reason of the immense distance of his body, is formed at the distance of half the *radius* from the *speculum*, is there apt to burn, that place is usually called the *focus* or *burning point*.

As the object approaches the *speculum*, the image recedes; for as in one and the same *speculum*, the *radius* is a standing quantity, it is manifest, that as D lessens, the proportion of DR to 2D—DR must increase, consequently, F, or the distance of the image from the *speculum* must do so too; and when the object has approached so near the *speculum* as to be at the center, the image will have receded so far as to be there likewise; for in this case, D being equal to R, 2D—R is equal to R, and of consequence, F is equal to D; so that the object and its image meet at the center of the *speculum*; upon the object's passing from the center towards the glass, the image is projected beyond the center, and when the object has approached so near the *speculum*, as to be distant from it but half the *radius*, the image is at an infinite distance; for in this case, D being equal to half the *radius*, 2D—R is nothing, consequently, F, that is the distance of the image, is infinite; or to speak more properly, the rays after reflexion proceed parallel; for which reason, if the flame of a candle be placed directly before a con-

cave

cave *speculum*, at the diftance of half the radius, the *speculum* will feem to be in flames, and the reflected light will be fo intenfe, as that by the help of it one may be able to read at a very confiderable diftance from the *speculum*. TAQUET afferts, that he has read at the diftance of no lefs than 400 feet; and to fay the truth, the diftance would be without limits, were it not for the atmofphere, whofe particles continually intercept the rays, and by fo doing, at length totally extinguifh the light. It fometimes happens, that when the flame of a candle is placed in the *focus* of a concave *speculum*, its image is projected on a diftant wall, which feems to invalidate the truth of what I juft now proved concerning the parallelifm of the rays after reflexion, but this is occafioned by the flames being too large to be contained totally within the *focus*, for were it fo fmall as to lie wholly within the *focus*, it would not project an image, but the rays after reflexion, would form a cylindrical body of light, which when projected on a diftant wall, would have a circular figure, of an equal circumference with the *speculum*.

When the diftance of the object from the *speculum* is lefs than half the *radius*, the image appears behind the *speculum*; for in this cafe, 2D—R is a negative quantity, and of confequence, fo is F, which fhews, that the diftance of the image which is denoted by F, muft be taken on the other fide of the *speculum*, with refpect to the object; as the object moves nearer to the *speculum* before, fo likewife does the image behind; and when the object is fo near as to touch the *speculum*, the image does the fame; for in this cafe, D being nothing, F, that is, the diftance of the image from the *speculum*, is likewife nothing.

As to the pofition of the images which are feen by reflexion from a concave *speculum*, thofe which
                                        appear

appear on the fame fide of the *fpeculum* with the object muft be inverted, and thofe which appear behind the *fpeculum* muft be erect. For the proof of which, let A B be an object placed before the concave *fpeculum* F G, at any diftance beyond the center C, in which cafe the image will be feen between the center and the *fpeculum*, fuppofe at D E ; from A and B, the extream points of the object, let the lines A H and B I be drawn perpendicular to the *fpeculum*, and of confequence crofling one another at the center ; this being done, fince the image is fuppofed to be at D E, and fince every point of an image is feen in the perpendicular drawn from the correfponding point in the object, it is manifeft, that D, the loweft point of the image, will correfpond to A, the higheft point of the object, and E, the higheft point of the image, will correfpond to B, the loweft point of the object, that is, the image will appear inverted. And by the fame way of reafoning, if D E be the object, fituated at fuch a diftance between the *fpeculum* and the center, as to have its image projected beyond the center at A B, the image muft appear inverted. On the other hand, where an object as D E, is placed between the *fpeculum* and the center, and confequently, projects an image behind the *fpeculum* ; for it muft be obferved, that the fame object D E, which when fituated between the center and the *fpeculum*, at a lefs diftance from the center than half the *radius*, projects an image as A B beyond the center, does likewife project another image as H I, behind the *fpeculum* ; and as the former image is vifible to an eye placed beyond it, fo the latter image is vifible to an eye placed between the object and the *fpeculum*, and it muft appear erect, inafmuch as the perpendicular C H, which paffes thro' D the higheft point of the object, does likewife pafs thro' H, the higheft point of the image.

A s

As to the magnitudes of an object and its image, they are to one another in the same proportion with the squares of their distances from the *speculum*; for if the line LCM be drawn thro' the center C, perpendicular to BA, DE, and HI, and consequently bisecting the angle at C, if BA be the length or breadth of an object, and DE the length or breadth of its image projected on this side the *speculum*, then LA and OD will be half the length or breadth of the object and its image, and the triangle CLA being similar to COD, LA is to OD, and consequently BA to ED, as LC to OC, that is, the length or breadth of the object, is to the length or breadth of its image, as the distance of the object from the center of the *speculum*, to the distance of

Pl. 10.
Fig. 13. its image from the same center; but it has been proved, that as AC, the distance of the object from the center, is to EC, the distance of the image from the center, so is AB, the distance of the object from the *speculum*, to EB, the distance of the image from the *speculum*; consequently, the length or breadth of an object, is to the length or breadth of its image, as the distance of the object from the *speculum*, to the distance of the image from the *speculum*; and forasmuch as similar surfaces are to one another, as the squares of their homologous sides, the magnitude of the object, is to the magnitude of the image, as the square of the object's

Pl. 10.
Fig. 14. distance from the *speculum*, to the square of the image's distance. And by the same method of arguing, if DE be an object whose image behind the *speculum* is HI, the magnitude of the former will be found to be to the magnitude of the latter, as the square of KO, to the square of KM. Hence it follows, that the object during its continuance beyond the center, must appear larger than its image, as being more distant from the *speculum*, and when it is in the center, where it meets the image, it

must

muſt appear equal to it, but being on the ſame ſide of the center with the *ſpeculum*, it muſt be leſs than its image, which in that caſe lies beyond the center, and conſequently, is at a greater diſtance from the *ſpeculum*.

It likewiſe follows, that the image which appears behind the *ſpeculum* is ever larger than the object; for ſince MK, the diſtance of the image behind the *ſpeculum*, is to OK, the diſtance of the object before the *ſpeculum*, as MC, the diſtance of the image from the center, to OC, the diſtance of the object from the center; and ſince in this caſe, the object is always leſs diſtant from the center than its image, during the appearance of the image behind the *ſpeculum*, it is evident, that the image muſt appear larger than the object; but then this is to be underſtood with reſpect to ſuch images only, as are projected by objects leſs diſtant than the center; for if an object be beyond the center, an eye being cloſe to the *ſpeculum*, ſhall ſee the image at the ſame diſtance, and of an equal magnitude with the object; and in this caſe, the ſeveral parts of the image do not appear in thoſe points where the perpendiculars from the correſponding points of the object meet with the reflected rays; the reaſon of all which ſeems to be this, the portion of the *ſpeculum* which the eye makes uſe of in this caſe is ſo exceedingly ſmall, that notwithſtanding the ſpherical figure of the *ſpeculum*, it may be looked upon as plane, and conſequently, the appearances muſt be the ſame as in other plain *ſpeculums*; that is, the image muſt appear as far behind the *ſpeculum* as the object is before it, and of the ſame magnitude with the object.

Pl. 10.
Fig. 14.

If an image formed on this ſide a concave *ſpeculum* be looked at with both eyes, it will appear double, provided the diſtance of the eyes from the image be but ſmall, and upon ſhutting either eye,

the

LECT. the contrary image will difappear; for fince the re-
XXIII. flected rays which form the feveral points of an
image meet and crofs one another at the image,
thofe which enter the right eye muft be reflected
from the left fide of the *speculum*, and thofe which
fall upon the left eye, muft be reflected from the
right fide of the *speculum*, and of confequence, one
and the fame point of the image muft appear to the
right eye, as fituated before the left fide of the *speculum*, and to the left eye, as fituated before the
right fide of the *speculum*; that is, it muft appear
double, and the right or left image muft vanifh up-
Pl. 10. on clofing the contrary eye. Thus, if the point
Fig. 15. C of the image A B, be looked at with both eyes,
one whereof is at O, and the other at Q, the eye
at O fhall fee it by means of the rays O N, which
are reflected from N, and of confequence, fhall fee
it as placed before N, but the eye at Q feeing it by
means of the rays Q M, which proceed from M,
fhall fee it as fituated before M, for which reafon,
the point C will appear double; and what has been
thus fhewn with refpect to the point C, may in the
fame manner be fhewn, with regard to all the other
points in the image, and therefore the whole image
muft appear double; as the eyes are more and more
removed from the images, they approach nearer
together, and at length coincide; the reafon of
which is plain, from the bare infpection of the figure; for fince the interval of the eyes continues
the fame, it is evident, that when they are farther
removed from the image, the rays whereby they fee
the point C muft be reflected from parts of the
speculum lefs diftant from one another than M and
N, and the diftance of the parts of the *speculum*
which reflect the rays to each eye, muft continually leffen as the eyes are more and more removed
from the image, and at certain diftances of the
eyes, muft become fo fmall as not to be fenfible.

5   And

And thus much concerning such spherical *specu-lums* as are concave; as to *convex speculums*, in order to determine the places of images formed by reflexion from them, let A be a radiating point, exposed directly before the *convex speculum* HK, whose center is C, AB a perpendicular from the radiating point to the *speculum*, which likewise denotes the distance of the radiating point from the *speculum*, AD a ray falling on the *speculum* at D, whose distance from B is indefinitely small, DE the reflected ray meeting the perpendicular in E, CD a *radius* drawn to the point of incidence, and of consequence bisecting the angle FDE; let the angle FCD be made equal to ECD, and let CF be continued till it meets AD produced; this being done, it is evident, that the angle at C in the triangle ACF, is bisected by the line CD, which cuts the opposite side, consequently, AC is to FC, as AD to DF; but forasmuch as D and B are supposed to be indefinitely near, AD is equal to AB, and DE to BE; and because the triangles CFD and CED are equal, DF is equal to DE, and FC is equal to CE; wherefore, AB is to BE, as AC to CE; that is, the distance of the radiating point from the *speculum*, is to the distance of the point E where the reflected ray cuts the perpendicular, which is called the *point of intersection*, as the sum of the distance of the radiating point and the *radius*, to the *radius* lessened by the distance of the point of intersection; that is, putting D for the distance of the radiating point, F for the distance of the point of intersection, and R for the *radius* as before, $D : F :: D+R : R—F$; consequently, reducing this analogy into an equation, and clearing F, F' will be found equal to $\dfrac{DR}{2D+R}$; that is, the distance of the point of intersection behind the *speculum*, and consequently the distance of an image

behind

behind the *speculum*, is equal to a rectangle under the distance of the object from the *speculum* and the *radius*, divided by the sum of twice the distance of the object added to the *radius*. Hence it follows, that if an object be placed so near a *convex speculum* as to touch it, its image will do so too; for in this case, D being nothing, F is likewise nothing; as the object recedes from the *speculum*, the image goes off behind; and when the object is removed to an infinite distance, the image appears behind in the midway between the *speculum* and its center; for in this case, D being infinite, 2D + R becomes 2D, and of consequence, F is equal to $\frac{R}{2}$, so that objects seen by reflexion from convex spherical *speculums*, appear constantly behind the *speculum*, within the limits of half the *radius*; and forasmuch as the images constantly appear on the same side of the center with the objects, they must be less than the objects; for if we suppose HI to be an object placed before the *convex speculum* FG, and projecting its

Pl. 10.
Fig. 14.
image at DE, it is manifest, that the image subtends the same angle at a smaller distance, than the object does at a larger distance, and consequently, must be less; and the disproportion between the object and its image, must increase as the object recedes, and decrease, as it approaches, because, as the object recedes from the center, the image approaches, and as that approaches, the image recedes; but as the image can never be more distant from the center than the object, it can in no case appear larger. The proportions which the magnitudes of the object and its image bear one to another, is the same with the squares of their distances from the *speculum*, as in the case of concave *speculums*; the proof of which being exactly the same with that made use of in the case of concaves, I shall not here repeat it.

As

As to the pofition of fuch images as are feen by L e c t. XXIII. reflexion from convex fpherical *fpeculums*, they muft always appear erect; for as they ever appear on the fame fide of the center with the objects, the perpendiculars which are drawn from the uppermoft parts of the objects, muft pafs thro' the uppermoft parts of the images; and thofe from the lower parts of the objects, muft likewife pafs thro' the lower parts of the images; thus, the perpendicular HC, which comes from H, the higheft point Pl. 10. in the object, paffes thro' D, the higheft point of Fig. 14. the image, and IC, which comes from I, the loweft point of the object, paffes thro' E, the loweft point of the image; and fo it is with regard to the perpendiculars which come from the intermediate points; fo that the feveral parts of the image have the fame fituation with the correfponding parts of the object, and of confequence the image appears erect.

APPEN

APPENDIX.

Of the COLLISION of NON-ELASTICK and ELASTICK BODIES

PROBLEM I.

IF *two bodies be either entirely void of. elasticity or perfectly elastick, and one strike the other directly; if* A *and* B *denote the quantities of matter or weights of the two bodies,* a *and* b *their velocities before the stroke; and if* A *be the swifter body when the bodies move the same way, the body which has the greater motion when they move contrary ways, and the moving body when one of them is at rest before the stroke; to determine the* ratio *of the bodies when their velocities before the stroke are given, or the* ratio *of their velocities before the stroke when the bodies are given; that is, to determine* $\frac{A}{B}$ *when* a *and* b *are given, or* $\frac{a}{b}$ *when* A *and* B *are given; so as that the motion of* A *before the stroke, shall be to its motion after the stroke, in the given* ratio *of* m *to* 1.

To give a solution of this *Problem,* it is neceſſary to know the motions of A before and after the ſtroke, both when the bodies are entirely void of elaſticity, and when they are perfectly elaſtick; and likewiſe to know the motion of A after the ſtroke, when the bodies move the ſame way, when they move contrary ways, and when B is at reſt before the ſtroke. The motion of A before the ſtroke, is Aa in all caſes. And from what has been delivered by

our

bf

is

ve

fe-

e-

o-

er

ve

re

nd

fix

of

af-

ay,

ye

ys,

we

ke,

A

O F

I

weig
fore
bodi
grea
movi
ſtrok
veloc
their

given

given

the n
after

T
to kn
both
and
to k
the
contr
ſtrok
in all

ſour Author, when the bodies are entirely void of elaſticity, the motion of A after the ſtroke, is $\frac{AAa + ABb}{A + B}$ when before the ſtroke the bodies move the ſame way, $\frac{AAa - ABb}{A + B}$ when they move different ways before the ſtroke, and $\frac{AAa}{A + B}$ when before the ſtroke B is quieſcent. And when the bodies are perfectly elaſtick, the motions of A after the ſtroke, when before the ſtroke the bodies move the ſame way, contrary ways, and B is quieſcent, are $\frac{2ABb + AAa - ABa}{A + B}$, $\frac{AAa - ABa - 2ABb}{A + B}$, and $\frac{AAa - ABa}{A + B}$. Hence, this *Problem* contains ſix *Caſes*, three when the bodies are entirely void of elaſticity, and three when they are perfectly elaſtick; which *Caſes* are thus ſolved.

When the bodies are entirely void of elaſticity. —

CASE I. If the bodies move the ſame way, Aa will be to $\frac{AAa + ABb}{A + B}$, as m to 1; whence we have $\frac{A}{B} = \frac{a - mb}{ma - a}$, and $\frac{a}{b} = \frac{mB}{A + B - mA}$.

CASE II. If the bodies move contrary ways, Aa will be to $\frac{AAa - ABb}{A + B}$, as m to 1; whence we have $\frac{A}{B} = \frac{mb + a}{ma - a}$, and $\frac{a}{b} = \frac{mB^1}{mA - A - B}$.

CASE III. If B be at reſt before the ſtroke, then will Aa be to $\frac{AAa}{A + B}$, as m to 1; whence we have

have $\frac{A}{B} = \frac{1}{m-1}$. In this case b is nothing, and consequently $\frac{a}{b}$ is infinite.

When the bodies are perfectly elastick.

CASE IV. If the bodies move the same way, Aa will be to $\frac{2ABb + AAa - ABa}{A + B}$, as m to 1; whence we have $\frac{A}{B} = \frac{ma + a - 2mb}{ma - a}$, and $\frac{a}{b} = \frac{2mB}{mB + A + B - mA}$.

CASE V. If the bodies move contrary ways, Aa will be to $\frac{AAa - ABa - 2ABb}{A + B}$, as m to 1; whence we have $\frac{A}{B} = \frac{2mb + ma + a}{A + B}$, and $\frac{a}{b} = \frac{2mB}{mA - A - B - mB}$.

CASE VI. If B be at rest before the stroke, Aa will be to $\frac{AAa - ABa}{A + B}$, as m to 1; whence we have $\frac{A}{B} = \frac{m+1}{m-1}$. In this case b is nothing, and consequently $\frac{a}{b}$ is infinite.

EXAMP. I. If the bodies be entirely void of elasticity, and move the same way, A with a velocity of 7, and B with a velocity of 3; and A lose half its motion by the stroke, or, which amounts to the same, if the motion of A before the stroke be to its motion after, as 2 to 1. In this case, a, b, m,

B, m, are 7, 3, 2 ; and $\frac{A}{B}$, which is equal to $\frac{a - mb}{ma - a}$ by *Cafe* 1, will be equal to $\frac{1}{7}$; fo that A and B will be as 1 and 7. Here Aa, the motion of A before the ftroke, is 7, and $\frac{AAa + ABb}{A + B}$, its motion after the ftroke is $3\frac{1}{2}$; but 7 is to $3\frac{1}{2}$, as 2 to 1.

EXAMP. II. If the bodies be entirely void of elafticity, and move the fame way, if A and B be as 1 and 4, and the motion of A before the ftroke be to its motion after, as 3 to 1, in which cafe m will be 3 ; then $\frac{a}{b}$, which is equal to $\frac{mB}{A + B - mA}$ by *Cafe* 1, will be equal to $\frac{6}{1}$; fo that a and b will be as 6 and 1. Here Aa, the motion of A before the ftroke, is 6, and $\frac{AAa + ABb}{A + B}$, its motion after the ftroke by *Cafe* 1, is 2 ; but 6 is to 2, as 3 to 1.

EXAMP. III. If B be at reft before the ftroke, and the motion of A before the ftroke, be to its motion after, as 10 to 1, in which cafe m will be 10 ; then will $\frac{A}{B}$ be $\frac{1}{9}$, or A and B will be as 1 and 9. If the velocity of A before the ftroke be expreffed by 1, that is, if a be 1, then will Aa be 1, and $\frac{AAa}{A + B}$ be $\frac{1}{10}$; but 1 is to $\frac{1}{10}$, as 10 to 1.

It is to be obferved, that A can never communicate all its motion to B, except when it is infinitely greater than B, in which cafe B will become nothing. For if A communicate all its motion to B, m will be 1 ; and $\frac{A}{B}$, which is as $\frac{1}{m - 1}$, will be as $\frac{1}{0}$; but $\frac{1}{0}$ is infinite ; and therefore A muft be infinitely

A a

finitely greater than B, to lose all its motion by the stroke.

EXAMP. IV. If the bodies be perfectly elastick; and move the same way with velocities which are as 3 and 2 ; and if the motion of A before the stroke be to its motion after, as 2 to 1 ; then will a, b, m, be 3, 2, 2; and $\frac{A}{B}$, which is as $\frac{ma + a - 2mb}{ma - a}$ by *Case* 4, will be $\frac{1}{3}$; so that A and B will be as 1 and 3. Here Aa, the motion of A before the stroke, is 3 ; and $\frac{2ABb + AAa - ABa}{A + B}$, its motion after the stroke, is $\frac{3}{2}$; but 3 is to $\frac{3}{2}$, as 2 to 1.

EXAMP. V. If the bodies be perfectly elastick, and move the same way, if A and B be as 4 and 5, and the motion of A before the stroke be to its motion after, as 3 to 1 ; then will A, B, m, be 4, 5, 3; and $\frac{a}{b}$, which is as $\frac{2mB}{mB + A + B - mA}$ by *Case* 4, will be $\frac{5}{2}$, so that a and b will be as 5 and 2. Here, Aa, the motion of A before the stroke, is 20 ; and $\frac{2ABb + AAa - ABa}{A + B}$, the motion of A after, is $\frac{60}{9}$; but 20 is to $\frac{60}{9}$, as 3 to 1.

EXAMP. VI. If A and B be perfectly elastick, and B be at rest before the stroke, if A move with a velocity of 4, and its motion before the stroke be to its motion after, as 3 to 1 ; then will a, b, m, be 4, 0, 3; and $\frac{A}{B}$, which is as $\frac{m + 1}{m - 1}$ by *Case* 6, will $\frac{4}{2} = \frac{2}{1}$; so that A and B will be 2 and 1. Here, Aa, the motion of A before the stroke, is 8 ; and $\frac{AAa - ABa}{A + B}$, its motion after the stroke, is $\frac{8}{3}$; but 8 is to $\frac{8}{3}$, as 3 to 1.

SCHOLIUM.

SCHOLIUM.

If it be required to know the motion of B after the ſtroke in the ſix *Caſes* before mentioned, that motion may be had, from what our Author has delivered, when the weights of the bodies, and their velocities before the ſtroke, are given.

If the bodies be intirely void of elaſticity; the motion of B after the ſtroke, when before the ſtroke, the bodies move the ſame way, when they move contrary ways, or when B is quieſcent, is

$$\frac{BAa + BBb}{A+B}, \quad \frac{BAa - BBb}{A+B}, \quad \text{or} \quad \frac{BAa}{A+B}.$$

And if the bodies be perfectly elaſtick; the motions of B after the ſtroke, when before the ſtroke the bodies move the ſame way, when they move contrary ways, or when B is quieſcent, is

$$\frac{2BAa + BBb - BAb}{A+B}, \quad \frac{2BAa - BBb + BAb}{A+B}, \quad \text{or} \quad \frac{2BAa}{A+B}.$$

PROB. II. *If two bodies A and B be given, and be perfectly elaſtick, if A be the leſſer body, and B be at reſt before the ſtroke; it is required to find an intermediate body of ſuch a weight or quantity of matter, which I ſhall denote by x, as that A ſtriking x at reſt, and x with the motion acquired by the ſtroke ſtriking B at reſt, the motion produced in B ſhall be greater than can be produced by an intermediate body of any other weight, or, in other words, that the motion in B ſhall be a* maximum.

The motion of x after it is ſtruck by A, is $\frac{2Aax}{A+x}$, and the motion of B after it is ſtruck by x, is $\frac{4ABax}{AB + Ax + Bx + xx}$, by *Schol. Prob.* I.

But

But by suppofition the motion of B is a *maximum*, and confequently its fluxion is nothing. The fluxion therefore of $\dfrac{4ABax}{AB + Ax + Bx + xx}$ is nothing;

that is, $\dfrac{4A^2B^2ax - 4ABxax^2}{\overline{AB + Ax + Bx + xx}|^2} = 0$. Confequently,

$4A^2B^2ax - 4ABxax^2 = 0$; and, by dividing by $4ABax$, $AP - x^2 = 0$; and $AB = x^2$; whence x is a mean proportional between A and B.

Our Author has given a clear folution of this *Problem*, but in a different manner.

Cor. I. If a number of bodies be in a continual geometrical progreffion, if the leaft of the bodies be A, the *ratio* of the increafe be e, and the number of bodies n; and if A ftrike the fecond body at reft, and the fecond with the motion acquired ftrike the third body at reft, and fo on to the laft; the bodies, their velocities and motions, will be thus expreffed.

Bodies - - A, eA, e^2A, e^3A &c. $\overline{e}|^{n-1}A$.

Velocities - - a, $\dfrac{2a}{1+e}$, $\dfrac{4a}{\overline{1+e}|^2}$, $\dfrac{8a}{\overline{1+e}|^3}$, &c. $\overline{\dfrac{2}{1+e}}|^{n-1}$ a.

Motions - - Aa, $\dfrac{2Aae}{1+e}$, $\dfrac{4Aae^2}{\overline{1+e}|^2}$, $\dfrac{8Aae^3}{\overline{1+e}|^3}$, &c. $\overline{\dfrac{2e}{1+e}}|^{n-1}$ Aa.

Examp. I. If the number of bodies increafing in geometrick proportion be 20, and the common *ratio* of the terms be 2, n will be 20, and e be 2. The laft body will be 524288 times greater than the firft; the velocity of the laft will be to the velocity of the firft, as 1 to $2216\frac{4}{7}$; and the motion of the laft will be about $236\frac{1}{7}$ times greater than the motion of the firft.

Examp.

Exam. II. If the number of bodies be 100, and he common *ratio* of the progreſſion be 2 ; then will 1 be 100, and e will be 2. In this caſe, the laſt body vill be above 6338253000000000000000000000000 times greater than the firſt, its velocity will be to the velocity of the firſt, as 1 to 271022000000000000 nearly ; and the motion of the laſt will be to the motion of the firſt, nearly as .2338480000000 to 1,

Cor. II. If the motion of the firſt body be to the motion of the laſt, as 1 to D, that is, if Aa be to $\overline{\frac{2e}{1+e}}\Big]^{n-1}$ Aa, as 1 to D, then will e be equal

to $\dfrac{D^{\frac{1}{n-1}}}{2-D^{\frac{1}{n-1}}}$

For example, if the number of bodies be 20, and the motion of the laſt be 100000 times greater than the motion of the firſt, n will be 20, D will be 100000, and e will be 10.9746 nearly ; ſo that each preceding body in the 20 bodies muſt be 10.9746 times greater than the body lying next behind it.

Cor. III: If the motion of the firſt body be to the motion of the laſt, as 1 to D, that is, if Aa be to $\overline{\frac{2e}{1+e}}\Big]^{n-1}$ Aa, as 1 to D, D will be equal to $\overline{\frac{2e}{1+e}}\Big]^{n-1}$, and putting R for $\dfrac{2e}{1+e}$, and L, for logarithm, we ſhall have $D = R^{n-1}$, and L, $D = \overline{n-1} \times$ L, R.

For example, if e be 4, and n be 25, $\dfrac{2e}{1+e}$ will

be

be $\frac{4}{7}$, the logarithm of which number is 0.2041199 $= L, R$; and $\overline{n-1} \times L, R = 4.8988795 = L, D$. The natural number of this logarithm is 79228 nearly; so that in this case the motion of the last body will be nearly 79228 times greater than the motion of the first.

COR. IV. If D and R be given, n may be found by being equal to $\dfrac{L, D + L, R}{L, R}$; for by the last *Corollary* $L, D = \overline{n-1} \times L, R$, and consequently $n = \dfrac{L, D + L, R}{L, R}$.

For example, if D be 100000, and e be 2, in which case R will be $\frac{4}{7}$; then will L, D, be 5.000000 and L, R 0.1249387; and $\dfrac{L, D + L, R}{L, R}$, will be 41.02 $= n$; so that more than 41 bodies will be necessary to make the motion of the last 100000 times greater than the motion of the first.

Of the Motion *of a* Globe *in a* Fluid Medium.

PROB. III. *If the diameter and density of a Globe moving in a fluid medium, if the density of the medium, the velocity with which the globe sets out, and the time of the motion, be all given; to determine the part of the velocity which is destroy'd by the resistance of the medium, the remaining part of the velocity, and the space described by the globe in the same time.*

Let D denote the diameter of the globe, d its density, ∂ the density of the fluid medium, V the velocity with which the globe sets out, t the time of the motion expressed in seconds, m the part of a diameter or number of diameters of the globe which it would describe with the velocity V in the time t,

and

and T the time in which the globe with the velocity V would *in vacuo* defcribe a fpace which is to $\frac{8D}{3}$ as d to δ; and then the part of the velocity deftroyed by the refiftance of the medium, will be $\frac{m\delta V}{\frac{2}{3}d + m\delta}$; the remaining part of the velocity will be $\frac{\frac{2}{3}dV}{\frac{2}{3}d + m\delta}$; and the fpace defcribed in the *medium* in the time t, will be $\frac{8Dd}{3\delta} \times$ Log. $\overline{1 + \frac{m\delta}{\frac{2}{3}d}}$ $\times$ 2. 302585093.

For Sir Isaac Newton has proved, that the part of the velocity which is deftroy'd by the refiftance of the medium in the time t, is $\frac{Vt}{T+t}$; that the remaining part, is $\frac{VT}{T+t}$; and that the fpace defcribed in the time t, is $TV \times$ Log. $\overline{\frac{T+t}{T}} \times$ 2.302585093. But by conftruction, T is as $\frac{8Dd}{3\delta V}$, and V is as $\frac{mD}{t}$. And therefore, by fubftituting $\frac{8Dd}{3\delta V}$ and $\frac{mD}{t}$ inftead of T and V in the foregoing expreffions, the part of the velocity deftroyed by the refiftance of the medium in the time t will be $\frac{m\delta V}{\frac{2}{3}d + m\delta}$, the remaining part of the velocity will be $\frac{\frac{2}{3}dV}{\frac{2}{3}d + m\delta}$, and the fpace defcribed in the time t will be $\frac{8dD}{3\delta} \times$ Log. $\overline{1. + \frac{m\delta}{\frac{2}{3}d}} \times$ 2.302585093.

Cor. 'I. If the denfity of the globe be equal to the denfity of the *medium*, that is, if d be equal to δ, the velocity deſtroy'd by the reſiſtance of the *medium* in the time t, will be $\dfrac{mV}{\frac{s}{7}+m}$.

This *Corollary* will obtain, if the globe and the *medium* be perfectly denſe or void of pores; for by being entirely void of pores, they will have equal denſities. And ſuch a globe moving in ſuch a *me-dium* the length of 3 times its diameter, will loſe above half its velocity; for if m be 3, $\dfrac{mV}{\frac{s}{7}+m}$ will be $\dfrac{9V}{17}$. And this will always be the velocity loſt in moving three times the length of the diameter, when the globe and the *medium* have equal denſities.

Cor. II. If a globe in moving through m times its diameter in a fluid *medium*, loſe the n part of its velocity; then will $n = \dfrac{m\delta}{\frac{s}{3}d + m\delta}$, $d =$ $\dfrac{\delta \times \overline{m - nm}}{\frac{s}{3}n}$, $\delta = \dfrac{\frac{s}{3}dn}{m - nm}$, and $m = \dfrac{8nd}{3\delta - 3n\delta}$. For $\dfrac{m\delta V}{\frac{s}{3}d + m\delta} = nV$; whence $n = \dfrac{m\delta}{\frac{s}{3}d + m\delta}$, $d = \dfrac{\delta \times \overline{m - nm}}{\frac{s}{3}n}$ $\delta = \dfrac{\frac{s}{3}dn}{m - nm}$, and $m =$ $\dfrac{8nd}{3\delta - 3n\delta}$.

Examp. I. If a globe loſe $\frac{1}{4}$ of its velocity in moving the length of 10 times its diameter in wa-ter, in which caſe n will be $\frac{1}{4}$, m will be 10, and δ will be 1; then d will be $\frac{5}{4}$, that is, the globe will be denſer than water in the proportion of 5 to 4.

Examp. II. If a globe 10 times as denſe as water, loſe $\frac{1}{5}$ths of its velocity in moving 10 times its diameter in a fluid; the denſity of that fluid will

will be 8 times as great as the denfity of water.
In this cafe d is 10, m is 10, and n is $\frac{1}{4}$; and
δ, which is equal to $\dfrac{\frac{3}{2}dn}{m - nm}$, is 8.

EXAMP. III. If a globe twice as denfe as water,
lofe $\frac{3}{4}$ths of its motion by moving in a fluid 14
times as denfe as water; it will fuffer this lofs of
velocity in moving the length of $1\frac{1}{7}$ D. For in
this cafe d, δ, n, are 2, 14, $\frac{3}{4}$; and m, which is
equal to $\dfrac{8nd}{3\delta - 3nd}$, will be $\frac{8}{7} = 1\frac{1}{7}$.

EXAMP. IV. If a perfectly folid globe move 24
times the length of its diameter in a perfectly folid
medium, it will lofe 9 parts in 10 of the velocity it
had at the beginning of the motion. For in this
cafe d is equal to δ, and m is 24; and n, which is
equal to $\dfrac{m}{\frac{8}{3} + m}$, will be equal to $\frac{72}{80} = \frac{9}{10}$.

EXAMP. V. If a globe of equal denfity with
water, move half the length of its diameter in air,
it will lofe the $\dfrac{1}{4587\frac{1}{2}}$ part of its velocity, on fuppofition that the denfity of water is to the denfity
of air, as 860 to 1. For in this cafe, d, δ, m, are
860, 1, $\frac{1}{4}$; and n, which is equal to $\dfrac{m\delta}{\frac{8}{3}d + m\delta}$, will
be $\dfrac{1}{4587\frac{1}{2}}$.

EXAMP. VI. If the earth moved round the
fun in a fluid *medium* of equal denfity with the air
at the furface of the earth, it would by the refiftance
of the *medium* lofe almoft all its motion in 10000
years, on fuppofition that the denfities of the earth,
of water, and of the *medium*, are 5, 1, $\frac{1}{850}$, or in
decimals

decimals 0.0011628. For the earth moves in its orbit with a velocity that carries it at the rate of 489393878279I miles, or 617142343 times the length of its own diameter in 10000 years, on suppofition that the fun's horizontal parallax is 10½ feconds. In this cafe therefore d, δ, m, are 5, 0.0011628, and 617142343; and confequently n, which is equal to $\frac{m\delta}{\frac{1}{3}d + m\delta}$, will be a $\frac{34771}{71711}$th part, which is nearly the whole, of its prefent velocity.

By the *French* meafures, a degree of a great circle of the earth contains 342366 *Paris* feet, or 365403.3158 *Englifh* feet, on fuppofition that a *Paris* foot is to an *Englifh* foot, as 1142 to 1070. And confequently the diameter of the earth, fuppofing the earth to be fpherical, will be 41870881 *Englifh* feet, or 7930 miles. The mean diftance of the fun from the earth, reckoning the parallax at 10½ feconds, is about 19644.2675 femidiameters of the earth, or 77889520.6375 miles; confequently the circumference of the earth's orbit is 489393878.2791 miles, which the earth defcribes in one year, or 29558161.6 feconds of time.

EXAMP. VII. If the earth move in an *Æther* 700000 rarer than the air at the furface of the earth, it will lofe about $\frac{1}{14}$th part of its prefent velocity in 10000 years; for in this cafe d, δ and m, are 5, 0.00000000166, and 617142343; and confequently n, which is equal to $\frac{m\delta}{\frac{1}{3}d + m\delta}$, will be equal to a $\frac{1.02445628938}{14.35778962271}$th part, that is $\frac{1}{14}$th part of the prefent velocity very nearly.

And if the earth moves 100000 years in this *Æther*, it will lofe almoft half of its prefent motion in that time.

EXAMP.

EXAMP. VIII. If we suppofe the earth to lofe the $\frac{1}{100}$th part of its prefent velocity by moving in an *ætherial medium* for 400000 years, in which time it will have defcribed 24685693680 times its diameter, the denfity of the *medium* will be above 200 millions of times lefs than the denfity of the air at the furface of the earth. For in this cafe d, n, m, are 5, 0.01, 24685693680, and confequently ♂, which is equal to $\frac{\frac{1}{3}dn}{m-nm}$, will be

$$\frac{0.1333333}{24438836743.2000000} = \frac{1}{183366000000}.$$ But the denfity of water being 1, the denfity of air is $\frac{1}{850}$; and confequently, the denfity of the air at the furface of the earth will be to the denfity of this *medium*, as above 213200000 to 1.

Of the Motion *of* Wheels *over* Obftacles.

PROB. IV. *If a wheel moving on an horizontal plane, meet with an immoveable obftacle in its way, over which it is to be drawn by a force fixed to its center; if the weight and diameter of the wheel, the height of the obftacle, and the direction of the force drawing the wheel, be all known; thence to determine the force that is fufficient to draw the wheel over the obftacle.*

Let GPME be the wheel, ND the horizontal plane on which it moves from N towards D, EF the obftacle over which it is to be drawn; let the wheel arrive at the obftacle, and touch its top E; and there let it be fuppofed to ftand preffing the horizontal plane at G with its whole weight. Draw OEK a tangent to the wheel in the point E, draw the diameter ACG perpendicular to the horizontal plane, and produce it till it meet the tangent in O; from

Pl. 11. Fig. 1.

from E draw the *radius* EC; draw EH perpendicular to AG; and mr, MC, perpendicular to EC, and consequently parallel to the tangent OK; and lastly, draw the *radius* Cm; if the whole weight of the wheel be expressed by CO, in the direction of which line that weight acts when the wheel is wholly supported by the horizontal plane at G, that weight may be resolved into two others CE and OE, acting according to the directions of those lines, the weight CE pressing against the top of the immoveable obstacle, and being wholly sustained by it, and the weight OE drawing the wheel down in a direction parallel to the tangent OEK. Let W denote the whole weight of the wheel, r its *radius*, h the height of the obstacle, and x the part of the whole weight which draws the wheel down in a direction parallel to OEK; and then we shall have this analogy; as x is to W, so is OE to CO, or HE to CE, from the similarity of the triangles CEO, and CEH; whence $x = \dfrac{W \times HE}{r}$; but HE from the nature of the circle, is equal to $\sqrt{AH \times HG}$, or to $\sqrt{AH \times EF}$, that is, in symbols, to $\sqrt{2rh - hh}$; and therefore $x = \dfrac{W \times \sqrt{2rh - hh}}{r}$. A force just equal to this weight, and acting in direct opposition to it, that is, drawing the wheel upward in the direction CM parallel to OK, will just be able to make the wheel rest on E the top of the obstacle, without suffering any part of its weight to rest on the horizontal plane at G. This force must be increased to produce the same effect, if it act in any other direction than that of CM. For let it draw the wheel in the direction Cm, m lying between E and M, and then the force acting in this direction may be resolved into two forces, which will be as Cr and rm, whereof Cr draws the wheel directly against E the

top

top of the obſtacle, and, ſo is loſt, and mr draws
it up in a direction parallel to, O K. But mr is leſs
than Cm or C M, and to become equal to it, and
conſequently, ſufficient to ſupport the wheel againſt
the top of the obſtacle without ſuffering any part of
its weight to reſt on the horizontal plane; it muſt be
increaſed in the *ratio* of Cm or C M to rm, that is,
putting s for the ſine of the angle which the direction of the force makes with C E, in the *ratio* of r to
s; but the force rm cannot be increaſed, but the
whole force C M muſt be increaſed in the ſame proportion. And therefore the force $\dfrac{W \times \sqrt{2rh - hh}}{r}$

muſt be increaſed in the proportion of r to s, and
then, putting F for the force, acting in the direction Cm, which is juſt ſufficient to ſupport the
wheel on the obſtacle without ſuffering it to preſs
on the plane N D, $F = \dfrac{W \times \sqrt{2rh - hh}}{s}$; and
the ſmalleſt addition to this force will make it draw
the wheel over the obſtacle.

Since the reſiſtance given by the obſtacle, is equal
to the force that is juſt ſufficient to make the wheel
reſt on the obſtacle without ſuffering any part of its
weight to preſs on the plane of the horizon, that
is, putting R for the reſiſtance given by the obſtacle, ſince R is equal to F; R will be equal to
$\dfrac{W \times \sqrt{2rh - hh}}{s}$.

It is to be obſerved, that the direction of the
force muſt lie between C E and C A; for if the
force draw the wheel in the direction C E it will be
wholly ſpent upon the obſtacle, and not in the leaſt
contribute to draw the wheel over it; and if it
draw the wheel directly upwards from C to A, it
will not make it to preſs againſt the obſtacle, and
conſe-

confequently, however great we may fuppofe it to be, can never draw it over it.

Cor. I. If the direction of the moving force change continually, paffing from CE to CM, and thence to CP, the fine of the angle which the line of direction makes with CE, will increafe in the paffage of that line from CE to CM, and decreafe in its paffage from CM to CP; but as s increafes or leffens, $\dfrac{W \times \sqrt{2rh - hh}}{s}$ will leffen or increafe, and confequently the force F will leffen in the paffage of the line of its direction from CE to CM, and thence increafe in the paffage of that line to CA. So that the force will be leaft when it acts in the direction CM, in which cafe the whole force will be employed in drawing the wheel over the obftacle; whereas in all other directions, part of the force will be loft by drawing directly againft the top of the obftacle. Hence the moft advantageous direction of the force, will be that which makes a right angle with CE, in which cafe s will be equal to r, and $F = \dfrac{W \times \sqrt{2rh - hh}}{r}$

Cor. II. If the height of the obftacle be given, in which cafe h will be as r, and the force draw the wheel in the direction CM parallel to OK; then F will be as $\dfrac{W \times \sqrt{2r - 1}}{r}$.

If the *radii* of four wheels be 1, 2, 3, 4, then will $\dfrac{\sqrt{2r - 1}}{r}$, be 1, $\dfrac{\sqrt{3}}{2}$, $\dfrac{\sqrt{5}}{3}$, $\dfrac{\sqrt{7}}{4}$, that is, as the numbers 1000, 866, 745, 661; and the forces requifite to fupport thefe wheels on the point E, fo as not to fuffer any part of their weight to reft on the horizontal plane, will be as their weights multiplied

plied into thefe numbers refpectively. The orce requifite to fupport the firft wheel, will be as its weight
multiplied into 1000, the force requifite to fupport
the fecond wheel as its weight multiplied into 866;
and fo of the reft. And if the weights of all the
wheels be equal, the forces neceffary to fupport
them, and confequently the refiftances given by
the obftacle to which thefe forces are equal, will be
as the numbers 1000, 866, 745, 661. So that
in wheels of a given weight, the leffer the wheel is,
the greater will be the refiftance which is given to
it by an obftacle of a given height.

Cor. III. If the height of the obftacle be indefinitely fmall and given, in which cafe the tangent OK will coincide with the horizontal plane
ND, and the point E coincide with the point G;
and if the force draw the wheel in a direction pa-
rallel to OK or ND; then will F be as $W \times \frac{\sqrt{2r}}{r}$,
or, becaufe 2 is a given quantity, as $\frac{W}{rr}$; and if
the weight of the wheel be given, F will be as
$\frac{1}{\sqrt{r}}$.
If the *radii* of four wheels of equal weights be 1,
2, 3, 4, and the wheels be drawn on a fmooth plane
parallel to the horizon, the forces neceffary to put
them in motion, when they draw in directions pa-
rallel to that plane, will be as $1, \frac{1}{\sqrt{2}}, \frac{1}{\sqrt{3}}, \frac{1}{\sqrt{4}}$;
that is, as the numbers 1000, 707, 577, 500.
And therefore, of wheels drawn on the plane of
the horizon by forces acting in directions parallel
to that plane, leffer wheels will require a greater
force to put them in motion than greater.

Cor. IV. If the height of the obstacle be proportional to the *radius* of the wheel, and if the force draw the wheel in a direction parallel to OK, that is, if h be as r, and F be as $\dfrac{W \times \sqrt{2rb - bb}}{r}$; then will the force, and consequently the resistance given by the obstacle, be as the weight of the wheel; for $\dfrac{\sqrt{2rh - hh}}{r}$ will be as $\dfrac{\sqrt{2rr - rr}}{r}$, that is, as 1; and therefore F will be as W.

Cor. V. If the direction of the force drawing the wheel be parallel to the horizontal plane, that is; if mC be parallel to ND; then will the force that is requisite to sustain the wheel on the point E, be $\dfrac{W \times \sqrt{2rh - hh}}{r - h}$. For in this case the angle mCE is equal to the angle CEH, and consequently, their sines are equal, that is, s is equal to CH, which in symbols is r — h. And therefore F, which universally is as $\dfrac{W \times \sqrt{2rh - hh}}{s}$, is in this case as $\dfrac{W \times \sqrt{2rh - hh}}{r - h}$.

If the height of the obstacle be given, in which case h will be as 1, then will F be as $\dfrac{W \times \sqrt{2r - 1}}{r - 1}$

If the *radii* of four wheels of equal weight, be 1, 2, 3, 4; then will F with respect to these four wheels, be as $\dfrac{1}{0}$, $\dfrac{\sqrt{3}}{1}$, $\dfrac{\sqrt{5}}{2}$, $\dfrac{\sqrt{7}}{3}$, that is, as infinite, 1732, 1128, 882. The height of the obstacle is equal to the *radius* of the first wheel, inasmuch as I have supposed them both to be as 1; and consequently the force must be infinite to make the wheel rest against E, and hinder any part of its

its weight from pressing on the horizontal plane at G.

Cor. VI. The force, is to the weight of the wheel, as the sine of the angle ECH, is to the sine of the angle which the line of direction of the force makes with EC; that is, $\dfrac{F}{W} = \dfrac{\sqrt{2rh - hh}}{r}$.

If the force be one half of the weight of the wheel, that is, if F be one half of W, $\sqrt{2rh - hh}$ will be one half of s; if P be equal to W, $\sqrt{2rh - hh}$ will be equal to s; and if F be as W, $\sqrt{2rh - hh}$ will be as s.

Of the Motion of Water through Orifices and Pipes.

Prob. V. *To determine the motion of water running out of a hole made in the bottom of a vessel.*

Sir Isaac Newton has given a general solution of this *Problem* in the following paragraph, which is contained in *prop.* 36: *prob.* 8. *lib.* 2.

" Sit ACDB vas cylindricum, AB ejus orificium
" superius, CD fundum horizonti parallelum, EF
" foramen circulare in medio fundi, G centrum fo-
" raminis, et GH axis cylindri horizonti perpendi-
" cularis. Et finge cylindrum glaciei APQB ejuſ-
" dem esse latitudinis cum cavitate vasis, et axem
" eundem habere, et uniformi cum motu perpetuo
" descendere, et partes ejus quam primum attingunt
" superficiem AB liquescere, et in aquam conversas
" gravitate suâ defluere in vas, et cataractam vel
" columnum aquae ABNFEM cadendo formare,
" et per foramen EF transire, idemque adæquate
" implere. Ea vero fit uniformis velocitas glaciei
" descendentis ut et aquæ contiguæ in circulo AB,

Pl. 11.
Fig. 2.

B b   " quam

" quam aqua cadendo et casu suo describendo alti-
" tudinem IH acquirere potest; et jaceant IH et
" HG in directum, et per punctum I ducatur recta
" KL horizonti parallela et lateribus glaciei occur-
" rens in K et L. Et velocitas aquæ effluentis
" per foramen EF ea erit quam aqua cadendo ab I
" et casu suo describendo altitudinem IG acquirere
" potest. Ideoque per theoremata GALILÆI erit
" IG ad IH in duplicata ratione velocitatis aquæ
" per foramen effluentis ad velocitatem aquæ in
" circulo AB, hoc est, in duplicata ratione circuli
" AB ad circulum EF; nam hi circuli sunt reci-
" proce ut velocitates aquarum quæ per ipsos ea-
" dem tempore et æquali quantitate, adæquate
" transeunt. De velocitate aquæ horizontem versus
" hic agitur. Et motus horizonti parallelus quo
" partes aquæ cadentis ad invicem accedunt, cum
" non oriatur a gravitate, nec motum horozonti
" perpendicularem a gravitate oriundum mutet,
" hic non consideratur. Supponimus quidem quod
" partes aquæ aliquantulum cohærent, et per co-
" hæsionem suam inter cadendum accedant ad in-
" vicem per motus horizonti parallelos, ut unicam
" tantum efforment cataractum et non in plures
" cataractas dividantur: sed motum horizonti pa-
" rallelum, a cohæsione illâ oriundum, hic non
" consideramus."

This *Theory* Sir ISAAC corrected by experiments,
proved it in six different cases, and drew several
corollaries from it. The reason why a correction
was necessary will be shewn in the *Scholium*. And
the truth of his and other corollaries flowing from
this theory, will more easily appear by expressing
the foregoing proportions of the velocities in symbols; to do which let A denote the *area* of the
circle AB, a the *area* of the hole EF, H the sum
HG, which is the perpendicular height of the water
in the vessel above the hole, x the height IH, from
which water or any other body must fall by the

force

force of gravity from a state of rest, to acquire the velocity of the water in AB, V the velocity of water in its passage through the hole EF, and v its velocity in the surface AB; and then the proportions will be thus expressed; $H + x : x :: V^2 . v^2 :: A^2 . a^2$; whence, $\sqrt{H + x} . \sqrt{x} :: V . v :: A . a$.

COR. I. The height from which a body must fall to acquire a velocity equal to the velocity of the water in the surface AB, is equal to $\frac{v^2 H}{V^2 - v^2}$, or $\frac{a^2 H}{A^2 - a^2}$. For by inversion and division of proportion, $x . H :: v^2 . V^2 - v^2 :: a^2 . A^2 - a^2$; whence $x = \frac{v^2 H}{V^2 - x^2} = \frac{a^2 H}{A^2 - a^2}$. But x denotes IH. And therefore $IH = \frac{v^2 H}{V^2 - v^2} = \frac{a^2 H}{A^2 - a^2}$.

COR. II. The perpendicular height of the water in the vessel, denoted by H, is equal to $\frac{IH \times \overline{V^2 - v^2}}{v^2}$, or $\frac{IH \times \overline{A^2 - a^2}}{a^2}$, by *Cor.* 1.

COR. III. The height from which a body must fall, to acquire a velocity equal to that with which the water flows through the hole, is equal to $\frac{V^2 H}{V^2 - v^2}$, or $\frac{A^2 H}{A^2 - a^2}$. For by division of proportion, $H + x = IG . H :: V^2 . V^2 :: A^2 . A^2 - a^2$, whence $IG = \frac{V^2 H}{V^2 - v^2} = \frac{A^2 H}{A^2 - a^2}$.

COR. IV. The perpendicular height of the water in the vessel, denoted by H, is equal to $\frac{IG \times \overline{V^2 - v^2}}{V^2}$, or to $\frac{IG \times \overline{A^2 - a^2}}{A^2}$, by *Cor.* 3.

Cor. V. If the *area* of the surface be equal to the *area* of the hole, H will be nothing in comparison of IH and IG which will be equal. For if A be equal to a, H will be nothing, by *Cor. 2,* and IH and IG will be equal and infinite, by *Cor. 2, and Cor.* 3.

The truth of this *Corollary* may likewise appear from the nature of gravity. For if A be equal to a, V must be equal to v. But V can never be equal to v while there is any acceleration of the motion of the water in its descent, thro' the vessel, as there will always be till H becomes nothing in comparison of the equal lines I H and I G, which in this case must be considered as infinite.

Cor. VI. If a be greater than A, in which case A^2-a^2 will be negative, H will be negative, by *Cor.* 4; and I G, and consequently V, will be affirmative, by *Cor.* 3. But a negative perpendicular height of the water in the vessel, and an affirmative velocity of the water flowing through the hole, require an inversion of the vessel or a turning of its bottom upwards; by which inversion the hole will become the upper orifice, and the upper orifice the hole; a will become A, and A become a; and the velocity will be affirmative, that is, the water will move downwards, as it ought to do from the nature of gravity. Farther, when a is greater than A, the vessel will be conical with its wider end downwards; but from the nature of gravity, water poured in at the top or narrower end of such a vessel, will descend in a cylindrical column, which will not fill the base, as the foregoing account of this motion requires; and therefore, to give this case the conditions required, there must be an inversion of the vessel.

Cor. VII. If the hole be small, and the surface of the water infinitely large, both a and v may

be confidered as o with refpect to A and V; confequently IH will be o, by *Cor.* 1. and IG will be
equal to H, by *Cor.* 3.

In this cafe, and this only, the fuperficial parts
of the water have no velocity at the very beginning
of the motion, but begin to defcend from a ftate
of reft, as quiefcent bodies do when the fupport is
taken away. In all other cafes, in which a and v
have fome magnitudes when compared with A and
V, the fuperficial parts of the water fet out with
fome velocity, and do not begin to defcend, on
the water's beginning to flow through the hole, as
heavy bodies near the furface of the earth begin to
defcend from a ftate of reft.

Cor. VIII. If the *ratio* of the furface to the
hole be given, as it will be when each of them continues the fame, or when both of them change in
the fame proportion; the velocity in the furface
will be proportional to the velocity through the
hole, and both will be proportional to the velocity
which would be acquired by a body in falling
through a height equal to the perpendicular height
of the water in the veffel. If $\frac{A}{a}$ be given, $\frac{V}{v}$ will
be given; and confequently v will be as V. And
fince $\frac{A}{a}$ is given, $\frac{a^2}{A-a^2}$, and $\frac{A^2}{A^2-a^2}$, will both
be given; and confequently both IH and IG will
be as H, by *Cor.* 1, and *Cor.* 3. But v and V are
as $\sqrt{IH}$ and $\sqrt{IG}$. And therefore, both v and V
will be as $\sqrt{H}$.

By this *Corollary*, when A and a continue invariable, and the heights of the water in the veffel
are 1, 4, and 16 feet; the velocities in AB and
EF will be as 1, 2, and 4. But bodies placed at
fmall diftances from the furface of the earth, do all
begin to defcend with the fame velocity very nearly,

B b 3 as

as has been proved by experiments. And therefore the superficial parts of the water, in this case,
begin to descend in a very different manner, or
with very different velocities from that with which
a heavy body placed at those heights, begins to
descend from a state of rest. The velocity in AB
is regulated by the velocity in EF, and the velocity in EF is always measured by $\sqrt{H}$, when $\frac{a}{A}$
is given.

COR. IX. The velocity of the water in the sur-
face AB is always the $\frac{a}{A}$ part of the velocity thro'
the hole, that is, v is the $\frac{a}{A}$ part of V, or in other
words, $v = \frac{aV}{A}$. When a is nothing in proportion
to A, as we may suppose it to be, when a is very
small, and A exceedingly great, then will v be no
sensible part of V, that is, it will be nothing; and
consequently, the superficial parts of the water will
in this case begin their motion, as heavy bodies do,
from a state of rest.

COR. X. The whole motion of the descending
column AMEFNB, is equal to the motion of a
cylinder of water, whose base is a, whose altitude
is H, and whose velocity is V, that is, to the motion aH × V. For Va is equal to vA, that is, the
motion of the water in EF is equal to its motion in
AB; and from the nature of the descending column, each of them is equal to the motion in any
section of the column parallel to EF or AB; and
consequently, the motion in all the sections, supposing them to be indefinitely many, that is, the whole
motion of the descending column, will be equal to
the motion in the hole multiplied into the number

of

of sections, that is, to Va.H, or aH $\times$ V. This property has been proved by Dr. Jurin.

Cor. XI. The force which can generate the whole motion of the water running out of the hole, is equal to the weight of a cylinder of water whose base is a, and altitude is 2IG, by *Cor.* 3; that is, equal to the weight of a cylinder of water, whose magnitude is 2aH $\times \frac{A^2}{A^2-a^2}$. For in the same time, in which the water running out is equal to this cylinder, this cylinder, by falling from the height IG by the force of its gravity, will acquire a velocity equal to that with which the water runs out. But when the quantities of matter and velocities of two bodies are equal, their motions, and consequently the forces which can generate those motions in equal times, will likewise be equal. And therefore the force which can generate the whole motion of the water running out of the hole, is equal to the weight of a cylinder of water, whose magnitude is 2aH $\times \frac{A^2}{A^2-a^2}$.

Cor. XII. The weight of the descending column AMEFNB is equal to the weight of a cylinder of water, whose base is a, and whose height is $\frac{2HA}{A+a}$, that is, whose magnitude is 2aH $\times \frac{A}{A+a}$. For let IO be a mean proportional between IH and IG, and then $\sqrt{IH}$. $\sqrt{IG}$ :: IH . IO :: IO . IG :: a . A; and, by division of proportion, HO . IH :: OG . IO; and by alternation and composition, HO + OG . 2HO :: IH + IO . 2IH :: a + A . 2a. But, by *Cor.* 11. in the time a drop of water falls by its own gravity from I to G, the quantity of water discharged by the hole will be equal to a $\times$ 2IG, or A $\times$ 2IO; and in the time the drop

descends

defcends from I to H, the quantity of water paffing through the furface AB, and difcharged by the hole, will be equal to $A \times 2IH$, and the difference of thefe quantities, namely $A \times 2HO$, will be the quantity difcharged in the time the falling drop defcends from H to G, which quantity is the column AMEFNB; for in the time the drop defcends from H to G, the fuperficial parts of the water, fetting out with the velocity of the drop at H, and defcending freely and without refiftance, will reach the hole. And therefore, all the water in the veffel will be to the water in the column AMEFNB, as $A \times H$ is to $A \times 2HO$, or as $H = HO + OG$ to $2HO$; or as $a + A$ to $2a$; whence, putting Q for the quantity of water in the defcending column, $A \times H . Q :: A + a . 2a$; and confequently,

$$Q = 2AH \times \frac{A}{A+a}.$$

This *Corollary* may be proved in another manner, thus. The cataract is the difference of the two hyperboloids KAMEFBL and KABL, fuppofing the affymptote KL to be infinitely extended both ways, and the *area* AB to be infinite; but by fluxions, as Dr. Jurin has fhewn, the hyperboloid KAMEFNBL is equal to $2a \times \overline{H + x}$, or to $\frac{2A^2 x}{a}$, becaufe H is equal to $\frac{A^2 x - a^2 x}{a^2}$ by *Cor.* 2; and the hyperboloid KABL, is equal to $2Ax$, and the difference of the two is $\frac{2A^2 x}{a} - 2Ax = \frac{2A^2 x - 2Aax}{a}$. All the water in the veffel is AH or, by fubftituting $\frac{A^2 x - a^2 x}{a^2}$ in the room of H, $\frac{A^3 x - Aa^2 x}{a^2}$; and confequently, the water in the veffel is to the water in the cataract, as

$A^3 x$

$\dfrac{A^2x - Aa^2x}{a^2}$ is to $\dfrac{2A^2x - 2Aax}{a}$, that is, after due reduction, as $A + a$ is to $2a$. Therefore $AH . Q$ :: $A + a . 2a$: whence, $Q = 2aH \times \dfrac{A}{A + a}$.

Cor. XIII. The weight of all the water in the veffel, is to the weight of that part of it which is fuftained by the bottom, as the fum of the circles AB and EF. is to their difference. For, fince $A \times H . Q :: A + a . 2a$, by *Cor.* 12. $A \times H$. $A \times H - Q :: A + a . A + a - 2a = A - a$ by divifion of proportion.

Cor. XIV. The weight of the water which the bottom fuftains is to the weight of the cataract, as the difference of the circles AB and EF, to twice the leffer circle EF. For $A \times H . Q :: A + a . 2a$, by *Cor.* 12. And by divifion of proportion, $A \times - Q . Q :: A + a - 2a = A - a . 2a$.

Cor. XV. The weight of water which the bottom fuftains, is to the weight of water perpendicularly incumbent thereon, as the circle AB, is to the fum of the circles AB and EF. For the weight of water which the bottom fuftains is $A \times H - Q$ $= AH - \dfrac{2aHA}{A + a}$, by *Cor.* 12. $= \dfrac{A^2H - aAH}{A + a}$; and the weight perpendicularly incumbent on the bottom is $A - a \times H = AH - aH$. But $\dfrac{A^2H - aAH}{A + a} . AH - aH :: A^2 - aA . A^2 - a^2$ :: $A . A + a$ by dividing by $A - a$.

Cor. XVI. The quantity of water in the defcending column is to the quantity perpendicularly incumbent on the hole, as twice the circle AB, is to the fum of the circles AB and EF. For the

quantity

quantity of water in the descending column is
$2aH \times \dfrac{A}{A+a}$. But $2aH \times \dfrac{A}{A+a}$. $aH :: \dfrac{2A}{A+a}$.
$1 :: 2A . A+a$.

Hence, when a is nothing, as we may suppose it to be when A is infinitely great, the descending column will be equal in magnitude to $2aH$, as Dr. Jurin has shewn it to be by determining its magnitude by fluxions.

Cor. XVII. The weight of the descending column, is to the weight of water which can generate the whole motion of the water running out of the hole, as the difference of the circles AB and EF, is to the greater circle AB. For, putting F for the force or weight which can generate the whole motion of the water running out of the hole, and supposing Q to denote the weight of the descending column, we shall have F equal to the weight of a quantity of water whose magnitude is $2aH \times \dfrac{A^2}{A^2-a^2}$, by *Cor.* 11. and Q equal to the weight of a quantity, whose magnitude is $2aH \times \dfrac{A}{A+a}$, by *Cor.* 12. And therefore, $Q . F :: 2aH \times \dfrac{A}{A+a} . 2aH \times \dfrac{A^2}{A^2-a^2} :: 1 . \dfrac{A}{A-a} :: A-a . A$.
Hence $Q = \dfrac{F \times A - a}{A}$, and $F = \dfrac{QA}{A-a}$; and consequently, the force which can generate the whole motion of the water running out of the hole, will always exceed the weight of the descending column, except when a becomes o, as we may suppose it to do, when it is very small, and A exceedingly great.

Cor. XVIII. The force which can generate the whole motion of the water running out of the hole,

hole, is to the weight of water perpendicularly incumbent on the hole, as twice the square of the
greater circle AB, to the difference of the squares
of the circles AB and EF. For the force which
can generate the whole motion of the water run-
ing out of the hole, is the weight of $2aH \times \frac{A^2}{A^2 = a^2}$
quantity of water, by *Cor*. 11. and the weight of
water perpendicularly incumbent on the hole, is the
weight of the cylinder aH. But $2aH \times \frac{A^2}{A^2 - a^2}$.
$aH :: 2A^2. A^2 - a^2$. In the same ratio is the
whole motion of the effluent water, to the motion
of the water in the cataract.

Cor. XIX. If in the middle of the hole be
placed a little circle PQ parallel to the horizon,
whose center is G, and if the *area* of this circle be
called o; the weight of water which it sustains during the efflux of the water through the ring sur-
rounding it, is to the weight of half the cylinder
oH, as a to a $- \frac{1}{4}$o; if R denote the weight sustained, R is to $\frac{oH}{2}$, as a to a $- \frac{1}{4}$o, and R is equal
to $\frac{aoH}{2a - o}$. For if we suppose A to be contracted
till it becomes equal to a, in which case IH will be
infinite, by *Cor*. 1. the water, notwithstanding this,
will descend about the column PQH which the
little circle sustains with velocities, which are every
where in the subduplicate *ratio* of the distance from
KL, and likewise in the reciprocal *ratio* of the several sections through which it passes; consequently, the cataract AEPHQFB, is equal to the difference of the two hyperboloids PEAKLBFQH
and AKLB. But the hyperboloid PEAKLBFQH
$= 2a - 2o \times \overline{H + x} = 2aH - 2oH + 2ax - 2ox$,
and the hyperboloid AKLB is 2ax; and the difference

Pl. 11.
Fig. 3.

rence of the two is $2aH - 2oH - 2ox$, which is the cataract AEPHQFB. The *ratio* of all the water in the veffel to this annular cataract, is

$$\frac{aH}{2aH - 2oH - 2ox}.$$ But from the nature of the motion of the descending water, a is to a — o, as $\sqrt{H + x} . \sqrt{x}$, whence $H = \frac{2aox - oox}{a^2 - 2ao + oo}$. The foregoing *ratio*, when this value of H is subftituted in its room, will, after due reduction, become $\frac{2a - o}{2a - 2o}$. Therefore aH, the whole quantity of water in the veffel, is to the annular cataract, as $2a - o$ to $2a - 2o$, whence the annular cataract is $\frac{2a^2H - 2aoH}{2a - o}$, which being fubducted from aH, leaves $\frac{aoH}{2a - o}$ for the quantity fuftained by the little circle o. Confequently, $R = \frac{aoH}{2a - o}$; and $R . \frac{oH}{2} :: a . a - \frac{1}{2}o$.

S C H O L I U M.

Upon examining this motion by experiments, Sir Isaac Newton found the velocity of the water in its paffage through the hole to be lefs than it ought to be, if the water in the veffel defcended from the furface to the hole freely and without refiftance, in the proportion of 1 to $\sqrt{2}$. For he obferved the vein of the effluent water, and found it to contract and grow narrower, to the diftance of about a diameter of the hole below it, at which place he meafured the diameter of the vein, and found it to be lefs than the diameter of the hole in the proportion of 21 to 25, and confequently, the *area* of a fection of the vein at that place to be lefs than the *area* of the hole, in the proportion of 441

to

to 625, that is, of ˙1 to $\sqrt{2}$. But as the vein con-
tracts the velocity increases. And therefore, at the
distance of a diameter of the hole below it, the velocity will be greater than in the hole in the proportion of $\sqrt{2}$ to 1. If IG be four feet or 48 in-
Pl. 11.
Fig. 2.
ches, and the diameter of the hole be 1 inch, 1 added to 48 will make the height from the place where
the velocity is greatest to be 49 inches ; and if the
velocities of the descending column in the hole and
that place, were truly measured by the subduplicate
ratios of those heights, as they would be if the water descended freely and without resistance, they
would be nearly equal, being as the numbers 69
and 70. And therefore, the velocity of the water
in the hole is less than it would be if it was proper-
tional to $\sqrt{IG}$, in the *ratio* of 1 to $\sqrt{2}$. This diminution of velocity can be owing to nothing but
the lateral motion of the descending water, retarding its perpendicular motion downwards, and making it less than it otherwise would be, in the said
ratio of 1 to $\sqrt{2}$. Hence, the velocity with which
the water flows through the hole is very nearly
equal to the velocity which a body, by falling
freely and without resistance from a state of rest at
I, would acquire in descending through $\frac{1}{4}$IG. For
the velocity acquired in falling through $\frac{1}{4}$IG, is to
the velocity acquired in falling through IG; as
1 to $\sqrt{2}$.

According to Sir Isaac Newton, a body falling *in vacuo* from a small height above the surface
of the earth, will describe $193\frac{1}{3}$ inches, or $16\frac{1}{9}$ feet
in one second minute of time, and will have acquir-
ed a velocity at the end of the fall, which being
continued uniform, would carry it through twice
that space, that is, $386\frac{2}{3}$ inches or $32\frac{2}{9}$ feet, in an
equal time. But uniform velocities are as the
spaces described by them in the same time, and the
velocities acquired by a body falling *in vacuo* through
the

$\sqrt{\dfrac{A^2H}{A^2-a^2}}$ inches, in a second minute of time. These expressions may be shortened, if A be considerably greater than a, for in all such cases $\dfrac{A^2H}{A^2-a^2}$ will be so nearly equal to H, that $\dfrac{A^2}{A^2-a^2}$ may be safely rejected; and then the foregoing measures of the velocity will become $5.6773196 \sqrt{H}$ feet, or $68.1278352 \sqrt{H}$ inches. To shew the truth of this by an example, let A be 100 square inches, and a 1 square inch, and then $\dfrac{A^2H}{A^2-a^2}$ will be $\dfrac{10000H}{9999}$; if H be four feet or 48 inches, $\dfrac{10000H}{9999}$ will be 48.0048 inches, which is only greater than 48 by 48 parts of an inch divided into 10000. The excess is so small, that it may be safely rejected.

Another true measure of the velocity of the water flowing through the hole, will be had by dividing the quantity of water discharged, by the area of the

hole

hole and time of the difcharge, taken together; the
quantity of water difcharged being expreffed in cubick inches, the *area* of the hole in fquare inches
or parts of a fquare inch, and the time of the difcharge in feconds. Let Q denote the quantity difcharged, d the diameter of the hole, and t the
time of the difcharge, and then V will be meafured
by $\frac{Q}{At} = \frac{Q}{0.78539816 d^2 t}$ inches, which will be the
fpace defcribed in one fecond of time.

This meafure is equal to the former, that is,
$\frac{Q}{0.78539816 d^2 t} = 68.1278352 \sqrt{H}$; and confequently, $Q = 53.5074764 d^2 t \sqrt{H}$ cubick inches;
or $13555.227 d^2 t \sqrt{H}$ grains; becaufe a cubick
inch of water weighs $253\frac{1}{7}$ grains. If W denote
the weight of water difcharged, then will $W = 13565.3\frac{1}{2} d^2 t \sqrt{H}$ grains.

In order to know, whether the velocities of
water flowing through circular holes of different
diameters, when placed at the fame perpendicular
diftance from the furface of the water, be all equal;
what relation the velocity of water flowing through
a hole, bears to the velocity of water flowing
through an horizontal pipe of an equal diameter,
inferted into the fide of a veffel at an equal perpendicular diftance from the furface of the water;
and under what circumftances the meafure of the
velocity laid down in my *Animal Œconomy* obtains;
I fay, in order to know thefe things, I caufed a
proper *apparatus* to be made, and from the experiments made with it, I compofed the following
Tables.

TABLE

TABLE I.

t	H	d	W	w	$\frac{w}{W}$
10	4	$\frac{1}{10}$	2711	2944	1086
		$\frac{4}{10}$	43377	47040	1084
		$\frac{6}{10}$	67776	72960	1076
		$\frac{8}{10}$	173507	178560	1029
	2	$\frac{1}{10}$	1917	2087	1088
		$\frac{4}{10}$	30672	33600	1095
		$\frac{6}{10}$	47925	51840	1082
		$\frac{8}{10}$	122688	128400	1046

TABLE II.

d	l	w	d	l	w	d	l	w
$\frac{2}{10}$	0	12736	$\frac{4}{10}$	0	47040	$\frac{8}{10}$	0	178560
	d	14385		d	54720		d	204720
	2d	14400		2d	56160		2d	224640
	3d	13792		3d	52800		3d	217440
	4d	13728		4d	52220		4d	212160
	5d	13663		5d	51600		5d	203520
	10d	12683		10d	47040		16d	188160
							23d	178560

The firſt Table contains, in the firſt column, under t, the time of the diſcharge in ſeconds; in the ſecond column, under H, the perpendicular heights of the water above the hole in *London* feet; in the third, the diameters of the hole in parts of an inch; in the fourth, under W, the weights of water in grains, which ought to have been diſcharged by the theory or foregoing rule; in the fifth, under w, the weights of water in grains which were diſcharged by experiment, each weight being a mean taken from five or ſix experiments; and in the ſixth column, under $\frac{w}{W}$, the *ratio* of the weight diſcharged by experiment,

periment to the weight which ought to have been difcharged by the theory.

The fecond Table confifts of three parts, and each part of three columns. The firft column of each part, contains the diameter of the pipe in parts of an inch; the fecond contains the lengths of the pipe in the terms of the diameter, beginning with the hole, which may be confidered as a pipe of an infinitely fmall length exprefled by 0; and the third column contains the weights in grains difcharged in ten feconds, each weight being a mean taken from particular experiments. The holes and pipes were all at the perpendicular diftance of four feet from the furface of the water, fo that here t was 10 feconds, and H four feet.

	TABLE III.								
d	H	l	W	w	$\frac{w}{W}$	H	W	w	$\frac{w}{W}$
$\frac{3}{16}$	2	1	2180	2180	1000	$\frac{1}{2}$	1090	982	901
		2	1541	2080	1349		770	922	1196
		3	1258	2057	1634		629	877	1393
		4	1090	1874	1719		545	762	1398
		5	980	1759	1804		490	720	1469
		6	890	1690	1899		445	665	1494
		7	824	1564	1898		412	620	1505
		8	770	1520	1972		385	585	1519
		9	727	1440	1982		363	553	1522
		10	689	1410	2045		344	525	1523
		12	629	1320	2098		314	470	1493
		14	582	1225	2102		291	430	1476
		16	545	1163	2134		272	383	1405
		18	514	1086	2113		257	350	1362
		20	487	1030	2113		243	320	1313
		24	445	866	1946		222	260	1168
		25	436	860	1972		218	253	1160
		28	412	844	2048		206	230	1116
		32	385	758	1967		192	202	1048
		36	363	659	1814		181	185	1018
		48	314	509	1618				
		60	281	421	1496				
		72	257	345	1342				

TABLE IV.

d	H	l	W	w	$\frac{w}{W}$	H	W	w	$\frac{w}{W}$
$\frac{4}{10}$	2	1	12332	10040	814	1	6166	5018	814
		2	8720	9270	1063		4360	4630	1063
		3	7120	8820	1238		3560	4400	1235
		4	6166	8570	1389		3083	4270	1385
		5	5515	8240	1494		2758	4040	1465
		6	5034	7840	1557		2517	3880	1541
		7	4661	7580	1626		2330	3766	1616
		8	4360	7360	1688		2180	3668	1682
		9	4111	7150	1739		2055	3570	1737
		10	3900	6950	1782		1950	3414	1751
		16	3083	5776	1873		1541	2955	1918
		25	2466	4785	1940		1233	2460	1995
		36	2055	4048	1970		1027	2120	2064
		49	1762	3480	1975		881	1730	1963
		64	1542	3250	2108		771	1326	1720
		81	1370	3062	2235		685	1120	1635
		97	1252	2700	2156		626	940	1502

The third and fourth Tables confift each of two
parts corresponding to different perpendicular heights
of the water in the veffel, and different diameters of
the pipes. In both Tables, H denotes the perpendicular height of the water in the veffel above
the pipe in feet; l the length of the pipe in inches;
W the weight in grains which ought to be difcharged by the firft *Propofition* of my *Animal Œconomy*; w
the weight in grains which was difcharged by experiment; and $\frac{w}{W}$ the *ratio* of the weight difcharged
by experiment to the weight which ought to have
been difcharged by that *Propofition*. The diameter
of all the pipes in the third Table was $\frac{3}{10}$ of an
inch, and of all the pipes in the fourth Table $\frac{4}{10}$

of

of an inch. And the time of the difcharge was 10 feconds in all the experiments of both Tables.

The quantity or weight of water which ought to be difcharged by the firft *Propofition* of the *Animal Œconomy*, may be thus found. I there proved, that the velocity of water flowing thro' a pipe, is as $\sqrt{\frac{F}{dl}}$. But if the force which can generate the motion of water flowing through a pipe lying parallel to the horizon, be equal to the force which can generate the motion of water flowing through a hole of an equal diameter with the pipe, when placed at an equal perpendicular diftance from the furface of the water; F, by *Cor.* 11. of this *Problem*, will be as $2d^2H$, on fuppofition that the *area* of the hole is extreamly fmall in comparifon of the *area* of the furface of the water. And therefore the velocity of water flowing through a pipe lying parallel to the horizon, is as $\sqrt{\frac{2dH}{l}}$. The weight of water difcharged, is as the orifice of the pipe, the time of the difcharge, and velocity, taken together; that is, as $d^2t\sqrt{\frac{2dH}{l}}$. And therefore, W is as $d^2t\sqrt{\frac{2dH}{l}}$.

A pipe of $\frac{1}{10}$ of an inch in diameter, and 1 inch in length, difcharged 2180 grains of water in 10 feconds, when it was inferted into the fide of the veffel at the perpendicular diftance of two feet from the furface. In this cafe therefore, d, t, H, l, were 0.1, 10, 2, 1; and $d^2t\sqrt{\frac{2dH}{l}}$ was equal to 0.06326. Hence we may find W in other cafes by this analogy; $2180 : 0.06325 :: W : d^2t\sqrt{\frac{2dH}{l}}$; whence $W = 48746.3 d^2t\sqrt{\frac{dH}{l}}$ grains.

In

In the firſt part of the third Table, W is $\frac{2180}{\sqrt{l}}$,

and in the firſt part of the fourth Table, $\frac{12332}{\sqrt{l}}$;

and W in the ſecond part of each Table is one half of W in the firſt part.

OBSERVATIONS *on the* TABLES.

OBS. I. By the firſt Table the diſcharges by ex-periment are nearly proportional to the diſcharges by the theory, that is, w is nearly proportional to W, or $\frac{w}{W}$ is nearly the ſame, whatever be the dia-meter of the hole, provided the time of the diſ-charge, and the perpendicular height of the water in the veſſel above the hole, be given. The diſ-charges by experiment were all ſomething larger than the diſcharges by the theory, which might be partly owing to the pouring in of the water at the top of the veſſel, in order to keep the veſſel con-ſtantly full during the time of the diſcharge ; for the pouring, tho' it was done gently, might a little increaſe the velocity wherewith the water ran out of the hole.

OBS. II. By the ſecond Table, the weight of water diſcharged, and conſequently the velocity, in-creaſes from the hole till the length of the pipe be-comes equal to about twice its diameter, that is, till l becomes equal to about 2d, and is greater there than at any other length of the pipe. The greateſt velocities in theſe pipes in proportion to the veloci-ties through their reſpective holes, are as the num-bers 1130, 1258 to 1000.

OBS. III. From the length of twice the diame-ter, that is from the length 2d, the velocity leſſens continually on increaſing the length of the pipe, and becomes equal to the velocity through the hole when the length of the pipe becomes equal to about 22.3657d$\sqrt{d}$ inches. For, by the ſecond Table,
the

the velocities of the water flowing through the pipes, were nearly equal to the velocities through their refpective holes, when the lengths of the pipes were 10d, 16d and 23d, that is 2 inches, 6.4 inches, and 18.4 inches. But 2, 6.4; and 18.4, are nearly as 1, 2.8, and 8, the fefquiplicate *ratios* of 1, 2 and 4, and 1, 2 and 4, are as the diameters $\frac{1}{10}$, $\frac{2}{10}$ and $\frac{4}{10}$. And therefore the velocities of the water flowing through the pipes, were nearly equal to the velocities through their refpective holes, when the lengths of the pipes were in the fefquiplicate *ratios* of their diameters. The diameter of the fmalleft pipe being $\frac{1}{10}$ of an inch, $d\sqrt{d}$ is 0.0894; and if d be of any other magnitude, and l be the length of a pipe of that diameter through which the water flows with a velocity equal to that with which it flows through its correfponding hole, we fhall have this proportion; as 2 is to 0.0894, fo is l to $d\sqrt{d}$, whence $l = 22.3657 d\sqrt{d}$.

OBS. IV. By the third and fourth Tables, the quantity of water difcharged by experiment in proportion to the quantity which ought to have been difcharged by the theory, that is $\frac{w}{W}$, increafes gradually till the pipe comes to be of a certain length, and after that it decreafes gradually on increafing the length of the pipe. In the two parts of the third Table this *ratio* was greateft, when the lengths of the pipes in inches were about 20 and 10, and it was greateft in the two parts of the fourth Table, when the lengths of the pipes were 81 and 36. But from the courfe of the numbers expreffing $\frac{w}{W}$ in the fecond part of the fourth Table, I think this *ratio* would have been greater in a ipe of 40 inches in length, than in the one I ufed pf 36, and therefore fhall fuppofe that it would have been greateft at the lengths of 81 and 40. Confequently, putting x for the length of the pipe in inches, at which

this

this *ratio* is greatest, x will be as $\sqrt{H}$ when d is given, and as d³ when H is given, and when neither d nor H is given, as d³$\sqrt{H}$. Hence we may form a rule for finding the length of the pipe, at which this *ratio* shall be a *maximum*; for it was a *maximum* in a pipe of $\frac{1}{10}$th of an inch in diameter, when its length was 20 inches, and the perpendicular height of the water in the vessel two feet. In this case therefore x, d and H, are 20, 0.1, and 2, and d³$\sqrt{H}$ is 0.01414; and in other cases, x may be found by this analogy; as 20 is to 0.01414, so is x to d³$\sqrt{H}$; whence x is equal to 1414d³$\sqrt{H}$. To see whether this rule be universal, and obtain in pipes of greater diameters, and at greater distances from the surface of the water, I shall suppose d and H to be 0.5 and 3, as in our Author's Table p. 227, and then 1414d³$\sqrt{H}$ will be about 600 inches or 50 feet, which length is twice as great as it was in reality; for the *ratio* was a *maximum* by that Table, when the length of the pipe was 25 feet; so that the value of x here determined seems to obtain only in pipes of small diameters.

OBS. V. By the third and fourth Tables, the quantity discharged by experiment in proportion to the quantity which ought to have been discharged by the theory, that is $\frac{w}{W}$, does not differ much in pipes whose lengths are within certain limits. $\frac{w}{W}$, in the pipes, whose lengths were 6 and 32 in the first part of the third Table, is less than in the pipe where this *ratio* is a *maximum*, in the proportions of 100 to 112 and 110, and the difference of $\frac{w}{W}$ and the *maximum* is still less in pipes of all other lengths between 6 and 32; so that in this part of the Table, 6 and 32 are the limits, at and within which there is a near agreement be-

tween theory and experiment. $\frac{w}{W}$ in the pipes
whose lengths are 4 and 16 in the second part of this
Table, is less than in the pipe where this *ratio* is a
maximum, in the proportion of about 100 to 109
and 108, and it is still less in pipes of all other
lengths within these limits. And $\frac{w}{W}$ in the pipes
whose lengths are 9 and 64 in the second part of the
fourth Table, for the pipes were not carried to
such lengths as were necessary to settle the limits in
the first part, is less than the *maximum* in the proportion of 100 to 118 and 120; and it is still less
in pipes of all other lengths within these limits.

Obs. VI. By the third and fourth Tables, the
quantity of water discharged by experiment always
exceeds the quantity which ought to be discharged
by the theory; it was near double within the limits
of the first part of the third Table and second part of
the fourth, and greater in the second part of the third
Table in the proportion of 5713 to 3858. If we suppose it to be double within the limits, in pipes of all
lengths, then will w be equal to 2W, or to 97492.6d^2
$\sqrt{\frac{dH}{l}}$ grains, W being equal to 48746.3$dv\sqrt{\frac{dH}{l}}$
grains, as was shewn, above.

Of the Foci *of* Optick Glasses.

PROB. VI. *If the distance of an object from a double
convex lens whose surfaces are spherical; if the radii of
both the spherical surfaces, the thickness of the lens, and
the sines of incidence and refraction, be all given; thence
to determine the distance behind the lens of the principal
focus or concourse of the rays issuing from the object and
falling perpendicularly, or very nearly so, on that surface of the lens which is turned towards the object.*

Let MN be a *lens*, E and e the centers of its Pl. 11.
spherical surfaces MCN and MDN, Q an object Fig. 5.
C c 4   placed

placed directly before the *lens*, Qq a line drawn from the object perpendicular to the surfaces of the *lens*, and consequently passing through the centers e and E; let the point A be indefinitely near to C, in which case QA and QC may be looked upon as equal; let q be the *focus* or concourse of the rays QA and QC after the first refraction by the surface MCN, and z their *focus* or concourse after the second refraction by the surface MDN. Put D for QC the distance of the object from the *lens*, r for the *radius* CE, ϱ for the *radius* eD, x for Dq, the distance of the *focus* behind the *lens*, after the first refraction, and z for Dz its distance behind the *lens* after the second refraction; and lastly, let I and R denote the sines of incidence and refraction of the rays passing out of air or any other *medium* into the first surface MCN, and consequently R and I the sines of incidence and refraction in their passage out of the second surface MDN into air or that other *medium*.

Pl. 11.
Fig. 4.
To determine z, we must first determine the measure of x in known terms, to do which draw AF perpendicular to Qq, EI perpendicular to QAI the incident ray produced, and ER perpendicular to the refracted ray Aq; and then, from the similarity of the two triangles QAF and QIE, and also of the triangles qAF and qER, and from QA being equal to QC, and qA equal to qC, we shall have AF equal to $\frac{QC \times EI}{QE}$, or, in symbols, to $\frac{DI}{D+r}$, by the two first triangles, and by the two last triangles, equal to $\frac{Cq \times ER}{Eq}$, or, in symbols, to $\frac{Rx}{x-r}$. Consequently, $\frac{DI}{D+r}$, is equal to $\frac{Rx}{x-r}$, and $x =$
$$\frac{DIr}{DI - DR - Rr}.$$

Pl. 11.
Fig. 5.
Having found the measure of x or Cq in known terms, z or Lz may be thus determined. For that measure,

meaſure, that is for $\frac{DIr}{DI - DR - Rr}$, put A, to za
and qa produced draw the perpendiculars eI and eR,
and draw am perpendicular to Qq. And then,
from the ſimilarity of the triangles qam and qRe, and
alſo of the triangles zam and zIe, and from qa being
equal to qD, and za equal to zD a m will be equal
to $\frac{eR \times qD}{qe}$ from the firſt triangles, and equal to
$\frac{eI \times zD}{ze}$ from the ſecond. eR is the ſine of incidence of the ray Aa falling on the ſecond ſurface
MDN, and eI the ſine of its refraction; and therefore eR will be I, and eI will be R. qD is equal
to qC — CD = A — t, putting t for CD the
thickneſs of the *lens*; and qe is equal to qC + eC
= qC + eD — CD = A + ℮ — t; conſequently
$\frac{R \times \overline{A - t}}{A + ℮ - t} = \frac{Iz}{z + ℮}$; and from this equation, A =
$\frac{I℮z + Rtz + Rt℮ - Itz}{Rz + R℮ - Iz}$. But A denotes $\frac{DIr}{DI - DR - Rr}$.
And therefore $\frac{I℮z + Rtz + Rt℮ - Itz}{Rz + R℮ - Iz} = \frac{DIr}{DI - DR - Rr}$.
By clearing z in this equation, we ſhall have z =
$$\frac{DIRr℮ + RRr℮t + DRR℮t - DIR℮t}{DII℮ + 2DIR - DIIt - DIR℮ - DRRt - IRr℮ - RRrt + IRrt - DIRr + DIIR}.$$
To give this equation a more ſimple form, di-
vide both numerator and denominator by I — R;
and then the numerator will become $\frac{R}{I - R} \times IDr℮$
$+ \frac{R}{I - R} \times Rr℮t — DR℮t$, or by putting B inſtead
of $\frac{R}{I - R}$, BIDr℮ + BRr℮t — DR℮t, and the denominator will become IDr + ID℮ — IDt + RDt +
Rrt — BIr℮; and the equation will be reduced to
another form, and ſtand thus;
$$z = \frac{BIDr℮ + BRr℮t — DR℮t}{IDr + ID℮ — IDt + RDt + Rrt — BIr℮}.$$

This

This is Dr. Halley's universal *Theorem* for finding the principal *focus* of rays falling diverging on a double *convex lens,* published in the *Philosophical Transactions.*

If the rays instead of falling diverging, fall parallel on a double convex *lens,* as they will nearly do, when the object is at an immense distance from the *lens,* D in this case may be considered as infinite ; and consequently, all the terms in which D is not found, may be thrown out of the equation, and then $z = \dfrac{BIDr_\rho - DR_\rho t}{1Dr + 1D_\rho - 1Dt + RI_t} =$

$\dfrac{BIr_\rho - R_\rho t}{Ir + I_\rho - It + Rt}.$

And lastly, if the rays fall converging on a double convex *lens,* the signs of all the terms in which D is found must be changed ; for when the rays fall converging, the point behind the *lens* to which they tend at their incidence, must be considered as the place of the object, which, from its being differently situated with respect to the *lens* from what it is when the rays fall diverging, requires the signs of all the terms in which D is found to be changed, which being done, we shall have

$z = \dfrac{DR_\rho t + BR_{r\rho}t - BIDr_\rho}{1Dt - 1Dr - 1D_\rho - RDt + Rrt - BIr_\rho}.$

These are the three general *Theorems* for finding the principal *focus* of rays falling, diverging, parallel, or converging on a double convex *lens.*

If the *lens* be made of glass, as *lenses* usually are, and the object be placed in air, then, since the sine of incidence of a ray passing out of air into glass, is to the sine of refraction, as 3 to 2, I, R and B will be 3, 2, and 2 ; and the foregoing general *Theorems* for finding the *foci* of rays falling diverging, parallel, and converging, on a double convex glass, will be $z = \dfrac{6Dr_\rho + 4r_\rho t - 2Dr_\rho t}{3Dr + 3D_\rho - 3Dt + 2Dt + 2rt - 6r_\rho},$

$z = \dfrac{6r_\rho - 2_\rho t}{3r + 3_\rho - 3t + 2t},$ and

$z =$

$$z = \frac{2Drt + 4rpt - 6Dtp}{3Dt - 3Dt - 3Dp - 2Dt + 2rt - 6rp}.$$

And if the *radii* be equal, and the thickness of the glass be neglected, or considered as o, then will these *Theorems* stand thus, $z = \frac{Dr}{D - r}$, $z = r$, and

$$z = \frac{-Dr}{-D - r}.$$

If the *lens* be a double concave glass, the *radii* of whose two spherical surfaces are equal, and if the thickness of the *lens* be considered as o, the *radii* will lie on different sides of the *lens* with respect to the object from what they did before, and consequently, the signs of the *radii* must be changed; and then the last *Theorems*, in which the *radii* were supposed to be equal, and the thickness of the glass was neglected or considered as o, will stand thus,

$$z = \frac{-Dr}{D + R}, \quad z = -r, \text{ and } z = \frac{Dr}{r - D}. \text{ By these}$$

Theorems, z is always negative when the rays fall upon the double concave, diverging, or parallel, and when they fall converging it is negative when D is greater than r. When z is negative, the *focus* falls on the same side of the glass with the object, contrary to what it does in all cases of a double convex *lens*, excepting that of diverging rays, when the distance of the object is less than the *radius*, or D is less than r. For in that case, z, which is equal to $\frac{Dr}{D - r}$, will be negative.

By this *Problem* we may determine how far a radiating point must be distant from the eye, to have the principal *focus* of the rays issuing from it placed in the *retina*, on supposition that the coats and humours of the eye are unchangeable as to their figures, magnitudes, and densities.

Let ABGz represent a human eye, in which ABG is the *cornea*, AMCNGB the cavity containing the aqueous humour, MCND the crystalline humour, and

Pl. 11. Fig. 6.

4

and AMDNGz the cavity containing the vitreous
humour. According to Doctor JURIN, the *radii*
of the ſpherical ſurfaces of the *cornea* and of the
cryſtalline humour, that is, of the ſpherical ſurfaces ABG, MCN, and MDN, are in 10th parts
of an inch, 3.3294, 3.3081, and 3.5056; and the
diſtance of the *cornea* from the anterior part of the
cryſtalline, the thickneſs of the cryſtalline, the diſtance of the poſterior part of the cryſtalline from
the *retina*, and the diſtance of the *cornea* from the
retina, are in the ſame parts of an inch, 1.0358,
1.8525, 6.2617 and 9.15. Let Q be the radiating
point, q the principal *focus* of the rays by the firſt
refraction of the aqueous humour, by virtue of
which refraction they fall converging on the cryſtalline, and let z be their *focus* after their refractions
by the cryſtalline and vitreous humours. By taking
the ſpecifick gravities of the humours of the eye,
I have found that the ſpecifick gravities of the
aqueous and vitreous humours are very nearly equal,
and each much the ſame with that of water; and
that the ſpecifick gravity of the cryſtalline is greater than the ſpecifick gravity of water, in the proportion of about 11 to 10. For the mean ſpecifick
gravities of five cryſtalline humours of oxen's eyes,
and of three cryſtalline humours of ſheep's eyes, were
11134 and 11033, the ſpecifick gravity of water
being 10000, and the mean of theſe two means, is
11083, which I ſhall ſuppoſe to be the ſpecifick
gravity of the cryſtalline humour of a human eye.
But the refractive power of the cryſtalline is very
nearly proportional to it's denſity, and the ſine of
incidence of rays paſſing out of the aqueous humour
into the cryſtalline, is to the ſine of refraction, very
nearly as 21 to 20, as I ſhall ſhew in the *Scholium*.
And conſequently, I will be 21, and R will be 20.
From theſe meaſures I now proceed to determine
the diſtance of a radiating point from the *cornea*,
that is, the diſtance of Q from B, ſo as that the

focus

focus of the rays, issuing from it and falling diverging on the *cornea,* may by the refractive powers of the aqueous, crystalline, and vitreous humours, be placed in the *retina* at z. By the refraction of the aqueous humour, the rays fall on the crystalline with such a degree of convergence as would make them unite at q. In the universal *Theorems* therefore for finding the principal *focus* of rays falling converging on a double convex *lens,* Cq is D, Dz equal to 6.2617 is z, the *radius* of MCN is r, the *radius* of MDN is ϱ, CD the thickness of the crystalline is t, and I and R are 21 and 20. And by clearing D in that *Theorem*, we shall have D =

$$\frac{BIr\varrho z + BRr\varrho t - Rrtz}{BIr\varrho + Itz - Ir z - I\varrho z - R\varrho t - Rtz} = 10.3102 = Cq.$$

And Cq + BC = 11.346 = Bq.

In the *Theorem* for finding x, Bq is x, QB is D, I is 4, R 3, the sine of incidence of rays passing out of air into water or into the aqueous humour, being to the sine of refraction, as 4 to 3, and the *radius* of the *cornea* is 3.3294 10th parts of an inch; consequently, D is 57.48, that is, about 5 inches and 3 quarters. So that supposing the eye to be unchangeable, a radiating point placed at the distance of $5\frac{1}{4}$ inches from it, will have its image placed in the *retina*.

SCHOLIUM.

Pl. 11.
Fig. 7.

" Let AB represent the refracting plane surface
" of any body, and IC a ray incident obliquely on
" the body at C, so that the angle ACI may be
" infinitely little, and let CR be the refracted ray.
" From a given point B perpendicular to the re-
" fracting surface erect BR meeting the refracted
" ray CR in R, and if CR represent the motion of the
" refracted ray, and this motion be distinguished in-
" to two motions CB and BR, whereof CB is parallel
" to the refracting plane, and BR perpendicular to
" it : CB shall represent the motion of the incident
" ray

" ray, and BR the motion generated by the refrac-
" tion." NEWT. *Opt. Prop.* 10. p. 245, 246. CBR is
equal to the angle of incidence, and CRB is equal to
the angle of refraction; consequently, if R be made
the center, and a circle be supposed to be drawn with
the *radius* CR, CR will be the fine of the angle of
incidence, and CB the fine of the angle of refraction; and, putting I and R for those fines, we shall
have this analogy, I . R :: CR . CB. Hence
$\frac{I^2 - R^2}{R^2} = \frac{CR^2 - CB^2}{CB^2}$, or $\frac{I^2}{R^2} - I = \frac{BR^2}{CB^2}$. But the
motion of the ray at its incidence represented by
CB, is given; and therefore, $\frac{I^2}{R^2} - I$ is as BR^2.
But by the aforesaid proposition BR^2 expresses the
refractive force, and is nearly as the density of the
body; as Sir I. NEWTON found, by computing BR^2
from the fines I and R in several bodies, and then
comparing it with their respective densities. And
consequently, putting D for the density of the
body, $\frac{I^2}{R^2} - I$ is as D, and $\frac{I}{R}$ as $\sqrt{D + I}$. In paf-
sing out of air into water $\frac{I}{R}$ is $\frac{4}{3}$, and, the density
of water being 10000, $\sqrt{D + I}$ is 10004: And in
passing out of air into the cryftalline, whose density
is to that of water as 11083 to 10000, $\sqrt{D + I}$ is
10528. Therefore in passing out of air into the
cryftalline $\frac{I}{R}$ will be $\frac{2}{7}$; for 10004 . 10528 :: $\frac{4}{3}$.
$\frac{42111}{10012} = \frac{7}{5}$ very nearly. $\frac{I}{R}$ in passing out of the
aqueous humour into the cryftalline, will be com-
pounded of the *ratio* $\frac{3}{4}$ and $\frac{7}{5}$ by the second *Theo-*
rem of the *Opticks*, p. 113; and therefore $\frac{I}{R}$ will
be equal to $\frac{21}{20}$; or I will be to R, as 21 to 20.

F I N I S.

11.

Fig. 4.

q

6.

BOOKS Printed for J. NOURSE, opposite Catherine-Street in the Strand, Bookseller to his MAJESTY.

1. THE Elements of Trigonometry, by W. Emerson. The Second Edition, with large Additions; together with the Tables of Sines, Logarithms, &c. 8vo. 1764. Price 7s.

2. The Elements of Geometry, by W. Emerson, 8vo. 1763. Price 5s.

3. A Treatise of Arithmetick, by W. Emerson, 8vo. 1763. Price 4s. 6d.

4. A Treatise of Algebra, by W. Emerson, 8vo. 1764. Price 7s.

5. Emerson's Navigation, 12mo. The Second Edition. 1764.

6. A New Method of Increments, by W. Emerson, 4to. 1763. Price 7s. 6d.

7. The Arithmetick of Infinites. Conick Sections. The Nature and Properties of curve Lines. By W. Emerson, 8vo. 1767. Price 7s. 6d.

8. A Treatise of Algebra, by T. Simpson. The Third Edition 8vo. 1766. Price 6s.

9. Essays on several curious and useful Subjects in speculative and mixed Mathematicks, by T. Simpson, 4to. 1740. Price 6s.

10. Mathematical Dissertations on a variety of Physical and Analytical Subjects, by T. Simpson, 4to. 1743. Price 7s.

11. Miscellaneous Tracts on some curious and useful Subjects, by T. Simpson, 4to. 1757. Price 7s.

12. Trigonometry, Plane and Spherical, with the Construction and Application of Logarithms, by T. Simpson. The Second Edition, 8vo. 1765. Price 1s. 6d.

13. The Doctrine of Annuities and Reversions, by T. Simpson, 8vo. 1742. Price 3s.

14. Appendix to Ditto. Price 6d.

15. The Elements of Geometry, by T. Simpson. The Second Edition, with large Alterations and Additions, 8vo. 1760. Price 5s.

16. Select

16. Select Exercises for young Proficients in the Mathematicks, by T. Simpson, 8vo. 1752. Price 5s. 6d.

17. The Doctrine and Application of Fluxions, by T. Simpson, 2 vol. 8vo. 1750. Price 12s.

18. The Elements of Euclid, by R. Simpson. The Second Edition. To which is added the Book of Euclid's Data, 8vo. 1762. Price 6s.

19. Introduction a L'Arithmetique vulgaire, 4to. 1752. Price 1s.

20. The Elements of Navigation, by J. Robertson. The Second Edition, 2 vol. 8vo. 1764. Price 18s.

21. The Elements of Astronomy, translated from the French of M. De La Caille, by J. Robertson, 8vo. 1750. Price 6s.

22. The Military Engineer; or a Treatise on the Attack and Defence of all kinds of fortified Places. In two Parts, 8vo. 1759. Price 9s.

23. The Accountant; or, The Method of Book-Keeping, by Dodson, 4to. 1750. Price 4s. 6d.

24. The Mathematical Repository, by J. Dodson, 3 vol. 12mo. 1748—1755. Price 12s.

25. An Easy Introduction to the Theory and Practice of Mechanicks, by S. Clark, 4to. 1764. Price 6s.

26. The British Gauger; or Trader and Officer's Instructor in the Royal Revenue of the Excise and Customs. Part I. Containing the necessary Rules of vulgar and decimal Arithmetick, and the whole Art of Practical Gauging, both by Pen and Rule; illustrated by a great Variety of curious and useful Examples. Part II. Containing an Historical and Succinct Account of all the Laws relating to the Excise; to which are added, Tables of the old and new Duties, Drawbacks, &c. on Beer, Ale, Spirits, Soap, Candles, &c. with a large and copious Index. By Samuel Clark, 12mo. 1765. Price 5s.

27. The Elements of Fortification, by J. Muller. The Second Edition, 8vo. 1756. Price 6s.

28. The Method of Fluxions and infinite Series, with its Application to the Geometry of curve Lines, by the Inventor Sir Isaac Newton. Translated from the Author's original Latin. To which is subjoined, a perpetual Comment upon the whole Work. by J. Colson, F. R. S. 4to. 1736. Price 15s.

Lightning Source UK Ltd.
Milton Keynes UK
UKHW020755081118
331957UK00010B/1159/P